Monte Verità Proceedings of the Centro Stefano Franscini
Ascona

Edited by H. Flühler, ETH Zürich

Conference on Recent Shifts in Vegetation Boundaries of Deciduous Forests, Especially Due to General Global Warming

Edited by:
F. Klötzli
G.-R. Walther

Springer Basel AG

Editors:

Prof. Dr. Frank Klötzli
Geobotanical Institute ETH Zürich
Zürichbergstrasse 38
CH-8044 Zürich
Tel.: +41-1-632 45 88
Fax: +41-1-632 12 15
E-mail: kloetzli@geobot.umnw.ethz.ch

Gian-Reto Walther
Geobotanical Institute ETH Zürich
Zürichbergstrasse 38
CH-8044 Zürich
Tel.: +41-1-632 45 88
Fax: +41-1-632 12 15
E-mail: walther@geobot.umnw.ethz.ch

A CIP catalogue record for this book is available from the Library of Congress, Washington D.C., USA

Die Deutsche Bibliothek Cataloging-in-Publication Data

Conference on Recent Shifts in Vegetation Boundaries of Deciduous Forests, Especially Due to General Global Warming <1998, Ascona>:
Conference on Recent Shifts in Vegetation Boundaries of Deciduous Forests, Especially Due to General Global Warming / ed. by F. Klötzli ; G.-R. Walther. – Basel ; Boston ; Berlin : Birkhäuser, 1999
(Monte Verità)
ISBN 978-3-0348-9743-3 ISBN 978-3-0348-8722-9 (eBook)
DOI 10.1007/978-3-0348-8722-9

Originally published by Birkhäuser Verlag in 1999
Softcover reprint of the hardcover 1st edition 1999
Camera-ready copy prepared by the editors.
Printed on acid-free paper produced from chlorine-free pulp. TCF ∞

9 8 7 6 5 4 3 2 1

Table of contents

Preface

When you arrive in the Ticino in winter, coming from the north, it is obviously warmer than the place you left, and also much greener. The presence of high palm trees give an exotic touch to the gardens. If you look at the shrubs and trees more closely, you realize that most of them are evergreen broad-leaved species from southeastern USA and from eastern Asia.
This late winter scene is framed by snow-clad high mountains which overtop lush-green lower valleys. However, the scene is not like that of the Mediterranean in winter but reminds you more of the Japanese main island Honshu or some forgotten valleys in the southern Alleghanies or western Caucasus.
These associations proved very relevant to our discussions of the spreading of evergreen broad-leaved plants in the Ticino and in similar areas – a process we have been calling "Laurophyllisation" (laurophyll, laurineous cf. Stocker 1963, laurel-like leaves, according to the definition in Klötzli 1988, excluding sclerophyllous plants from the mediterranean type climate regions).
Obviously there have been warmer years in the past. However, the recent warming has lasted longer than before and there is no evidence of such a dramatic spread of laurophyllous species before our times.
Thus the spreading of laurophyllous ornamental garden species into the surrounding forests appears to be a new process in our region.

Acknowledgements

The meeting took place at the Centro Stefano Franscini, Monte Verità, Ascona. It was financed by the Swiss Federal Institute of Technology (ETH) Zürich. We would like to thank the responsible members of the ETH, especially Prof. Hannes Flühler, and the staff of the Centro Stefano Franscini, Mrs. Katia Bastianelli and Brigitte Pichler, for their support in organising the meeting.
Special thanks go to State Councillor Dr. Marco Borradori, forest engineer Ivo Ceschi, head forester of the canton of Ticino, and Dr. Benedetto Lepori, from the university of the italian part of Switzerland, and Prof. Valerio Sala, director of the Botanical Garden of the Brissago Islands. We also thank Dr. Aldo Antonietti, vice-director of the Federal Office for Environment, Forest and Landscape (BUWAL), who made the public presentation in the Palazzo "Sopracenerina" in Locarno, made available to us by the Sopracenerina corporation.
Our heartfelt thanks go especially to our colleagues of the DIONEA S.A., Locarno, for their support in organising the conference and the excursions in particular, where we could enjoy the great hospitality of the Ticino. Special thanks go to Mr. Sandro Canonica and his friends for their kind welcome in the beautiful surrounding of their vineyard in Lamone where we could enjoy the real "insubrian spirit"...
We thank also the staff of the Geobotanical Institute involved in the conference. First of all the head of the institute, Prof. Peter Edwards, for his support in advance and during the conference, Mrs. Anne Pickhardt for her skillful assistance in completing the proceedings and Mr. Kaspar Pflugshaupt and Mr. Roger Stupf for their technical support in their compilation.

We are pleased to present this book and hope that it will provide the reader with an overview of recent findings and ongoing activities in the context of recent shifts in vegetation boundaries of deciduous forests, especially due to general global warming.

The editors

Frank Klötzli
Gian-Reto Walther

A winter picture of a Swiss forest rich in laurophyllous species
(Photo: G.-R. WALTHER)

List of participants

ANTONIETTI A., Dr.
Bundesamt für Umwelt, Wald und Landschaft (BUWAL), CH-3003 Bern

BODMER H.C., Dr.
Dept. Wald- und Holzforschung, ETH Zürich, CH-8092 Zürich

BORHIDI A, Prof. Dr.
Institut of Ecology and Botany, H-2163 Vàcràtòt, Hungary

BRUELHEIDE H., PD Dr.
Systematisch-Geobotanisches Institut, Univiversity of Göttingen, D-37073 Göttingen

CARRARO G., Dipl. Ing. for.
DIONEA SA, CH-6600 Locarno

CONEDERA M., Dr.
FNP Sottostazione Sud delle Alpi, CH-6501 Bellinzona

DOYLE G., Dr.
Faculty of Sciences, University College Dublin, Dublin 4, Ireland

EDWARDS P.J., Prof. Dr.
Geobotanisches Institut ETH Zürich, CH-8044 Zürich

FUJIWARA K., Prof. Dr.
Institute of Environmental Science & Technology, Yokohama National University, Yokohama, Japan

GAGNIDZE R., Prof. Dr.
Institute of Botany, Georgian Academy of Science, 38007 Tbilisi, Georgia

GIANONI G., Dipl. Ing. for.
DIONEA SA, CH-6600 Locarno

GRABNER S., Dr.
Institut für Botanik, Universität Innsbruck, A-6020 Innsbruck

HEISELMAYER P., Prof. Dr.
Institut für Botanik, Universität Salzburg, A-5020 Salzburg

HOLDENRIEDER O., Prof. Dr.
Dept. Wald- und Holzforschung, ETH Zürich, CH-8092 Zürich

HÜBL E., Dr.
Botanisches Institut, Universität für Bodenkultur, A-1180 Wien

HÜBL E., Prof. Dr.
Botanisches Institut, Universität für Bodenkultur, A-1180 Wien

KIENAST F., PD Dr.
WSL, CH-8903 Birmensdorf

KLÖTZLI F., Prof. Dr.
Geobotanisches Institut ETH Zürich, CH-8044 Zürich

KRÄUCHI N., Dr.
WSL, CH-8903 Birmensdorf

NAKHUTSRISHVILI G., Prof. Dr.
Institute of Botany, Georgian Academy of Science, 38007 Tbilisi, Georgia

PAUL L., Dr.
Dept. Wald- und Holzforschung, ETH Zürich, CH-8092 Zürich

PILLAR V., Dr.
Dept. de Botânica, Univ. Federal do Rio Grande do Sul, Porto Alegre, Brasil

POLI MARCHESE E., Prof. Dr.
Istituto di Biologia ed Ecologia Vegetale, Università di Catania, I-95128 Catania

SALA V., Prof.
Dir. Parco Botanico Isole di Brissago, CH-6612 Ascona

SCHULZ Ch.
Systematisch-Geobotanisches Institut, Univiversity of Göttingen, D-37073 Göttingen

VOSER P., Dr.
FORNAT, CH-8708 Männedorf

WALDER M.
CH-6653 Verscio

WALTHER G.-R., Dipl. natw. ETH
Geobotanisches Institut ETH Zürich, CH-8044 Zürich

ZAZANASHVILI N., Dr.
Institute of Botany, Georgian Academy of Science, 38007 Tbilisi, Georgia

Recent shifts in vegetation boundaries of deciduous forests,
especially due to general global warming, pp. 1–13
edited by F. KLÖTZLI and G.-R. WALTHER

Shaping the Mountain of Truth: an introduction to the Monte Verità

Hans-Caspar BODMER[1], Lioba PAUL[2] and Ottmar HOLDENRIEDER[2],
[1] Sect. for Nature & Landscape Conservation, ETH-Zürich,
[2] Sect. of Forest Pathology & Dendrology, ETH-Zürich,
CH-8092 Zürich, Switzerland

Abstract

The Monte Verità, currently used as the congress centre of the Federal Institute of Technology, is an estate located in Ascona, Switzerland, at the southern edge of the Alps. The location is known as a centre of European counterculture from the beginning of the 20th century. We present an overview on its history, site conditions, land use and park management problems. The woody flora of the park, comprising many exotic species, is listed.

The natural background

It is a pleasure for us to introduce you to the Mountain of Truth or the Monte Verità, the location at which this congress on vegetation shifts takes place. The Monte Verità is a small hill (321m a.s.l.) which belongs to the municipality of Ascona. It is situated at the southern edge of the Alps (8°46' E, 46°9' N) close to the lake Lago Maggiore (193m a.s.l.).

The Alps are relatively young, their origin is in the cretaceous and tertiary period. Their formation is not yet finished and shaping by growth and erosion is still going on. The Swiss Alps form the highest part of the Alps between Nice and Vienna, reaching up to 4634m. a.s.l. Ascona is located just south of the Insubrian line, a narrow tectonic disturbance zone, which separates the southern Alps from the northern tectonical complexes. It received its name from a celtic tribe, the Insubrians, who lived in this region 2000 years ago. The southern Alps are geologically very complex. Different types of rocks alternate within short distances and create varying site conditions (Koenig 1978). The surroundings of Ascona are formed by metamorphic crystalline rocks and the

Monte Verità is composed of amphibolite, accompanied by biotite-diorite (E.D. Müller, pers. comm.). It has been shaped by erosion during the tertiary period and later on by glacial action during ice ages (Hantke 1991).
This, however, also happened to surrounding landscapes and is not very special. Nevertheless there is one very special geological feature of this hill: It shows a pronounced magnetic anomaly. It has been speculated that this might explain the attractiveness of this mountain for unusual people (Szeemann 1980:6).

However, another probably more significant factor for the attraction of special people as well as special plants is Ascona's climate (see Figure 1). The average rainfall amounts to nearly 2000 mm. The summer is warm and wet, heavy thunderstorms are common. The central area of Ascona has been flooded repeatedly during recent years. In winter the average monthly temperature never falls below freezing point and the winter season is very dry. This leads to frequent forest fires during the winter.

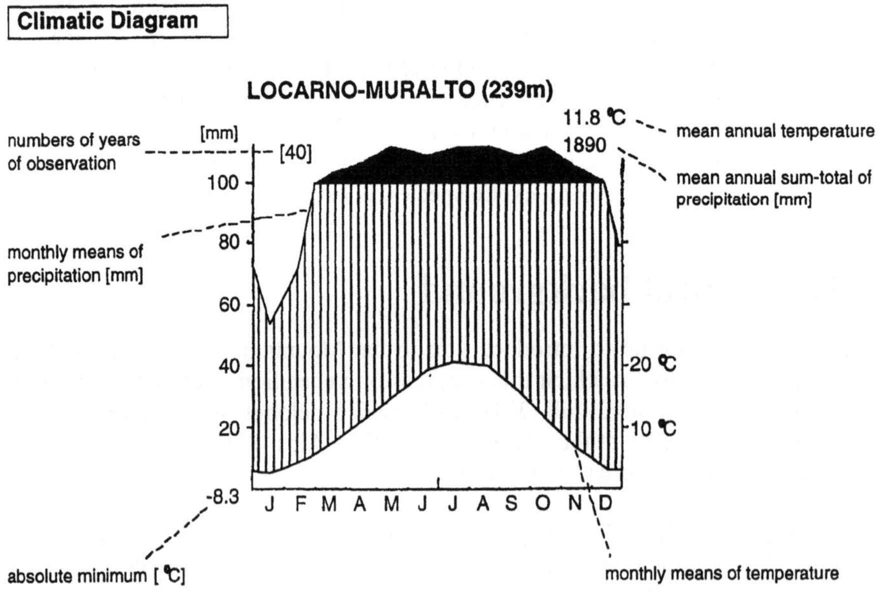

Figure 1: Climatic diagram of Locarno-Muralto (one kilometre east of Ascona)

The extraordinary mild and humid summer climate makes it even possible for some subtropical plants to survive without shelter at a northern latitude of over 46 °N. You will note this richness of garden flora, with palms, *Magnolia*,

Camellia, *Araucaria* and *Camphora* trees, which is much more diverse and more colourful than the indigenous vegetation (see Oberdorfer 1964).
Parks and gardens are a special feature of the Insubrian region and the Monte Verità too has its park shaped by exotic plants as well as by non-indigenous people. Parks need shaping, without human activity they would disappear. Parks are man-made paradises, parks are safe sites for dreams and recreation. To approach such an object, first we have to understand its history and to answer the following questions: "Who are you?" , "Where are you coming from?" and "Where are you going to?".

Social background

In Europe the beginning of this century was characterized by fast development in the industrial sector. Many people felt a strong desire to go 'back to nature' (see Figure 2). People tried to escape from cities willing to live a more primitive life with greater promise of happiness.

Figure 2: "Der Liebe Macht" (power of love).
Oil painting of Gustav Arthur Gräser, 1899 (Museum Casa Anatta, Monte Verità). Men are leaving the dark industrial world (on the right side) and seek the paradise (left).

In his book about the Monte Verità, Martin Green remarks: "At the end of the nineteenth century, intellectual Europe became preoccupied with the problem of its own unhappiness, malaise, or to use Freud's word - Unbehagen. The favorites of this rich and powerful civilization felt themselves to be more unhappy than more primitive people." (Green 1986). Another aspect is that many people were suffering from poor environmental conditions in towns and they were forced to find a better place to live. Last but not least travelling to the south became more and more fashionable. People were seeking the southern ambiente but they were afraid of the different culture of southern countries. A specific counterculture developed in Ascona and the centre of it was the Monte Verità.

It was difficult to approach the history of this location from a scientific point of view. The available sources for the history of land use in this area are poor. Most sources are personal stories, even tales at times. Finally, the following picture took shape out of the mist:
The Monte Verità was a part of the *patriziato* of Ascona. The municipality of Ascona was not the same in the last century as it is nowadays, it was a poor community, as were most of the other villages in this region. The common land was used by all the citizens of Ascona for grazing sheep or goats, for supplying firewood and of course for collecting chestnuts and mushrooms. An old regulation of the *patriziato* allowed everybody to use these common goods, poor or rich, but only once a day. This was the traditional multifunctional land use for centuries.
Towards the second part of the last century the Monescia, as the Monte Verità was called at that time, was shaped for the first time more seriously by man: People built terraces to use the sunny slopes for grape production. A system of small terraces in the forest adjacent to the park still reminds one of the famous "nostrano", a local wine, which was produced in Ascona. The vineyards were given up in the final decade of the last century, because of an insect (*Viteus vitifoliae*), which had been introduced from America, killing the grapes (Kobelt, Wädenswil, pers. comm.). With the beginning of this century the Monescia was degraded and abandoned, and time was just right for a change.
Looking at the demographic development of the region we made an interesting observation: At the beginning of this century, the proportion of the "stranieri", foreigners, who were living in Ascona, was climbing up to over 45 % (Figure 3). That means nearly every second registered person was not from here. Before 1900 and after 1920 the distribution is less spectacular (Ufficio di Statistica 1993).

The people at Monte Verità

In 1900 Henry von Oedenkoven, the son of a rich Belgian industrialist, arrived from Germany together with a group of seven young people and decided to settle here and to form a commune. These people came from very different industrialized sites in Europe and were searching for a new sense of life. Oedenkoven, 24 years old, was motivated by suffering from a serious disease, probably gout. About her motivation, his partner, Ida Hofmann wrote: "... there is no need to tell you about the wonderful climate and the fantastic surroundings; most people know the magical quality of this area" (Hofmann-

Oedenkoven 1906). However, as city dwellers with an idealized view of nature they did not recognize the poor agricultural quality of the site.

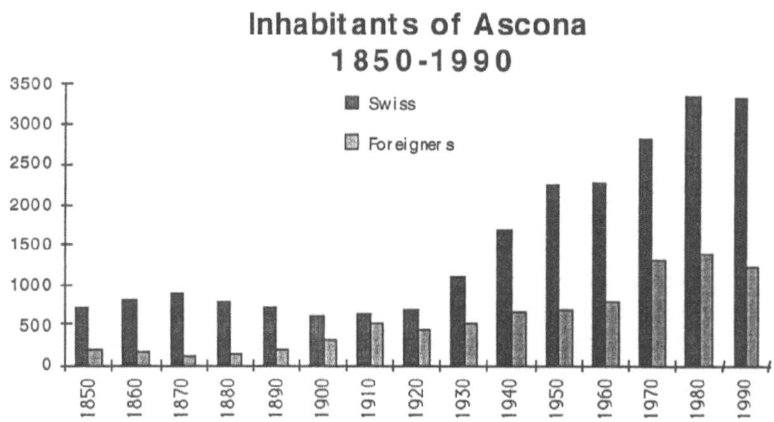

Figure 3: Swiss and foreign inhabitants of Ascona from 1850 to 1990 (Ufficio cantonale di statistica 1993)

Anyway, they gave the name 'Monte Verità' , Mountain of Truth, to the hill and started to shape it. A sanatorium was built which soon became internationally recognized. It was rigorously managed by Oedenkoven, who was an orthodox vegetarian. Even spices, fat, salt and wool were forbidden. Leather was permitted only for sandals. Several members of the community did not agree with these ideals and escaped from time to time to local restaurants. Many anecdotes curl around such distracted 'searches for truth'. Within a few years the Monte Verità became a point of attraction not only for prophets of a 'natural' lifestyle and other remarkable people, but also for tourists, coming with special boats from Italy to see the nude vegetarians.

During that period important ideas were developed and fascinating life experiments were performed on the Monte Verità. These people shaped our society with feminism, pacifism, socialism, psychoanalysis, dadaism, surrealism, modern dance and literature. There is probably no place in Europe where more 'visions' were developed than here in Ascona. For example the following personalities were active on the Monte Verità :

Gustav 'Gusto' Arthur Gräser (1879-1958), philosopher and painter, with a very individual lifestyle, lived temporarly in caves; Rudolf von Laban (1879-1958), founder of the expressive dancing, 'principles of dance and movement notation'; Carl Gustav Jung (1874-1961), psychoanalist, founder of the 'complex psychology'; Hermann Hesse (1877-1962), writer, Nobel prize-winner 1946; Elisar von Kupffer (1872-1942), built the Sanctuarium Artis Elisarion, a temple with the famous painting 'Clear World'; Leopold Wölfling, (1868-1935) archduke, member

of the royal house of Habsburg; Erich Mühsam (1878-1934), writer and revolutionary, member of the 'Zentralrat der bayerischen Räterepublik', who spent 7 years in prison. Some of them are meeting in Figure 4.
Comprehensive information on their activities as well as those of many other temporary inhabitants of the Monte Verità can be found in Dadò (1980), Green (1986) and Landmann (1931, 1973).

The decline of the commune came with the end of World War I, partly because of political changes in the home countries of the members, partly because of personal problems. In 1923 three years after Oedenkoven emigrated via Spain to Brasil, the Monte Verità was purchased by a group of Germans from Berlin. One of them was Werner Ackermann, who later on wrote a book about the mountain (Landmann 1931), which now is an important source of information for us. They continued the hotel business in a more traditional manner and organized large parties in the park. Among other park shaping activities a tennis court was constructed during this period. But soon mismanagement led to the failure of the enterprise and the mountain of truth was advertised for sale.
From 1926 the new owner was Baron Eduard von der Heydt, a rich German banker, who became - like many rich foreigners - a Swiss citizen eleven years later. Originally, he did not like the Monte Verità and he offered only half the asked price for it. Nevertheless the object came into his possession, the 'Mountain of Truth just fell into his lap'. He built the hotel, where we are just now, in 1927. It was designed by Emil Fahrenkamp in the "Bauhaus-Style". As one of the early flat roof buildings, the feelings of the people ran high about it. Von der Heydt tried to renew the tradition of Oedenkoven by managing the hotel at least partly in the vegetarian style. He also attracted extraordinary people, among them many artists and politicians (von der Heydt & von Rheinbaben 1958) and he was very involved in collecting East Asian art. He left his world-famous collection to the City of Zürich, where you can see it in the Rietberg Museum. So the Monte Verità again became a focal point for creating and exchanging philanthropic ideas.

Legend to Figure 4:

<table>
<tr><td colspan="2">Sanatorium Monte Verità in 1906</td><td>Baron Eduard von der Heydt on the roof of Casa Anatta</td></tr>
<tr><td>Gustav 'Gusto' Gräser in his old age as hermit in a fork of a tree</td><td colspan="2">The initiators of the Monte Verità exercising their dew-walk with Henri Oedenkoven, Ida Hofmann, Anni Pracht, Raphael Friedberg and Corneli Gabes Gouba</td></tr>
<tr><td colspan="2">Sunbathing and fango therapy in 1906 on the Monte Verità</td><td>The famous dancer Rudolf von Laban with Betty Baaron in 1914</td></tr>
</table>

Figure 4: Captions see p. 6

With the death of von der Heydt in 1964 the Monte Verità became the property of the Republic and Canton of Ticino. According to the last will of the deceased the Monte Verità should become a site where contemporary cultural events take place. A period with changing activities followed, a few international meetings took place and also the great exhibition of Harald Szeemann on the history of the Monte Verità. Two books on the Monte Verità were published (Dadò 1980, Green 1986) and a museum was installed in the Casa Anatta, the house of the first owner von Oedenkoven. Casa Selma, one of the small buildings in the park was renovated for an exhibition and the "temple" (Elisarion) with the wall painting 'Clear World of the Blessed' by Elisar von Kupffer was transferred to the Monte Verità. Fahrenkamps building was still used as hotel, which however never became profitable. The vegetation took over large parts of the park reminiscent of Sleeping Beauty's castle.
In 1989 the ownership was transferred to a trust and a contract was signed with the Swiss Federal Institute of Technology of Zürich to establish a congress centre at Monte Verità, named after Stefano Franscini (1796-1857), a member of the Swiss Federal Council who was engaged in the foundation of the ETH in 1854. From 1990 onwards the hotel was renovated, a lecture hall was constructed and several scientific workshops took place. The mountain again became a focal point for people searching for truth, although from a very different point of view compared to the founding fathers. Your stay here shows, that the magic forces of this particular site are still active.

The park

All those already mentioned people contributed to the shaping of the mountain ot truth. The park reflected their ideas and their attitudes towards nature. At the conference about the common ground of wild and cultivated plants Evans said: "Land is not shaped as a by-product of our activity: to shape land, to organize our space in nature, to build with and into growing things is the very essence of our being" (Evans 1994). We feel responsibility for this object, its assemblage of plants and architecture and the ideas of which it reminds us. We don't want to lose the traces of this small paradise and therefore we have to shape it. Otherwise the very dynamic vegetation would take it over quickly, eventually leading to an oak dominated forest (Gianoni et al. 1988).
Unfortunatly, historic sources concerning the park are very scarce, just the backgrounds of certain old pictures give some information. We see small lodges in a sparse forest, free views to the mountains and to the lake, people growing vegetables, people bathing in the sun with a dense hedge in the background. Gusto Gräser sitting on a tree and finally Baron von der Heydt, as proud owner of a professionally shaped hotel park.
We know only that von der Heydt employed 'renowned landscape architects' to organize the park (Landman 1931: 278), but the source does not mention any

names. Von der Heydt was a collector, not a creator. We know from literature that he did not want to touch the parts of the area which were covered by forest (Landman1931:278).
Nevertheless we get some insight into the development of the park during the last 50 years by a series of aerial photographs. According to that time series the development of canopy cover could be reconstructed. The canopy cover was continuously increasing up to 75 % by the end of the 90ies.
From a terrestric point of view, human activities have increased during recent years. New exotic trees were planted, fallen leaves were removed and dumped on other sites below trees, locally soil was filled up and compacted, car parks were enlarged and the landscape was furnished. Thereby slowly the park is losing its face and well-planned shaping becomes a challenge. Where do we want this park to go?

We do not have a definitive answer yet. Just some thoughts and ideas. The park is dominated by woody plants, most of which are non indigenous. We recorded more than 80 species of woody plants, from which approximately 70 % are exotic (Table 1). Parks and arboreta are traditional sources of plants, which invade the surrounding area later on .

Table 1: Species of woody plants in the park of Monte Verità. Abbreviations: + indigenous species, (+) indigenous in the region, but not at Mt. Verità, - non indigenous species. (Note: Because the list was compiled during winter, not all plants could be determined to species.)

Species	**Family**		**Provenance**
Abies alba	Pinaceae	(+)	Europe
Acacia dealbata	Mimosaceae	-	Australia, Tasmania
Acer pseudoplatanus	Aceraceae	+	Europe
Ailanthus altissima	Simaroubaceae	-	China
Albizia julibrissin	Mimosaceae	-	Asia
Aucuba japonica	Cornaceae	-	Japan
Berberis thunbergii	Berberidaceae	-	Japan
Betula pendula	Betulaceae	+	Europe, Asia Minor
Buddleja davidii	Buddlejaceae	-	China
Buxus microphylla	Buxaceae	-	Japan
Buxus sempervirens	Buxaceae	+	Southern Europe
Camellia cf. *japonica*	Theaceae	-	Japan
Castanea sativa	Fagaceae	-	S-Europe, W-Asia
Cedrus deodara	Pinaceae	-	W-Himalaya
Cephalotaxus fortunei	Cephalotaxaceae	-	China
Chamaecyparis lawsoniana	Cupressaceae	-	NW-California, SW-Oregon
Cinnamomum glanduliferum	Oleaceae	-	Southeast Asia
Cistus sp.	Cistaceae	-	(Mediterranean area)
Corokia cotoneaster	Cornaceae	-	New Zealand

Corylus avellana	Betulaceae	+	Europe, Asia Minor
Cotoneaster spp.	Rosaceae	-	(Eastern Asia)
Choisya ternata	Rutaceae	-	Mexico
Cryptomeria japonica	Taxodiaceae	-	China, Japan
Cunninghamia lanceolata	Taxodiaceae	-	Middle and Southern China
Cupressus lusitanica	Cupressaceae	-	Mexico
Cupressus sempervirens	Cupressaceae	-	Mediterranean area
Deutzia sp.	Hydrangeaceae	-	(Eastern Asia)
Diospyros kaki	Ebenaceae	-	Japan, China
Elaeagnus pungens	Elaeagnaceae	-	Japan
Eriobotrya japonica	Rosaceae	-	Japan
Fagus sylvatica	Fagaceae	+	Europe
Forsythia x intermedia	Oleaceae	-	(Eastern Asia)
Fraxinus excelsior	Oleaceae	+	Europe
Hebe spp.	Scrophulariaceae	-	Australia, New Zealand
Hedera helix	Araliaceae	+	Europe
Hydrangea sp.	Hydrangeaceae	-	
Ilex aquifolium	Aquifoliaceae	+	Europe
Jasminum cf. *officinale*	Oleaceae	-	Persia, Himalaya, China
Laurus nobilis	Lauraceae	-	Mediterranean area
Ligustrum lucidum	Oleaceae	-	China, Japan
Lonicera japonica	Caprifoliaceae	-	China, Japan
Lonicera pileata	Caprifoliaceae	-	SW-China
Lonicera fragrantissima	Caprifoliaceae	-	Eastern China
Magnolia spp.	Magnoliaceae	-	
Mahonia aquifolium	Berberidaceae	-	Western N-America
Nerium oleander	Apocynaceae	-	Mediterranean area
Olea europaea	Oleaceae	-	Mediterranean area
Parthenocissus tricuspitata	Vitaceae	-	Japan, China
Photinia serrulata	Rosaceae	-	Southern China
Phyllostachys sp.	Gramineae	-	(Eastern Asia, Himalaya)
Picea abies	Pinaceae	(+)	Europe
Picea orientalis	Pinaceae	-	Caucasia
Pinus nigra	Pinaceae	-	Southeast Europe
Pinus parviflora	Pinaceae	-	Japan
Pinus pinea	Pinaceae	-	S-Europe to Western Asia
Pinus strobus	Pinaceae	-	Eastern N-America
Pinus sylvestris	Pinaceae	(+)	Europe to Middle Sibiria
Pittosporum tobira	Pittosporaceae	-	China, Japan
Platanus x hispanica	Platanaceae	-	
Prunus avium	Rosaceae	+	Europe, N-Africa, Asia Minor
Prunus laurocerasus	Rosaceae	-	SE-Europe to Caucasia
Pseudotsuga menziesii	Pinaceae	-	Western. N-America
Pueraria lobata	Papilionaceae	-	Japan, China

Quercus robur	Fagaceae	+	Europe
Quercus rubra	Fagaceae	-	Eastern N-America
Quercus petraea	Fagaceae	+	Europe
Quercus pubescens	Fagaceae	+	Southern Europe
Rhododendron spp.	Ericaceae	-	(Eastern Asia)
Robinia pseudoacacia	Papilionaceae	-	Eastern USA
Rosmarinus officinalis	Lamiaceae	-	Mediterranean area
Rubus sp.	Rosaceae	+	Europe
Salix caprea	Salicaceae	+	Europe
Sambucus nigra	Caprifoliaceae	+	Europe
Sarothamnus scoparius	Papilionaceae	+	Europe
Sorbus aria	Rosaceae	+	Europe
Spirea sp.	Rosaceae	-	
Taxus baccata	Taxaceae	+	Europe, Asia Minor
Thuja occidentalis	Cupressaceae	-	E-Canada, USA
Thuja plicata	Cupressaceae	-	Western N-America
Thujopsis dolabrata	Cupressaceae	-	Japan
Tilia cordata	Tiliaceae	+	Europe
Trachycarpus fortunei	Palmaceae	-	S-China, Japan
Vaccinium myrtillus	Ericaceae	+	Europe, Asia
Viburnum lantana	Caprifoliaceae	+	Europe
Viburnum tinus	Caprifoliaceae	-	Mediterranean area

A striking example is the palm *Trachycarpus fortunei*. This hardy palm from eastern Asia was originally planted here to create the 'ambiente' of the south. Since 1923 a palm has been the emblem of the Monte Verità . However, nowadays the palm is perceived by many local people as a symbol of foreign colonisation and treated as a weed. *T. fortunei* is regenerating very vigorously in the park and we can observe here an example of a recent plant invasion. Christ (1882: 27-68) in his inventory of Insubrian flora did not list *T. fortunei*. In 1921 Venner described the occurrence of palm seedlings on the island of Brissago, Hegi (1932) also mentioned the occurrence of seedlings, which 'hardly ever develop further'. In 1950, Anliker reported the first invasions of *T. fortunei* into natural vegetation.

We have to find an appropriate way of considering this plant in the park management plan. Establishing areas where the seedling demography and natural succession will be studied is one option. Here in the park with its surrounding forest we have the chance to observe initial stages of vegetation shifts. This opportunity is restricted to a very few sites, because almost all land in this extraordinary mild 'hyperinsubric' zone is in private hands. The price of 1 square metre is over US$ 850 (more than 1200.- SFr.). There is probably no vegetation scientist, who could afford such an expensive experimental site.

Despite the fact that the park of the Monte Verità does not represent a definite epoch, it has a specific character which should certainly be preserved. This has to be done in an economical way, avoiding overmanagement and using natural processes. Optical calming without losing historic traces appears feasible.

A park is a focal point where different problems meet and where we can test new planning tools, which might be useful for solving larger scale problems later on. One example of such a tool is the plant growth generator AMAP. It relies on the principles of quantitative botany designed by Hallé and Oldemann (1970). Today the growth of several hundreds of plants has been analyzed and mathematically expressed. Each plant consists of thousands of polygons and allows a more realistic picture of the growth process to be shown. All we will need is the digital terrain model where we can "plant" the desired trees and let them grow to get an impression of how the park will look in ten, twenty or even fifty years (Lecoustre & de Reffye 1993).
We do not know yet whether we really need such tools to substitute our imagination or to communicate our dreams. But we know that if we cease to create dreams and if we cease to shape the park, we will lose a paradise.

References

Anliker J. (1950): Die im Freien aushaltenden Palmen der Südschweiz. Schweizer Beiträge zur Dendrologie, 2, 33-51.

Christ H. (1882): Das Pflanzenleben der Schweiz. Verlag Friedrich Schulthess, Zürich.

Dadò A.(ed.) (1980): Monte Verità. Berg der Wahrheit. Electa Editrice, Milano.

Evans P. (1994): A common ground for wild and cultivated plants. In: A. R. Perry & R.G. Ellis (eds): The common ground of wild and cultivated plants: introductions, invasions, control, and conservation. pp. 1-6. National Museum of Wales, Cardiff.

Gianoni G., Carraro G. & Klötzli F. (1988): Thermophile, an laurophyllen Pflanzenarten reiche Waldgesellschaften im hyperinsubrischen Seenbereich des Tessins. Ber. Geobot. Inst. ETH, Stiftung Rübel, Zürich, 54, 164-180.

Green M. (1986): Mountain of truth. The counterculture begins. Ascona, 1900-1920. University Press New England, Hanover and London.

Hallé F. & Oldemann R.A.A. (1970): Essai sur l'architecture et la dynamique de croissance des arbres tropicaux, Masson & Cie, Paris.

Hantke R. (1991): Landschaftsgeschichte der Schweiz und ihrer Nachbargebiete: Erd-, Klima- und Vegetationsgeschichte der letzten 30 Millionen Jahre. Ott Verlag, Thun.

Hegi G. (1932): Fremdlinge der Schweizer Flora. Schweizer Blätter für Naturschutz, 7, 37-59.

Hofmann-Oedenkofen I. (1906):Monte Verità. Wahrheit ohne Dichtung. Verlag Karl Rohm, Lorch (Württemberg).

Koenig M.A. (1978): Kleine Geologie der Schweiz. 3. Auflage. Ott Verlag, Thun.

Landmann R. (1931): Monte Verità. Die Geschichte eines Berges. Berlin 1931.

Landmann R. (1973): Ascona - Monte Verità. Von Ursula von der Wiese überarbeitete und ergänzte Ausgabe. Benziger Verlag, Zürich, Köln.

Lecoustre R. & de Reffye Ph. (1993): AMAP, un modeleur de végétaux, un ensemble de logiciels de CAO/DAO à l'usage des professionnels de l'amenagement et des paysagistes. Revue Horticole Suisse, 66, 142-146.

Oberdorfer E. (1964): Der insubrische Vegetationskomplex, seine Struktur und Abgrenzung gegen die submediterrane Vegetation in Oberitalien und in der Südschweiz. Beitr. naturk. Forsch. SW-Deutschland, 23 (2), 141-187.

Szeemann H. (1980): Monte Verità. Berg der Wahrheit. In: Dadò A. (ed): Monte Verità. Berg der Wahrheit. Electa Editrice, Milano.

Ufficio di Statistica (1993): Le popolazioni: Definizioni per l'uso, Documenti statistici nr. 33, Bellinzona.

Venner J. V. (1921): Der Tessin. Ein Führer durch das ennetbirgische Land. 2. Auflage. Verlag Frobenius, Basel.

von Heydt E. & von Rheinbaben W. (1958): Auf dem Monte Verità. Erinnerungen und Gedanken über Menschen, Kunst und Politik. Atlantis Verlag, Zürich.

Recent shifts in vegetation boundaries of deciduous forests,
especially due to general global warming, pp. 15–29
edited by F. KLÖTZLI and G.-R. WALTHER

Recent vegetation shifts in Switzerland

Frank KLÖTZLI & Gian-Reto WALTHER
Geobotanical Institute, ETH Zürich, CH-8044 Zürich, Switzerland

Abstract

Recent climatic shifts, based on mild winters and more hot summers have released vegetation shifts whereby laurophyllous species from surrounding gardens are spreading in mesic thermophilous forests up to elevations of 400m above the valley floors (> 200m a.s.l.). Since their introduction in the 17th century these species have never been seen in the forests, except as occasional small saplings. Now there is a distinct trend from deciduous to evergreen broad-leaved forest.

Introduction

In a recent paper it was demonstrated that undisturbed forests always contain herb layers which may fluctuate considerably (Klötzli 1995). More intense forest management may even bring changes to the physiognomic and floristic structure of a herb layer provoking shifts to other forest plant communities (e.g. more mesic or/and nutrient-rich types). However, nowhere changes were that relevant as to bring a shift from one biome to another.
In a given pair of relevés, an ancient (>25 ys) one and a recent one of the same locality, it is very difficult to assess the direction of such changes. The same difficulties arise in short time series as well (Klötzli et al. 1996). But for a whole set of relevés, >100 pairs, using principal coordinate analysis the difference in the position of the two clusters can be seen and analysed as to the directing ecological factors. For a large set of paired relevés of the southern foot of the Alps (Ticino, part of Insubria) the same procedure was applied. This time the two clusters moved apart, depicting major changes. The extent and causes of these changes and shifts and their general trend will be taken up in the following chapters (compare also Gianoni et al. 1988; Carraro et al. 1998 on numerical data on shifts and their interpretations, see also Klötzli et al. 1996 on changes in general site factors depicted in principal coordinate analysis).

General shifts in relevés of the Canton Ticino

Comparing relevés of the Canton Ticino of two different periods - but produced at exactly the same site and locality - give results which lead to far clearer interpretations and analysis than from the sets taken in northern Switzerland (Klötzli et al. 1996).

The following differences are obvious:

1) The difference in species is larger than for the north of Switzerland. Whereas in the north the change is around 50%, it amounts to 60% and more in the south. Of the missing 50% resp. 40% only 35% or 25% resp. are replaced by species growing under the same site conditions, i.e. with similar ecological demands.
2) There are a number of plot species showing tendencies to spread (detailed evaluation in prep.). A large amount of species augmenting their abundance/dominance values often have not the same indicator values. They may show no common trends, and their movement are often chaotic, probably due to intrinsic very local changes in the site conditions or in organismic relationships.

The number of papers treating species shifts and fluctuations is growing. Most of them are based on the evaluation of permanent plots, and they come to similar conclusions, though not always in accordance with shifts in environmental influences. Considering some of the latest papers (Herben et al. 1997, van der Maarel & Sykes 1997, Wilson et al. 1996) we may draw the following conclusions:

- shifts are species-specific and they differ in their mobility rate
- there are no associations betweeen two species changing positively, but there are relationships between species augmenting & retreating simultaneously, i.e. they occupy the same ecological niche (e.g. *Deschampsia flexuosa*, *Festuca rubra*, *Nardus stricta*, but not *Anthoxanthum odoratum*); however,
- the overall number of significant results is not higher than expected on a random basis (compare below: "mobility" sensu van der Maarel & Sykes 1997)
- the number is higher when species are associated, one shifting to lower, one to higher numbers, i.e. an internal restructuring is taking place
- "mobility" is not significantly different from random expectations; the "carousel model" does not signify that species are moving evenly or that species have random mobility.
- generally mobility is quite high – around 50% in one year – expressed as an increase in "cumulated species frequency"
- trying to assess assembly rules Wilson et al. 1996 stress the fact that these restrictions on coexistence are mostly due to species interactions, but certain species (groups) combine non-randomly.

- considering the structure of plant-communities vegetation may vary discretely and not only continuously, whereby a community organises itself in a deterministic and less in a stochastic way
- a plant communities' structure and composition is therefore (to a certain extent) predictable, and random effects are rather small, a capacity that is "inherited" with each possible, partly random change in the community due to shifts in the effects of site factors or then due to intrinsic events in the ecosystem (e.g. competitive or pathogenic effects).

Taking all shifts in our forests and possible influences into general consideration the shifts are pointing to differences in average temperatures, nutrient fluxes, a drop in the pH values and also in the average light-conditions.

The ornamental species have mostly been introduced in the 17th (Isole Borromee, 1670) and 18th century (Ridolfi 1926, Schröter 1936). Since then they have shown certain attemps to "leave the gardens", but, besides developing small saplings, dying mostly after 2-3 years, have not been successful. The first true installations go back to the late seventies of this century. As will be proven there are obvious correlations with climatic factors. Presently, a group of exotic species is assembling in Southern valleys.

Changes in climatic factors

Several sets of climatic factors have been analysed, especially temperature fluctuations and extremes. Winter values were taken from new readings in small meteorological stations being on a gradient in areas with and without exotic or other thermophilous species (data provided by my colleagues and coworkers from DIONEA S.A. in Locarno, working in the same scientific group, further evaluations from meteo-data in old time series, especially extremes, by Walther 1995, M.Sc.-Thesis).

Comparing the old average data of the sixties based on the so far collected meteorological data the following shifts are clearly visible; especially since the early seventies:

1) A general rise in the number of extreme warm days (daily maxima >30°C) (Figure 1)
2) less pronounced minima and higher maxima during high winter, again especially connected to the occurrence of exotic species (see Figure 2); generally, winters are less cold (compare the data of two stations Solduno & Golino in Klötzli et al. 1996, one station (Solduno) in areas with regeneration of laurophyllous species and the other one without such shifts under less warm conditions of a shadow slope).

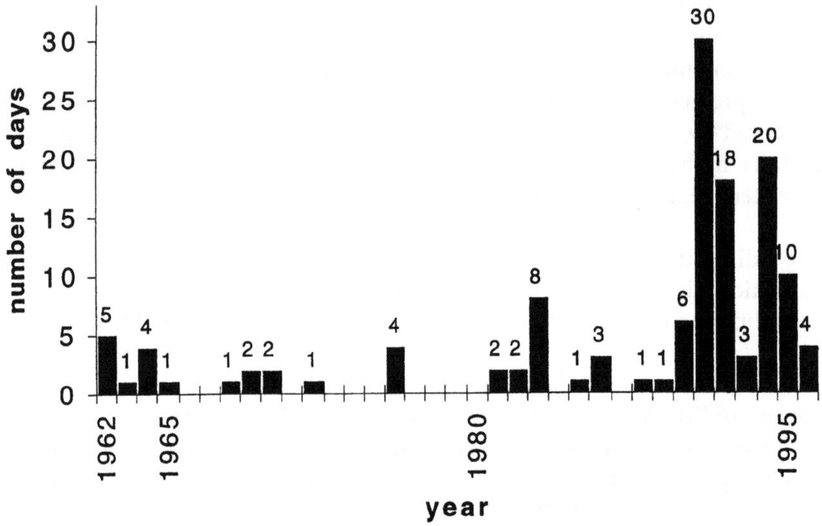

Figure 1: Number of extreme warm days (1962 - 1996)
(from Walther 1995, modified)

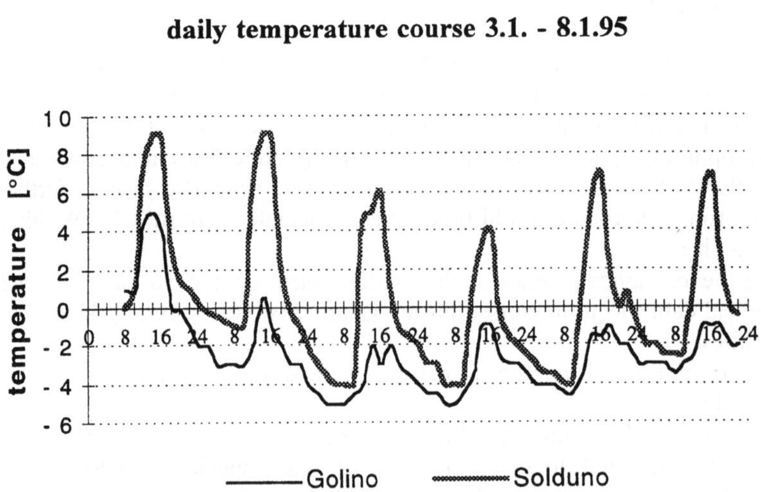

Figure 2: Daily temperature course from sites with exotic species (Solduno) and without exotic species (Golino)

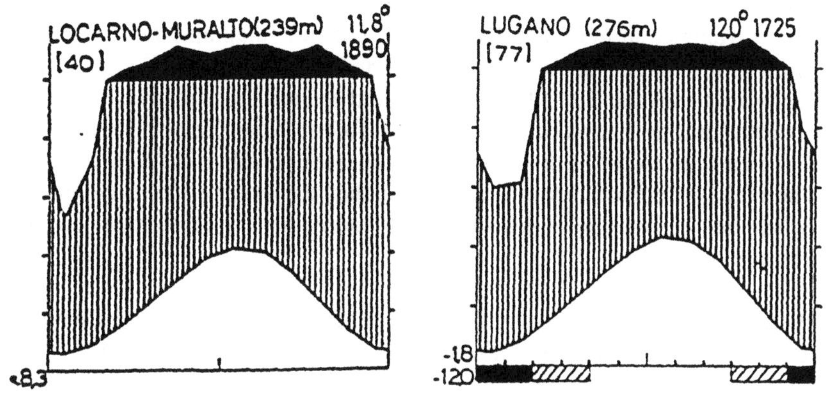

Figure 3: Climate diagrams of Locarno, Lugano (Switzerland)

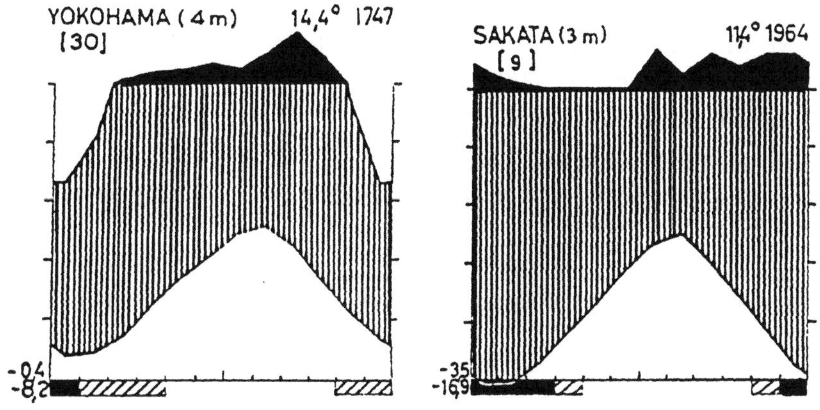

Figure 3 (cont.): Climate diagrams of Yokohama and Sakata (Japan)

Bringing the climate time series together with vegetation shifts we may conclude:

- The strong shifts in vegetation structures including the herb layer were highly triggered by the onset of mild winters and warmer (hot) summers.
- The recent climatic data (since the late seventies) lead to an obvious and clear difference in the climate diagrams. The new diagrams now resemble those from stations south of the northern limits of laurophyllous trees in Japan (e.g. Sakata, 39° NL, where the old values are already suitable to depict a limiting station of laurophyllous forest; New Zealand stations are less clear due to the lack of deciduous species) (Walter & Lieth 1960-67). The main difference lies in the level of the temperature curve (Figure 3) which has undergone a lift of half to more than a centigrade on the average.

Thus there is a twofold sponsoring of exotic species: Their saplings suffer less from low winter temperatures and take more advantage of more damp and warm summer conditions. Such combinations did not appear in the old climate data.

The position of the biome shift

Shifts are only known from the most thermophilous mesic areas on south-facing slopes around the insubrian lakes, especially near the shore near the villages of e.g. Gandria, Melide, Morcote (Lake Lugano) and Brissago up to Locarno (Lake Major) (see Figure 4). Last outliers may be found in areas north of Bellinzona. According to Walther (1995) the following sets may be distinguished:

A) areas with subspontaneously occurring exotic trees
(e.g. *Cinnamomum glanduliferum*, *Laurus nobilis*, *Ligustrum lucidum*)

B) areas with dominant exotic shrubs and lower trees
(e.g. *Eleagnus pungens*, *Prunus laurocerasus* etc.)

C) areas with some exotic shrubs

D) areas with indigenous evergreen shrubs
(e.g. *Ruscus aculeatus*, *Ilex aquifolium*, *Hedera helix* (climbing on trees) and *Buxus sempervirens*)

Their positions are mostly not very far away from the lakes, reaching up to 600m on southern aspects, and 400m on northerly aspects (valley bottom near 200ma.s.l.).

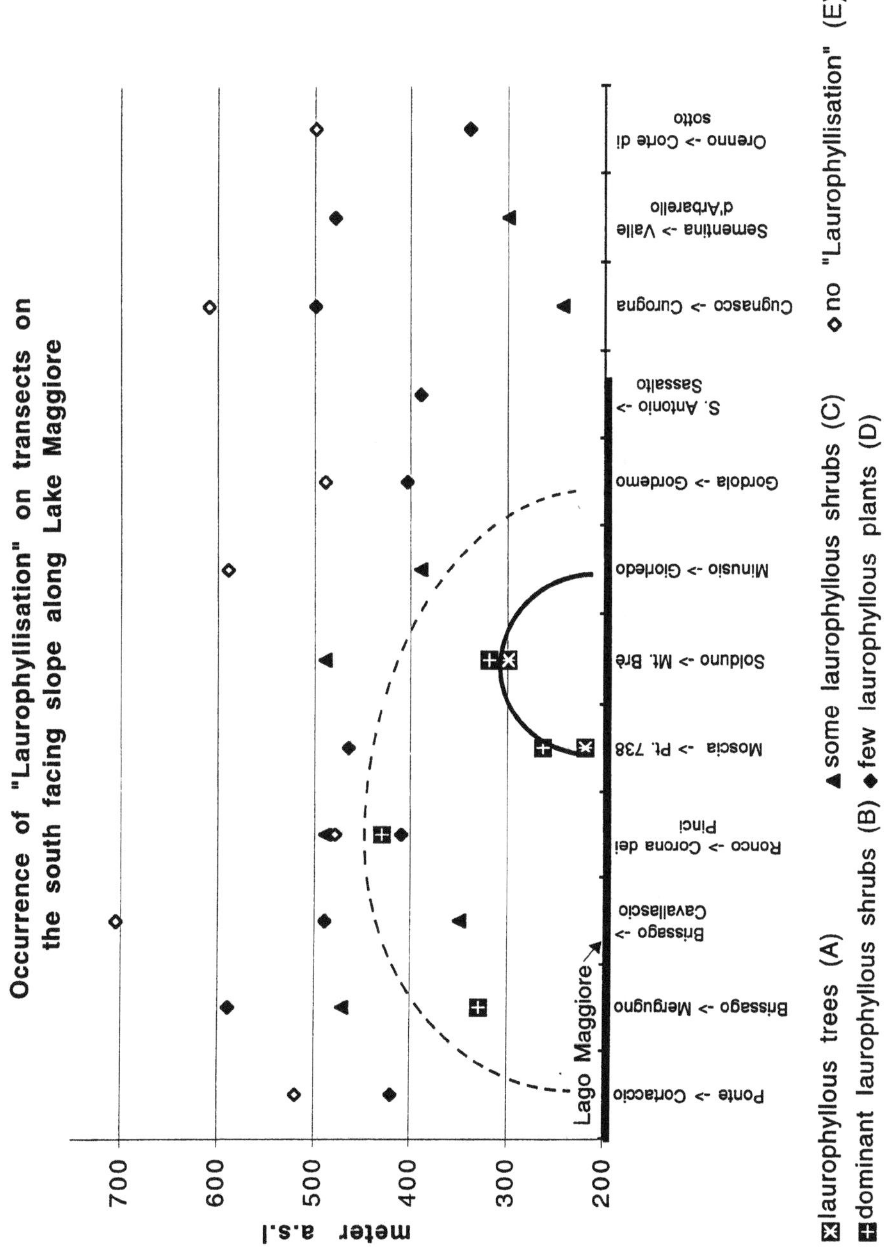

Figure 4: Occurrence of laurophyllisation on the south-facing slope above Lake Major

Table 1: Sensitivity of evergreen plants to low temperature

Frostdamaged leaves expected with temperatures below:	Plantspecies		Foliage	Damages occurred in Arco in harsh winters 1955/56 and 1962/63
-2°to-4°C	*Passiflora* spp. *Bougainvillea spectabilis* *Eucalyptus globulus*	● ● ●	* ▣ ■	completely killed
-5°to-6°C	*Cycas revoluta* *Phoenix dactylifera, Ph. jubaea* *Washingtonia filifera* *Cinnamomum camphora* *Citrus aurantium* *Acacia dealbata* *Eucalyptus viminalis*	● ● ● ● ● ⦿ ●	* ■	foliage completely killed by frost, moderate wood-damage. Plants recovered
-7°to-8°C	*Jubaea chilensis* *Chamaerops humilis* *Quercus suber* *Cinnamomum glanduliferum* *Myrtus communis* *Ligustrum japonicum,* *Ligustrum lucidum* *Nerium oleander*	● ● ● ⦿ ● ● ● ●	▣ ■	frostdamaged leaves, bulbs and sproutaxes healthy. In Spring normal sprouting
-9°to-10°C	*Laurus nobilis* *Pistacia lentiscus* *Olea europaea* *Rosmarinus officinalis*	⦿ ● ⦿ ●	▣ ■	1/10 to 1/3 of leafmass killed by frost
-11°to-12°C	*Araucaria araucana* *Pinus pinea* *Trachycarpus fortunei* *Magnolia grandiflora*	● ● ⦿ ●	▣ ■	less than 1/10 of leafmass damaged

Frostdamaged leaves expected with temperatures below:	Plantspecies		Foliage	Damages occurred in Arco in harsh winters 1955/56 and 1962/63
-12°to-15°C	*Cupressus sempervirens* *Quercus ilex* *Pittosporum tobira*	● ● ◉	□ ■	No visible frostdamage
< -15°C	*Sequoia gigantea* *Cryptomeria japonica* *Taxus baccata* *Prunus laurocerasus* *Buxus sempervirens* *Ilex aquifolium*	● ● ◉ ◉ ◉ ◉	■	No visible frostdamage

Legend:

- * partly subtropical plants
- ■ laurophyllous or then mostly in the area of the biome of the laurophyllous forests
- □ sclerophyllous or then mostly in the area of the biome of the sclerophyllous forests
- ● often as ornamental trees or shrubs in Insubria
- ◉ partly regenerating naturally or partly escaped from gardens

Adaptations of exotic species to lower winter temperatures

Although many exotic species may subsist at temperatures as low as -10°C (or even lower) (see Table 1) for one or two nights (compare also Larcher 1978), they were not able to spread before in these forests, not even the most hardy ones, because their saplings are more sensitive to low winter-temperatures or late frosts (similarly for a number of more hardy species in the area of Zürich).

The reason may also be given by the general change in winter- and summer-temperatures, saplings being also sponsored by the high humidity in warmer summers, getting a better chance than under conditions of drought.

Discussion of the actual development in Insubria

To get an idea on the weight of those shifts it is best to compare climatic conditions where those shifts depicting "laurophyllisation" are going on directly with the position of laurineous forests in e.g. the classical countries Japan and China.

A clear insight is provided by the (very schematic) diagram by Stocker (1963), modified by Klötzli (1988), compare also Walter & Breckle (1986). The position of the insubrian climate (see Figure 5) is then very near to the domaine of the laurineous forests, especially the northwestern japanese one (e.g. Sakata). But the moment the average meteo data of the last 20 years are taken into account the data of the insubrian stations move into the cooler part of the japanese stations in the laurineous area.

Thus it is not really astonishing that such shifts had to appear, especially under the given double positive effects of milder winters and warm damp summers. Why slopes above 600m a.s.l. are not touched is also obvious: If an average difference of 0.6°C/100m is assumed then the difference to the valley bottom would amount to 4 times 0,6°C, i.e. ca. 2.5°C, just a difference that keeps the station out of the area of laurineous forests.

Furthermore, the exotic species are less stressed than in the older days, fires being more under control, real drought periods (passing 2-3 weeeks during the growing season) and harsh winter-temperatures being much less frequent or absent.

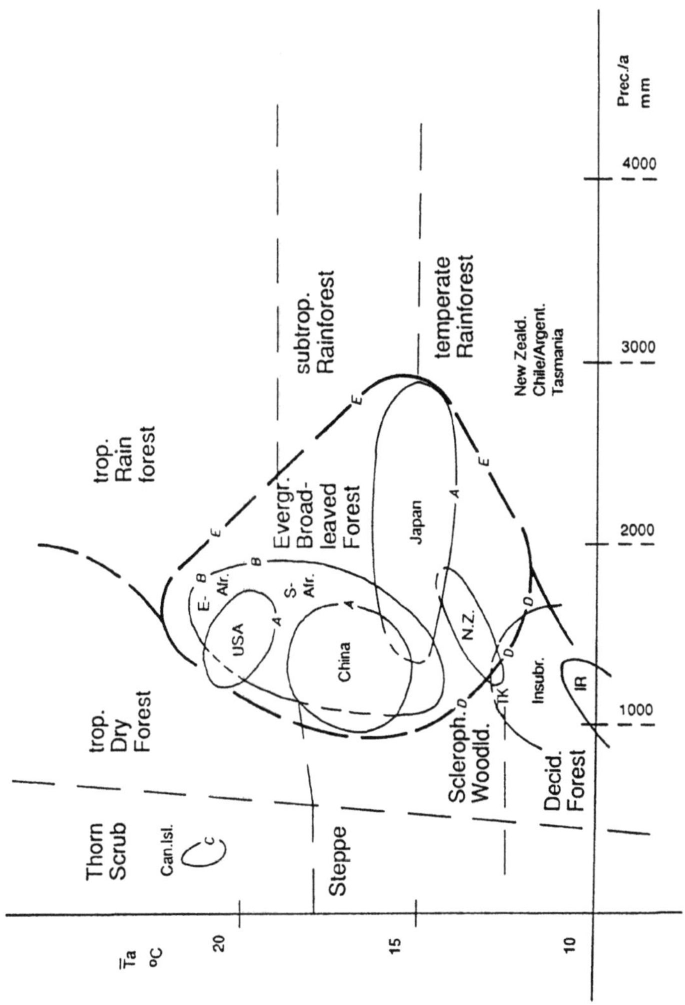

Figure 5: The position of the evergreen broad-leaved forest

Also the old exploitation intensity in these formerly coppiced forests is less marked and gardens with exotic species have spread into hardly inhabited areas from where e.g. Blackbirds transport mostly dark-blue berries into the surrounding forests. Thus „vicinism“ has also changed for these exotic species and initiating the arrival of new isles of laurophyllous undergrowth (compare e.g. Zonneveld, 1995, van der Maarel, 1996, van der Maarel & Sykes, 1993).
In some areas thick layers of shrubs have built up under *Quercus*, *Castanea* or *Ostrya*, e.g. with *Eleagnus pungens* (set B). In the warmest areas a considerable amount of laurineous trees have reached the tree layer (*Laurus nobilis*, *Cinnamomum glanduliferum*, *Prunus laurocerasus*, *Ligustrum lucidum*). Also *Trachycarpus* is already moving from shrub to tree layer. Other adjacent but cooler areas have less visible signs of "laurophyllisation“ (i.e. the spreading of laurophyllous species). Thus different degrees of pressure of exotic species in different layers of deciduous forests occur simultaneously on a given thermophilous slope: Slope quality is governing the degree of laurophyllisation. The highest degree is clearly connected to the warmest spot of Switzerland, i.e. the Brissago Islands, where also *Pittosporum tobira* and *Acacia dealbata* is spreading spontaneously. The neighbouring slopes have the highest density of laurineous trees in the upper tree layers and a compulsory regeneration of young palms (*Trachycarpus fortunei*, more rare *Phoenix canariensis*; compare also Walther 1995).

Today the situation is such, that trends are pointing to a gradual change in forest structure, allowing only few exotics in the upper tree layer, more in the lower tree layer, thickets of exotics in the shrub layer and a remarkable regeneration of saplings (incl. palms) in the shrub layer. But so far indigenous evergreen herbs have not suffered by competition.
In the end a forest will build up not quite in the way of a true laurineous forest but quite similar to the ecotonal types between the laurineous and deciduous types on an elevational gradient (e.g. gradients in the Japanese Alps (Miyawaki 1984), or the Chilean Andes near Puerto Montt (Oberdorfer 1960)). Deciduous Fagaceae will remain dominant. Should temperature rise higher than expected (average $T_a > 18°C$, with mild winters) then the way is open to an evergreen upper tree layer with laurineous Lauraceae and eventually Fagaceae.
I suggest to consider these facts when modelling possible shifts in all parts of Swiss vegetation due to climatic change (see e.g. Kräuchi 1994, Bugmann 1994). Then the new environmental situation we are analysing will give a new picture to the Swiss forests.

Acknowlegdements

Without the cooperation of our collegues from DIONEA S.A. in Locarno, Gabriele Carraro & Giuseppe Gianoni, both dipl. ing. forest. SPF, and André Grundmann, dipl. natw. ETH, the extent of this work would not have been possible. Also PD Dr. Otto Wildi gave his advice in statistical problems. Professor Valerio Sala, director of the Botanical Garden on the Brissago Islands enabled us to work on these islands. To all of them we owe our heartfelt Thanks.

Field-work was initiated by a grant from the National Research Programme (NRP 31, Climate Changes and Natural Disasters), carried by the Swiss Fund for Scientific Research. We thank the responsibles for backing our project.

References

Bugmann H.K.M. (1994): On the Ecology of Mountainous Forests in a Changing Climate: A Simulation Study. Diss. ETHZ: 258p.

Carraro G., Gianoni P., Klötzli F., Mossi R. & Walther G.-R. (1998): Observed changes in vegetation in relation to climate warming. Final report NRP 31, vdf Zürich, in print.

Gianoni G., Carraro G. & Klötzli F. (1988): Thermophile, an laurophyllen Pflanzenarten reiche Waldgesellschaften im hyperinsubrischen Seenbereich des Tessins. Ber. Geobot. Inst. ETH, Stiftung Rübel, Zürich 54:164-180.

Herben T., Krahulec F., Hadincová V., Pecháckova S. & Kovárová M. (1997): Fine-scale spatio-temporal patterns in a mountain grassland: do species replace each other in a regular fashion? J. Veg. Sci. 8: 217-224.

Klötzli F. (1988): On the global position of the evergreen broad-leaved non-ombrophilous forest in the subtropical and temperate zones. - Veröff. Geobot. Inst. ETH, Stiftung Rübel, Zürich 98: 169-196.

Klötzli F. (1995): Projected and chaotic changes in forest and grassland plant communities. Preliminary notes and theses. Annali di Botanica 53: 225-231

Klötzli F., Walther G.R., Carraro G., Grundmann, A. (1996): Anlaufender Biomwandel in Insubrien. Verh. Ges. Ökol., 26: 537-550.

Kräuchi N. (1994): Modelling forest succession as influenced by a changing environment. Mitt. Eidg. Forschungsanstalt WSL 69(2): 143-271.

Larcher W. (1978): Klima und Pflanzenwelt in Arco. Kulturelle Veröffentlichungen der Kurverwaltung Arco (Trento : Italien), 163p.

Miyawaki A. (1984): International Excursion & Symposium on Vegetation Science. Handbook for the International Excursion August 1-16, 1984, Japan. Dep. Vegetation Science, Inst. Environmental and Technology, Yokohama Nat. Univ. 208p.

Oberdorfer E. (1960): Pflanzensoziologische Studien in Chile. In: Tüxen R. (ed.) Flora et Vegetatio Mundi, Vol. II, 208p.

Ridolfi R. (1926): Alberi ed arbusti del parco civico e dei giardini pubblici di Lugano. Tipografia Luganese - Sanvito & C, Lugano, 71p.

Schröter C. (1936): Flora des Südens. Rascher, Zürich, 151p.

Stocker O. (1963): Das dreidimensionale Schema der Vegetationsverteilung auf der Erde. Ber. dtsch. bot. Ges. 76: 168-178.

Van der Maarel E. & Sykes M.T. (1993): Small-scale plant species turnover in a limestone grassland: the carousel model and some comments on the niche concept. J. Veg. Sci. 4: 179-188.

Van der Maarel E. (1995): Vicinism and mass effect in a historical perspective. J. Veg. Sci. 6: 445-446.

Van der Maarel E. & Sykes M.T. (1997): Rates of small-scale species mobility in alvar limestone grassland. J. Veg. Sci. 8: 199-208.

Walter H. & Breckle S.-W. (1986): Ökologie der Erde (Band 3). Spezielle Ökologie der gemässigten und arktischen Zonen Euro-Nordasiens (Zonobiom VI-IX). Gustav Fischer, Stuttgart. 587p.

Walter H. & Lieth H. (1960-1967): Klimadiagramm-Weltatlas. Gustav Fischer, Jena.

Walther G.-R. (1995): Ausbreitung und Grenzen laurophyller Arten im Südtessin. M. Sc.-Thesis, ETH Zürich: 37p. + Annex (Polycopy)

Wilson J.B., Ullmann I. & Bannister P. (1996): Do species assemblages ever recur? J. Ecol. 84: 471-474.

Zonneveld I.S. (1995): Vicinism and mass effect. J. Veg. Sci. 5: 441-444.

Plenary discussion

Question: H. BRUELHEIDE

My question refers to the climatic data of the weather station in Zurich. Is the weather station located inside the city and could the observed warming effect be due to intensified urbanisation? Could the increase in the city's area have caused the warming? According to urban ecological studies, increases in heating and traffic and reductions in evaporating surfaces can contribute to an increase in annual temperature from 0.5 to 1.5 °C (von Stülpnagel A., Horbert M., Sukopp H. 1990: The importance of vegetation for the urban climate. In: Sukopp H. (ed.): Urban ecology: 175-193. Academic Publ., The Hague). Is there any evidence that the warming effect also occurs in small towns without an urban climate?

Answer: F. KLÖTZLI

The climate station of Zurich is located on the edge towards the forested hill sites. But considering the number of climatic stations evaluated in the report of the

Swiss Meteorological Institute of about 300 there is over all an increase of 0.1°C in the sixties, 0.4°C in the seventies and 1.5°C in the eighties. That's the medium value for all the stations, not only Zurich. Zurich is not too bad in a way as its meteo station is not really in town, it is in a large garden on Zurichberg. It is under more or less forest edge conditions and not actually down town near the lake where the difference is higher of course. If you are experiencing real damp conditions you can experience it near the lake in the city. Going up the hill you feel how the temperature and the dampness is also changing.

Recent shifts in vegetation boundaries of deciduous forests,
especially due to general global warming, pp. 31–51
edited by F. KLÖTZLI and G.-R. WALTHER

Climatic influence on vegetation changes: a verification on regional scale of the Laurophyllisation

Gabriele CARRARO, Pippo GIANONI and Roberto MOSSI
DIONEA S.A., CH-6600 Locarno, Switzerland

Abstract

The authors analyse the changes in forest vegetation in the north and south of the Alps by comparing 130 vegetation relevés of the sixties with more recent relevés of the nineties.

The resulting changes in species compositions and vegetation structure can be explained with various factors one of which is the milder winter climate of the last three decades. A verification of the vegetation changes on regional scale shows that the milder winters have a strong effect.

With this example, the connection between similar bioclimates and the concept of the "chorological tension", some prognoses of the laurophyllisation in western Eurasia and their consequence for the practical implementation and for scientific conclusions are explained.

Preface

The investigation of a probable reference value of thermophilous forest structures, especially focussing on potential climatic changes, is subject of this project.

The research project is mainly based on three different topics:

- the comparison of old relevés from 1940/60 with the present situation on the northern and the southern side of the Alps
- the verification of the phenomenon on a regional and local scale (Locarnese)
- the comparison with similar ecosystems and the evaluation of literature

The comparison of typical climatic data of specific similar, just slightly warmer bioclimates with the Centraleuropean data - as presented by Klötzli (1988) -

evidently shows the importance of the mean temperature and the winter temperature as limiting factors for some vegetation types.
On both sides of the Alps, the northern side and Insubria, a fair amount of precipitation is guaranteed even in summer. Therefore, the influence of the precipitation is rather limited and can be neglected.
The topography of the surface has probably a higher influence (depression - ridge) in this context. The analyses were concentrated on zonal sites, generally avoiding the driest sites and the wettest depressions.

Changes that occurred in recent decades

It is well known how forest ecosystems have changed rapidly according to the type of management on the northern side of the Alps as well as on the southern side.
Especially during and shortly after the world wars forests were very intensively harvested. This resulted in the following characteristics which did change significantly later on:

- reduced forest surface, clearing widely distributed
- management of "Selva" or so called "palina" (= coppice system) (South of the Alps), management of so called "Mittelwald" (= coppice system with old trees) (North of the Alps) in favour of few particular species (*Castanea*, *Quercus*, *Carpinus*)
- intense harvesting of fruits and seeds
- regular gathering of litter and parts of the humus-layer
- low pressure by game grazing (roe deer, red deer, etc.)
- commonly practised bird-catching (blackbird)
- dispersed pasturing with small cattle (goat) and large cattle
- less frequent forest fires with much larger burned surface (Conedera et al. 1996)
- small and compact village structures, small garden areas, not directly adjacent to forest areas
- small input of nutrients by air pollution

The radical change, if not a complete overthrow of all of these phenomena, means a strong modification of factors which determine the composition of the vegetation.
In fact, the last 40 years represent the period with the most extensive management in forest ecosystems within the last four centuries. In addition, new conflicts like air pollution, game density, etc. threaten these ecosystems.
Although various authors of the past, whose relevés have been used, choose adult forests with a closed canopy and little human influence, it is obvious that relevés from the forties and sixties mostly show the influence of traditional

management. This is especially expressed by authors who choose areas with a high species diversity intending to show representative examples.
Despite the criteria for the selection of the relevés defined by the authors (zones close to the climax stage, little disturbance, none or extensive forest management) the forest history of the single stand cannot be reconstructed as precisely as by other authors (Kissling et al. 1988), since generally a diffuse influence of all factors must be presumed.

Evident climatic changes

Apart from the variation of the aforementioned factors only the climate changed slightly within the last 20-30 years. Those variations are small and could be explained by climatic cycles, such as the effect of climate warming around expanding agglomerations, or they could be linked to the global warming of the atmosphere.
The research on eventually modified climatic parameters is focused on factors that seem to limit the distribution of thermophilous species on the colline belt, like minimum-temperatures, extreme-temperatures and days with frost.
For various plant species one relevant factor may be the number of days with frost, which means days with temperatures below 0°C. Such temperatures lead to a change in the physical state of water which affects plant physiology (Larcher 1995; Klötzli et al. 1996).
Since the beginning of this century, the number of days with frost decreased in numerous meteorological stations (in urban and rural areas) all over Switzerland, if not on a global scale (Lauscher 1981).
In a typical meteorological station with Centraleuropean suboceanic climate like Zurich (yearly mean temperature 8.5°C, mean temperature in July 17.6°C, mean temperature in January -1.1°C, precipitation 1090mm/year) 130 to 60 days with frost were recorded between the beginning of this century and 1970. From 1970 onwards, the number of days with frost has oscillated between 100 and 50 days per year.
In this context a more spectacular variation - probably favoured by climate modification due to urbanization - is provided by the typical western Insubrian meteorological station of Lugano (annual mean temperature 11.7°C, mean temperature in July 21.4°C, mean temperature in January -1.9°C, annual precipitation 1750mm/year). Here, the mean value of days with frost from the beginning of this century until the end of the sixties dropped from 64.1 to 30.7 per year (mean values 1968-1993) representing a net reduction of more than 50%.
Another interesting fact in this context is a modification of extreme values; both, the colder and the warmer years show an increasing trend towards warmer conditions.

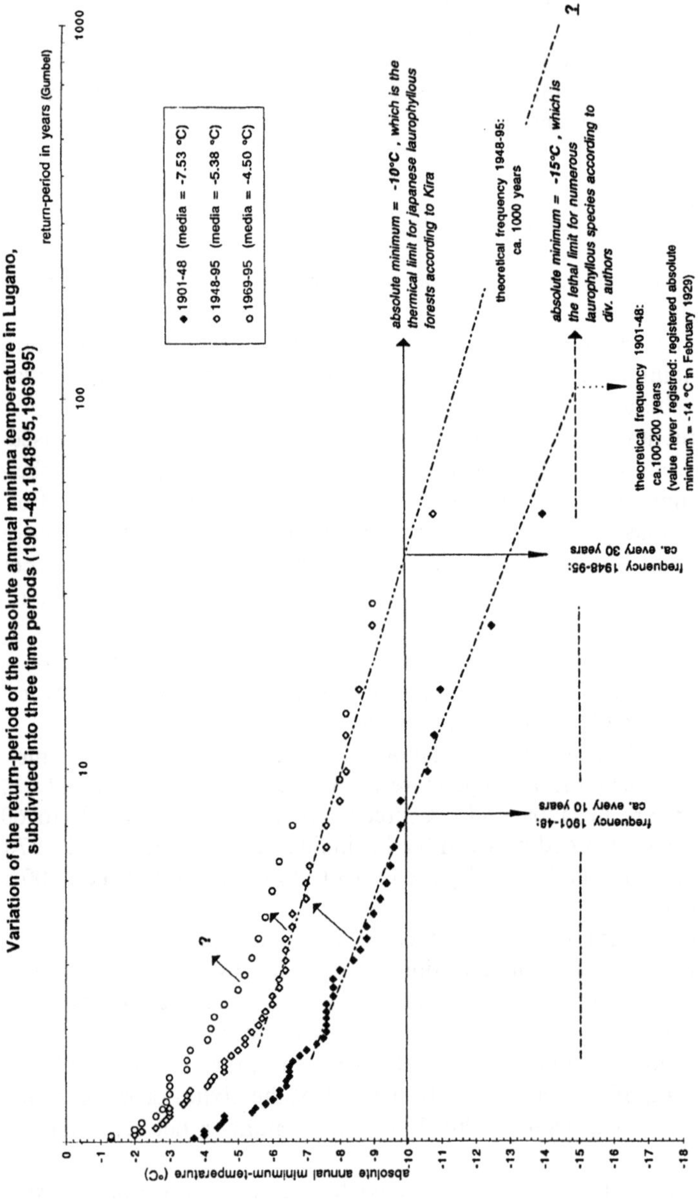

Figure 1: Variation of the return-period of the absolute annual minima temperature in Lugano

The variation of the mean temperatures of the coldest month and of the absolute minimum-temperatures may be better understood if they are analysed like other extreme events in nature.
Hydrology for example considers extreme events usually in connection with their probability of occurrence, calculated according to the law of return-periods (in Figure 1 expressed in logarithmic scale).
The Gumbel-rule shows that the return-period of the monthly mean values in January and the absolute minimum-temperatures have substantially changed during the second half of this century.
This means, that the occurrence of extreme events, lethal for sensitive species and limiting for some plant communities, are presently less likely though not impossible to occur (see Kira 1977, Walder 1985, Klötzli et al. 1996). During the last 20 to 30 years this tendency seemed even more pronounced; numerous other stations controlled in the same aspect show an analogous trend.
Temperatures of the coldest month of the year have risen in the second half of this century by 0.5 – 2°C on average and the absolute mean temperature has increased by 1.5 – 2.5°C.
Summer temperatures and days with maxima exceeding 30°C have indicated a slight increase in return rate (see Klötzli et al. 1996) however, these aspects need to be verified.

Modification in forest vegetation

Other research projects in which old and new relevés were compared, have shown that the following evidences are typical for adult forest populations (Kissling et al. 1988; Kissling 1989; Kuhn et al. 1987, 1990; Werdenberg & Hainard 1989):

- the number of species per relevé and the total species number was reduced by increased competition: this process of a simplification of the biocoenosis has occurred especially in populations that were influenced by man and therefore artificially enriched with species
- a process of recuperation of soil-fertility and of primitive vegetation-conditions is indicated
- the forest vegetation is enriched by indicators of nitrification with possible link to the atmospheric input (air pollution)

On the southern side of the Alps, where the cultivation of an exotic species (*Castanea*) has a long tradition, 40% of the species were persistent, though in different proportions, 60% disappeared and were replaced by ca. 30%, resulting in a netto loss of 30%.
Kuhn (1990) and Werdenberg & Hainard (1989) explain this reduction with increased competition in populations that used to be more enlighted and open

and with the reconstitution of the soil fertility in previously overharvested areas.
An aerial nitrogen input is probabely the reason for these changes. However, definite prove is still missing. This factor plays a dominant role by the natural recuperation of the fertility of forest-soils.
For the comprehension of coherent changes in the modification of the species, principal coordinate analysis according to the statistical tools by Wildi & Orloci (1996) has been used.
The above results concern the 176 relevés of the northern side of the Alps and show single points representing one relevé each. They move coherently in space in the same direction. In the South of the Alps the same shift could be observed.
The principal coordinate analysis indicates a general trend. For its specification it is necessary to proceed comparing the presence of species, directly taking the ecological attitude into account and considering, whether they are part of the climax forest communities of the region investigated.
The individual species have been divided into two categories:

- persistent species (+/– 10% presence) or species increasing in number
- species decreasing in occurrence (with variance in presence > -10%)

This subdivision slightly favours species with a decreasing tendency. It is justified by the fact that the analysis is restricted to mature forest populations and that the species that endure a light diminution can be preserved in the forest population and thus confirm their particular strategy.
In contrast, species with remarkable reduction must be regarded as lost.
With the botanical knowledge, the ecological values (Landolt 1977), the sociobotanical valence (Oberdorfer 1964) and the vegetation studies, the category of persistent or increasing species has been subdivived into the four following subcategories and differenciated into a northern and southern part of the Alps:

(1) laurophylloid thermophilous species (with persistent and plain leaves, Gianoni et al. 1988)
(2) non laurophyllous neophyte species
(3) species belonging to the climatic climax and/or being nitrophilous (based on observations carried out in populations of mature reference forests, occurring in analogous positions)
(4) non nitrophilous climax forest species

Symmetrically to the previous group the species with clear diminution in presence have been subdivided into the following categories:

(5) typical species of the submontane and montane belt
(6) mesophilous species, shade tolerant species and species that are characteristic for smaller forest gaps
(7) species occurring on nutrient poor soils, eventually being heliophilous
(8) heliophilous species, eventually occurring on nutrient poor soils

Adding importance to one species one may take into account its presence in general in the whole series of the relevés, the wideness of the variance between the variance of the median of cover and the hierarchical variance of the various layers, and the importance of other known factors (forest economies, parasites). This empirical method has the advantage of permitting an approximate interpretation of different phenomena in the forest which can be synthesized as follows:

- the distinct reduction of the mean number of species per relevé (-15% on the northern side of the Alps, -30% on the southern side) and the mean cover value of the single species
- general and marked increase of species which constitute the synusia of laurophyllous species of group 1 on the southern side of the Alps: *Hedera helix*, *Prunus laurocerasus*, *Ilex aquifolium* (incl. *I. a.* ssp. *plathyphylloides* – Christ 1903), *Taxus baccata*, *Trachycarpus fortunei*, *Laurus nobilis*, *Lonicera japonica*, *Asplenium adiantum nigrum* (s.l. incl. *A. onopteris*) is defined as "Laurophyllisation" (Klötzli et al., 1996). (The same phenomenon occurs, to a minor extent, on the northern side of the Alps with *Hedera helix*, *Ilex aquifolium*, *Taxus baccata*, *Cotoneaster horizontalis* and others and - rarely - *Prunus laurocerasus*.).
- the increase of summergreen neophytes (group 2) particularly on the southern side of the Alps, is less marked on the northern side: *Robinia pseudoacacia*, *Ailanthus altissima*, *Prunus serotina*, *Impatiens parviflora*: under particular conditions even mature forests are not immun to the penetration by exotic species
- the consolidation of numerous nitrophilous species typical for closed forest-consortia (group 3) on both the southern and the northern side of the Alps (*Acer*, *Rubus*, *Lamium*, *Geum*, *Carex pendula*, etc.), apart from the consolidation or the redemption of species typical for forest-climax (group 4: *Fagus*, *Luzula*, *Calamagrostis*, etc.) some of the species are typical for areas with mild climate (*Tamus communis*, *Evonymus latifolia*)
- the retreat of some species characteristic for a fresher climate or higher vegetation belts (group 5) which occur on the northern side of the Alps (*Abies alba*, *Sambucus racemosa*, *Lonicera alpigena*, *Galium rotundifolium*, *Knautia silvatica*, *Veronica latifolia*) and on the southern side of the Alps (*Lonicera xylosteum*, *Rhododendron ferrugineum*, *Saxifraga cuneifolia*, *Astrantia minor*, *Valeriana tripteris*)
- a strong reduction of the remaining groups 6, 7, 8 caused the major change in the principal coordinate analysis

In a general, simplified picture only few new species succeeded with installation, representing at least one of the following strategies: shadowtolerance, nitrophily, thermophily.

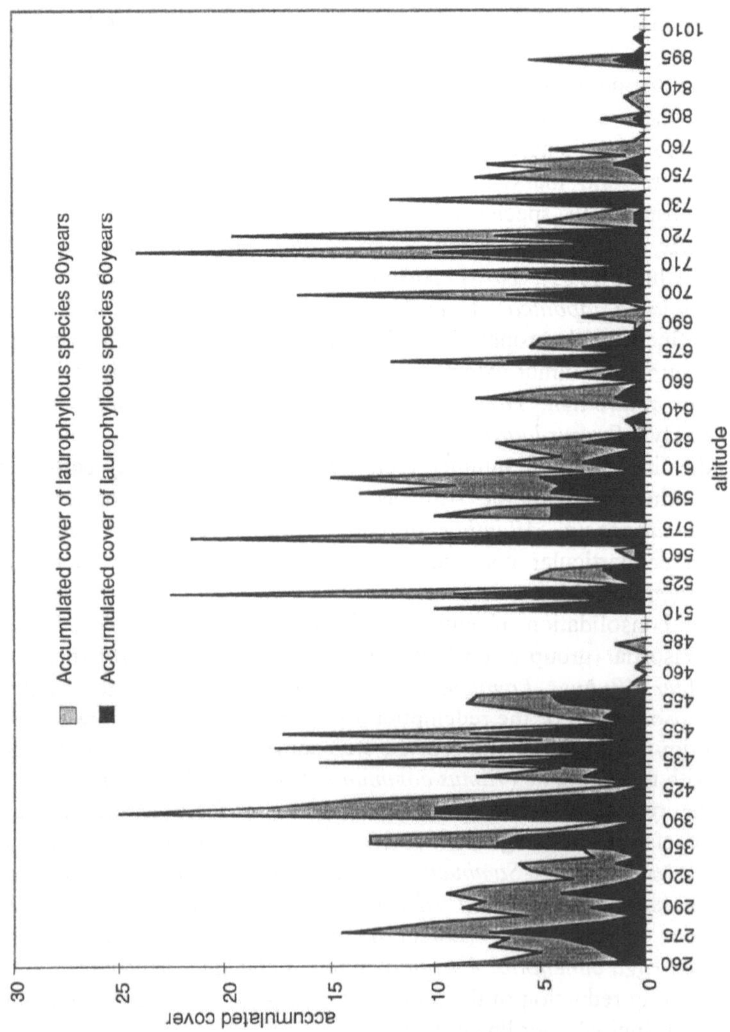

Figure 2: Cover variation of laurineous species on the southern side of the Alps (*Hedera*, *Ilex*, *Daphne laureola*, *Ruscus*, *Vinca*, *Asplenium adiantum-nigrum*, *Laurus*, *Ligustrum lucidum*, *Cotoneaster*, *Trachycarpus*, *Lonicera japonica*, *Prunus laurocerasus*, *Viburnum*)

Forest soils seem to become more fertile and appear evidently less eluminated. Apart from the climaxspecies and the nitrophilous species all thermophilous elements, that are able to survive in mature, little eluminated forest environment, and to conduct photosynthesis during the winter period have been confirmed to occur or to increase. Not a single exception on both the southern and the northern side of the Alps up to the limit of the colline belt (group 1 = exotic laurophyllous species) occurred, no matter how far from the settlements. However, the reduction of group 5 (species occurring in colder climate) is less marked, which could indicate a vegetation shift towards higher belts so that in this case the phenomenon shows characteristics of a change due to the climate.

The exotic species of group 1 and group 2 swarm out from the gardens and the settlements establishing within the structure of populations together with native species of the same type within the first 200-300 meters.
This belt probably corresponds with areas in which the activity of Blackbirds (*Turdus merula*) is most intense and efficent. Based on personal observations confirmed by literature (Glutz v. Blotzheim 1988) this species can be considered the principal vector for the dispersal of laurophyllous species.

The same species is active and sometimes dominant in the closest laurisilvae (Macaronesia) and in the ecotonal area of summergreen to evergreen forests in the Far-East, where a major dependence for zoochory phenomena with respect to the heart of the biome has been noted.

Interpretation of data

The changes presented in the previous chapter including the limitations mentioned above, may be regarded as a matter of fact, while the explanation of the diverse phenomena includes some suggestive factors.
The process of the interpretation of data, which is part of this chapter, has been undertaken without simplification and considering many different causal correlations of the complex system.
Nevertheless, some significant tendencies can be shown, in particular with data of the formation of species groups which respond to specific fundamental strategies: the definition of such groups is on one hand based on the existent phytosociological knowledge, considering distribution and growth of a species on a regional scale, and on the other hand on longterm observations in nature.

Analogical bioclimates

Climax vegetation represents an attraction-pole in the evolution of forest populations. The Centraleuropean vegetation-complex has been strongly influenced by man since the Neolithicum, therefore, it is far away from the climaticclimax.
Due to the last glaciations, the investigated vegetation is - in phylogenetical terms - very young and therefore little consolidated and less stable as expected. The most frequent explanation consists of the particular chorology in Central-Europe, of the isolated position of European summergreen forests with respect to other types of thermo-hygrophilous vegetation, due to the separation from the Mediterranean and the Sahara. In contrast, the glaciation combined with the damming effect of the mountain chains should have had a smaller effect than the biogeographical isolation of Europe. (Meusel et al. 1965, 1978).
From the comparison of relevés a functional need is apparent: a layer of evergreen plant species that are able to photosynthesize throughout the year.
In this context it is useful to compare the climaxvegetation of an analoguous bioclimate as it exists in Basque Provinces and in Japan in more consolidated situations from a more chorological point of view.
The potential vegetation of Insubria is to a large extent ascribable to an *Ilex-Fagus*-forest, which has not been described so far. In the Basque Provinces adjacent to France natural and semi-natural forest populations rich in laurophyllous species were discovered.
The parallelism of the structure of mixed broad-leaved forest with laurophyllous species can help to understand the phenomena observed.
This is even more obvious if diverse parameters and indices for the climate defining the zone of laurophyllous species are taken into account.
Insubrian stations are situated at the border to, or in the climatic zone, in which climax forests are either of the Laurisilvae-type or summergreen with laurophyllous species.

Climatic influence: a verification on regional scale

Is there a correlation between the dissemination of these species and the recent mildness of the climate, particularly in the winter season?
Is it possible to explain the above mentioned spreading solely by returning to the postulates mentioned above (recolonisation of a more mature forest state closer to the climaxby a typical synusia; presence of hemeroby effects) or is there a fundamental contribution by climatic variation to the better vitality of laurophyllous species in the seventies and eighties?
To answer these questions the distribution of those elements in relation to the variation of the climate on a regional scale was investigated more closely. The

aim was to compare situations, that were identical in all aspects apart from the climatic aspect.
Analyzing various sites in the Locarnese with identical or at least very similar zonal conditions were found, with respect to:

- the forest association (Luzulo-Carpinion; Querco-Fraxinetum, EK 34-33)
- the type of tree coverage (mature "palina" of *Castanea*)
- the altitude and the inclination
- the absence of fires in recent decades
- the distance to the nearest settlements is included in the circle of expansion starting from the gardens (50-200m)

All the stations presented mesoclimate, topographical conditions and a picture of homogenous vegetation for a circle of at least 50 - 150m.
This comparison combined with the empirical findings of Walther (1995) confirm the presence of a thermical altitudinal gradient with respect to the vitality of laurophyllous species (indicated limit in south exposition 600m a.s.l., 400m a.s.l. on north-exposed slopes). The analysis carried out in the investigation area during winter 1994/95 – in addition to phytosociological informations from the past and the present situation – help to understand to what extent the spread and the vitality of these species are correlated with climatic winter conditions, which seem to be crucial in this context.
In particular, a difference in the number of days with frost of about 50%, a 1-3°C higher mean temperature in January, the absolute minimum temperature with the difference of 2-4°C clearly correlate with the vitality of laurophyllous species in the natural environment.
The variation of these data corresponds with the general climate variation that occurred in recent decades as presented in the chapter "evident climatic changes".
In this sense the following statements can be made:

- the mildness of the winters in recent decades have contributed to the better vitality and distribution of these species
- as a consequence, under the investigated conditions, the synusia of laurophyllous species can be defined as thermosensitive

Laurophyllisation in western Eurasia

The spontaneous Laurisilvae closest to the European continent are limited to few areas in Macaronesia. Other Laurisilvae of interest occur in the Far-East.
The sequences of vegetation presented in Insubria are similar to those of the colline belt of the Pyrenean-Cantabrian area with (hyper-)humid climate, which is comparable to the hyperinsubrian climate. The sequences are considered an extremely indicative parallelism.

Analogically, the Caucasian-Hyrcanic complex can be included in possible prognoses and considerations about evolutionary tendencies of the vegetation based on an interregional comparison of ecological strategies of the various species groups (synusia), considering both, the present climate and the projection of a scenario with warmer climate.

In this sense the following predictions should be verified:

- Due to structural and climatic analogies with the Insubrian area, the phenomenon of the spread of laurophyllous species is predicted for European areas with oceanic climate and sufficient summer precipitation, i.e. July mean temperature > 18/20°C, absolute minima > -10/-15°C, January mean temperature > 0/2°C, summer mean precipitation (June, July, August) >250/350mm.
  Included are parts of Istria, eventually parts of the Apenninian slopes between La Spezia and Carrara, Cantabria, the coast of the Bay of Biscay, southwestern Caucasus to the northeastern slopes of the Pontic Mountains: these areas can be considered as potential transitional biomes between deciduous and evergreen broad-leaved forest. A warming of the atmosphere - particularly referring to winter temperature - retaining the same amount of precipitation during the growing season favours the competitiveness of autochthonal and exotic laurophyllous species.
- Bretagne, Wales and Southern Ireland (July mean temp. 14-17°C) represent transitional situations with mild winters and humid but relatively fresh summers: these areas are potential transition-biomes from deciduous to cold evergreen broad-leaved forests.
  A warming of the atmosphere - particularly referring to summer temperatures - would lead to an increase in the competitiveness of autochthonal and exotic laurophyllous species and to an increase in the number of potential species.
- In regions with more severe transitional climate (January mean temp. -2/0°C), sufficient summer precipitation (>150mm) and absolute minimum temperatures between -15 and -20°C (at the foot of Jura, the borders of Lake Leman, parts of the Swiss midlands and central-western Europe) a potential for various evergreen but more frost-tolerant shrubs exists, proportionally to the mitigation of the winter temperatures. In areas with sparse precipitation a warming could modify the equilibrium of the ecotone *Quercus/Carpinus* <> *Fagus/Abies* towards the first pole (Carpinion) (see Kienast et al. 1996); in fact there are some sectors of the midlands showing that the factor $FT_{July} / P_{year} \times 1000$ points to a precarious equilibrium T2030 F=20-24; T2100 F=24-30 in which the major summer evapotranspiration would permit zonal formation of Carpinion.
- In the Mediterranean basin relicts of the laurocerasus belt already exist either in azonal positions (gullies, springs) and/or exposed to an (oro-) mediterranean humid climate (Madeira, Tras - Os - Montes, Algarve, Nebrodi (Di Martino et al. 1977), Albania, western Greece-Istringia, Pontic Mountains,

Taurus (Schmidt 1969) and the northern slopes of the Hyrcanic Mountains: all situations with the same absolute minimum temperature > -10/-15°C, summer precipitation > 50-100mm). The natural limiting factor is not temperature, but the summer precipitation which seems to be crucial for the variation of transitional formations from sclerophyllous to laurophyllous species. Because of the centenarian to millarian degradation due to human action few areas remain with a clear distribution potential, suitable for the constitution and ecology of such relicts. An increased warming, meaning an aridification during summer, would compromise even more the existence of such relicts.

In the present stage only few parts of Insubria allow the constitution of Laurisilvae, particularly due to the absolute minimum temperatures. However, the hyperoceanic zones show a deficit in summer warmth, the Mediterranean zones a hydric deficit (Laurisilvae only in azonal sites possible) and the Centraleuropean climate too severe winters.

In case of a progressive temperature rise it could be assumed that the Centraleuropean zones will reach a situation similar to the transitional zone with cold winter climate.

The same holds for the temperate forests near the Atlantic Ocean, which also show a tendency to transitional forms with fresh summers.

The Mediterranean zone will lose parts of the extra-zonal vegetation with laurophyllous hydrolabile plants, whereas Insubria seems to become a region, in which Laurisilvae is the potential climatic climax vegetation.

Durability of vegetational systems: the chorological tension as an estimation factor.

The comparison of the present situation in the Pontic zone, the Far-East and North America already animated various authors to formulate the hypothesis that the elements of the Centraleuropean forest landscape could be species-deficient, presenting open or suboccupied ecological niches ("offene Stellenpläne" in Hübl 1988). In other terms a vital space has become evident from a climatic, structural and chorological point of view - analogous to the formation of other continents - not being completely saturated by the autochthonous flora.

Apart from the knowledge used so far (ecophysiology of the species and the vegetation formations) it is necessary to consider an ulterior integrative dimension - as estimation factor for a determined vegetation - which indicates the tendency to converge not only towards more mature (tendency to evolve towards the proper climax), but also more complete formations (convergence towards typical functional structures of the bioclimate).

The distance between the present situation "in situ" and these attraction-poles determines a tension which we call "chorological tension", substantially a term which quantifies the difference in the structure with respect to the climax vegetation found in the particular bioclimate.

The two following situations characterize a chorological tension:

- examples of present or former forest climaxes in the investigation area
- differences between different states of climax vegetation and other formations found in the same hemisphere, in situations characterized by a more complete chorological heritage.

A sensitive level of chorological tension may either be already present in a particular region - Insubria or in minor degree Western Europe - or it may increase depending on numerous factors e.g. the climate.

The anticipated modifications of a few degrees Celsius have to be considered as extremely important along the ecotonal belt of vegetation. In other words wherever a rapid warming of the atmosphere occurs new ecological niches may be procured and with them an altered potential for the distribution of other species.

Insubria - and in different modes other European regions - were historically kept as climatic islands, either by the differences with respect to the Mediterranean area, or through a floristic position, which is separated from the different biogeographical refuges (Caucasus / Portugal-Cantabria).

The invasion of exotic species may therefore seriously threaten the local biodiversity, as shown in the abundant literature.

Operational plans

On the level of scientific efforts the following steps are indicated:

- the amelioration and refining of methods to enlarge vegetation science in different parallel fields (pollen analysis, role of fires, pedology, zoology, phytopathology, mycology, etc.)
- the inclusion of quantitative characters referring to the population dynamics (biological values, state, critical mass, dynamic of diaspores, etc.) and on this basis developing criteria for the estimation of the quality and presence of biological bridges for particular formations and species (of strategic to invasive value)
- parallel studies in other European regions and other continents, in comparable ecotonal (cold-warm) or divergent positions (cold-humid versus dry-warm) introducing the relative terms of structural - functional analogy and of "chorological tension"; verifying the opportunity to introduce antagonists of invasive species from the area of origin (biological control)
- collecting knowledge on existing equilibria in the principle ecotonal zones of the continent of laurophyllous-deciduous forests and laurophyllous-sclerophyllous forests and ameliorating the knowledge on the occurrence and the genetics of populations of some Centraleuropean species and the peri- and oromediterranean distribution (thermo-hygrophilous ecotypes)

- favouring the position of Geographic-Information-Systems in relation to the connection between mesoclimate and vegetation for a better understanding of the present climate-depending ecotones in minor scale including the relative aspects of the hemeroby-degrees (presence of man, fire, zooanthropogenic pressure, etc.)
- setting up and collecting a network of observations and informations on particular species which spread "aggressively" or are "invasive" in analogous climatic situations on an international scale
- setting up a longterm observation network and considering as many scopes of a biomonitoring as possible, that seem to be of secondary interest at present, but could be useful in the future

Concerning forest ecosystems the following suggestions of important issues can be made:

- the conservation and protection of mature stages and the active enfavouring of autochthonous laurophyllous species with thermophilous ecotypes
- the conservation of the genetic variability of populations, the amelioration of the biological network if damaged; the conservation of unbroken forest populations with reduced proportion of external forest edges
- the anticipation of the suppression of the spread of aggressiv neophytes and reglementing the transcontinental introduction of plant species and living material.
- the conservation of the actual population of exotic laurophyllous species
- the promotion of European laurophyllous species in sylviculture and landscape
- the promotion of the conservation of laurophyllous biomes already mentioned in this report as strategic areas for the biodiversity of western Eurasia (Caucasus, Balcan, Pyrenees, extra-zonal relicts in the Mediterranean basin incl. the remnants of laurophyllous forests in Macaronesia, especially on the Azores).

References

Alther E.W. (1977): Entwicklung von Böden und Pflanzenbeständen auf Brachland und ihre Dynamik während 150 Jahren. Zeitschrift "Das wirtschaftseigene Futter" 23(3/4): 144-167.

Antonietti A. (1968): Le associazioni forestali dell'orizzonte submontano del Cantone Ticino su substrati pedogenetici ricchi di carbonati. Mitt. Eidg. Forschungsanstalt WSL 44(2): 85-226.

Brzeziecki B., Kienast F., Wildi O. (1995): Modelling potential impacts of climate changes on the spatial distribution of zonal forest communities in Switzerland. Journal of Vegetation Science 6: 257-268.

Carraro G. & Gianoni G. (1987): Studio sulla presenza delle diverse specie di quercia nostrana in Ticino e della loro importanza per la selvicoltura a dipendenza della stazione e tenendo conto del fenomeno d'ibridazione. Lavoro di diploma ETHZ: 198p. Polycopie

Carraro G. (1995): Lo spazio forestale fra consumo e rispetto: aspettative e conflitti. Forestaviva 15: 55-60.

Ceschi I. (1960): Analisi fitosociologica, pedologica e strutturale di un ontaneto naturale nella zona insubrica ticinese. Lavoro di diploma ETHZ. 25p. n.p.

Ceschi I. (1992): La diffusione del prugnolo tardivo (*Prunus serotina* Ehrhart) nel Ticino. Boll. Soc. Sci. Natur - 80(2): 97-105.

Chenevard P. (1910): Catalogue des plantes vasculaires du Tessin. Mem. de l'Institut National Genevois.

Christ H. (1903): Zur Flora des oberen Lago Maggiore. Ber. Schweiz. Bot. Ges. 13:154-156

Conedera M. et al. (1996): Incendi boschivi al Sud delle Alpi: passato, presente e possibili sviluppi futuri. vdf Zürich.

Cronk Q. & Fuller J. (1995): Plant Invaders. Chapman & Hall, London.

Delarze R., Caldelari D., Hainard P. (1992): Effects of fire on forest dynamics in southern Switzerland. Journal of Vegetation Science 3: 55-60.

Di Martino A. et al. (1977): Sintesi sugli studi condotti sulla vegetazione delle Madonie. Giorn. Bot. Ital. 111: 370-371.

Ellenberg H. & Klötzli F. (1972): Waldgesellschaften und Waldstandorte der Schweiz. Mitt. Eidg. Forschungsanstalt WSL 48(4): 589-930.

Ellenberg H. & Rehder H. (1962): Natürliche aufzuforstenden Kastanienflächen im Tessin 113(3/4): 128-142.

Felbermeier B. (1994): Die klimatische Belastbarkeit der Buche. Forstwiss. Cbl. 113:152-174.

Freuler B. (1904): Forstliche Vegetationsbilder aus dem südlichen Tessin. Verh. Schweiz. Naturforsch. Ges. 86: 269-292.

Glutz v. Blotzheim (Ed.) (1988): Handbuch der Vögel Mitteleuropas 11(2): 920-924. Aula-Verlag, Wiesbaden.

Gianoni G., Carraro G. & Klötzli F. (1988): Thermophile, an laurophyllen Pflanzenarten reiche Waldgesellschaften im hyperinsubrischen Seenbereich des Tessins. Ber. Geobot. Inst. ETH, Stiftung Rübel, Zürich, 54:164-180.

Guidi M., Piussi P. & Lasen C. (1994): Linee di tipologia forestale per il territorio prealpino friulano. Annali Acc. Ital. Sci. For. (43): 221-285.

Hübl E. (1988a): Die sommergrünen Wälder Japans und Westeurasiens, ein floristisch-klimatographischer Vergleich. Veröff. Geobot. Inst. ETH, Stiftung Rübel, Zürich 98: 225-298.

Hübl E. (1988b): Lorberwälder und Hartlaubwälder (Ostasien, Mediterraneis und Makaronesien). Düsseld. Geobot. Kolloq. 5:3-26.

Jäger E. (1968): Die klimatischen Bedingungen des Areals der Dunklen Taiga und der sommergrünen Breitlaubwälder. Ber. Dtsch. Bot. Ges. 81: 397-408.

Jäger E.J. (1975): Wo liegen die Grenzen der Kulturareale von Pflanzen? Wiss. Beitr. Univ. Halle-Wittenberg 6 (4): 101-113.

Kienast F., Brzeziecki B, Wildi O. (1996): Long-term adaptation potential of central european mountain forests to climate-change - A GIS-assisted sensitivity assesment. Forest Ecology and Management. 80(1-3): 133-153.

Kissling P. (1989): Changement floristique depuis 1950 dans les forêts des Alpes suisses. Bot. helv. 99(1): 27-43.

Kissling P., Kuhn N. & Wildi O. (1988): Le relevé mérocénotique et son application à l'étude du changement floristique en forêt. Bot. Helv. 98(1): 39-75.

Kira T. (1977): A climatological interpretation of Japanese vegetation zones. In: Miyawaki A. (Ed.): Vegetation science and environmental protection. Maruzen, Tokyo. 21-30.

Klötzli F. (1976): Grenzen von Laubwäldern in Europa. Ber. Deutsch. Bot. Ges. 89:371-380.

Klötzli F. (1988): On the global position of evergreen broad-leaved non-ombrophilous forest in the subtropical and temperate zones. Veröff. Geobot. Inst. ETH, Stiftung Rübel, Zürich, 98: 169-196.

Klötzli F., Walther G.-R., Carraro G. & Grundmann A. (1996): Anlaufender Biomwandel in Insubrien. Verh. Ges. Ökol. 26: 537-550.

Kowarik I. & Sukopp H. (1986): Unerwartete Auswirkungen neu eingeführter Pflanzenarten. Universitas 41(483): 828-845.

Kowarik I. (1985): Zum Begriff "Wildpflanzen" und zu den Bedingungen und Auswirkungen der Einbürgerung hemerochorer Arten. Publ. Naturhist. Ges. Limburg 35(3-4): 8-25.

Kräuchi N. (1994): Modelling forest succession as influenced by changing environment. Mitt. Eidg. Forschungsanstalt WSL 69(2): 143-271.

Kuhn N., Amiet R. & Hufschmid N. (1987): Veränderung in der Waldvegetation der Schweiz infolge Nährstoffanreicherungen aus der Atmosphäre. Ber. EAFV 295.

Kuhn N. (1990): Veränderung von Waldstandorten - Ergebnisse, Erfahrungen und Konsequenzen, mit einem Konzept für die Dauerbeobachtung von Waldbeständen. Berichte EAFV / WSL 319.

Kuhn N. (1993): Ursachen floristischer und ökologischer Vorgänge in Waldbestände. Schweiz. Z. Forstwes. 144(5): 347-367.

Landolt E. (1977): Ökologische Zeigerwerte zur Schweizer Flora. Veröff. Geobot. Inst. ETH, Stiftung Rübel, Zürich, 64.

Larcher W. (1995): Physiological Plant Ecology - Ecophysiology and stress Physiology of functional Groups. 3rd Ed., Springer, Berlin.

Lauscher F. (1981): Säkuläre Schwankungen der Dezennienmittel und extreme Jahreswerte der Temperatur in allen Erdteilen. Zentralanstalt für Meteorologie und Geodynamik in Wien. Nr. 252.

Meusel H., Jäger E. & Weinert E. (1965): Vergleichende Chorologie der Zentraleuropäischen Flora. Fischer, Jena. 1.

Meusel H., Jäger E., Rauschert S. & Weinert E. (1978): Vergleichende Chorologie der Zentraleuopäischen Flora. Fischer, Jena. 2.

Miyawaki A. & Suzuki K. (1980): Natürliche und ihre sommergrünen Sekundärwälder in Mitteljapan - Über die pflanzensoziologische Stellung der Quercetalia serrato-grosseserratae. Phytocoenologia 7: 492-506.

Oberdorfer E. (1964): Der insubrische Vegetationskomplex, seine Struktur und Abgrenzung gegen die submediterrane Vegetation in Oberitalien und in der Südschweiz. Beitr. z. naturk. Forsch. Südwestdeutsch. 23(2): 141-187.

Ortubay Fuentes A. (1994): Relacíon Clima-Vegetacíon en la Comunidad Autonoma del Pais Vasco. Tesis Doctoral. Universitad del Pais Vasco. n.p.

Ovington J.D. (Ed.) et al. (1983): Temperate broadleaved evergreen forests. Elsevier (Amsterdam).

Ozenda P. & Borel J.L. (1990): The possible responses of vegetation to a global climatic change (Scenarios for Western Europe, with special reference to the Alps) IOS Press. Amsterdam: 221-249.

Pignatti S. (1979): I piani di vegetazione in Italia. Giorn. Bot. Ital. 113: 411-428.

Rampazzi,F., G. Carraro, P. Gianoni, A. Focarile, B. Jann & N. Patocchi (1993): Studio naturalistico del fondovalle valmaggese. Memorie della Soc. tic. sci. nat., vol. 3.

Rivas-Martinez, Loidi J., Canto P., Sancho L.G. & Sanchez-Mata D. (1984): Datas sobre la vegetacíon del Rio Bidasoa (España). Lazaroa 6: 127-150.

Scheeder Th. (1994): Die Eibe (*Taxus baccata* L.) - Hoffnung für ein fast verschwundenes Waldvolk. IHW-Verlag.

Schmidt G. (1969): Vegetationsgeographie auf ökologischer Grundlage. Leipzig.

Thiébaut B. et al. (1978): Étude des hêtraies de l'arc montagneux périméditerranéen de la vallée du Rhone à celle de l'Ebre (II-III). Oekologia Plantarum 13(4).

Vaio V. (1993/94): La riserva naturale del Balbianello (Lago di Como) - Aspeti botanici - Tesi sperimentale di laurea. Pavia. n.p.

Walder M. (1985): Die Auswirkungen des Kälteeinbruchs im Januar 1985 auf wärmeliebende Pflanzen in einem Tessiner Garten. Schweiz. Beiträge zur Dendrologie 35: 189-199.

Walter H. & Lieth H. (1960-67): Klimadiagramm-Weltatlas. Gustav Fischer, Jena.

Walter H. (1976): Die ökologischen Systeme der Kontinente (Biogeosphäre) Prinzipien der Gliederung mit Beispielen. Stuttgart.

Walther G.-R. (1995): Ausbreitung und Grenzen laurophyller Arten im Südtessin. M. Sc.-Thesis, ETH Zürich: 37p. + Annex (Polycopy)

Werdenberg K & Hainard P. (1989): Modifications de la composition floristique dans la forêt genevoise et pollution atmosphérique par l'azote. Le lien est-il réel? Une réponse à l'Insitut fédéral de recherches forestières. Saussurea 19: 57-66.

Wildi O. & Orlóci L. (1996): Numerical exploration of community patterns: a guide to the use of MULVA-5. SPB Academic Publishing - Amsterdam.

Zoller H. (1961): Die kulturbedingte Entwicklung der insubrischen Kastanienregion seit den Anfängen des Ackerbaus im Neolithikum. Ber. Geobot. Inst. ETH, Stiftung Rübel, Zürich, 32: 263-279.

Plenary discussion

Question: G. Nakhutsrishvili

Was ist ihre Quelle für kaukasische Daten über Vegetation, im speziellen über Syntaxa?

Answer: G. Carraro

Wir haben ein Büchlein von Korotkov erhalten, "Prodromus of the Vegetation Syntaxa of the USSR". Daraus haben wir einige Ideen über die Situation im Kaukasus erhalten. Jedoch muss ich zugeben, dass wir viel Mühe hatten, Informationen über die Verhältnisse z.B. in ihrer Heimatregion zu erhalten, welche für uns interessant wären. Die andere Aufnahme über Kastanienwälder stammt aus einer Arbeit von Lüdi, und er hatte durch eine Freund zufällig ein paar Aufnahmen erhalten, aber darüber hinaus kennen wir in vielen Fällen zuwenig, bzw. es gibt im Moment kein Informationsnetz, um an solche Informationen zu gelangen.

Question: R. Gagnidze

I think your report is very interesting also for caucasian specialists. You noted that *Abies* is characteristic to the mesolithic period. It means this period should be characteristic of moist conditions. How many years did the mesolithic period exist and how long did it last in Europe?

Answer: G. Carraro

7000-8000 years before present; *Abies* disappeared at the beginning of the local Neolithicum (6000-5000 B.P. = beginning of agriculture and keeping domestic animals)

Question: R. GAGNIDZE

The mesolithic period coincides with the Atlantic period of the Holocene, which is characteristic of a "climate optimum". What position did the evergreen broad-leaved species have in the ecosystem of the Atlantic period? Did the evergreen species predominate in quantity the deciduous ones?

Answer: G. CARRARO

We don't have any information on other species. For example it is difficult to understand why *Ilex aquifolium* is not to see and even today we don't see *Ilex aquifolium* in pollen diagrams.

Question: R. GAGNIZE

The pollen grains of *Ilex* and *Rhododendron* are badly reflected in Holocene deposits of the Caucasian high-mountainous region. How are they reflected in Holocene deposits of Europe? My second question: Is the anatomical structure of the leaves of *Rhododendron* and *Ilex* species studied and do their structures differ in the populations of dry and mesolithic period conditions?

Answer: G. CARRARO

As I showed you in the pictures, you could see larger not so hard leaves, that means laurophyllous ones and the others. But there are a lot of intermediate forms. There are already studies on the relation between the area in cm^2 and the weight in gram of the ashes. So it could be one way to become a physiological measure about a level. That means, thin and large is laurophyllous, thicker and harder is sclerophyllous. There are some studies of Prof. Larcher.

Question: E. POLI

Erster Punkt: *Castanea sativa*-Wälder werden von verschiedenen Verfassern als natürliche Wälder angesehen, ganz einfach darum, weil sie diesselbe Artenkombination wie natürliche Wälder aufweisen. Nur nimmt *Castanea* einfach den Platz von den natürlichen lokalen Holzarten z.B. *Quercus pubescens*, *Quercus ilex* oder anderen Arten ein.

Zweitens: *Ilex aquifolium* hier im Tessin wie auch andere Laurisilvae-Arten haben eine andere Bedeutung als in manchen mediterranen Gebieten wie z.B. in Sizilien, wo eher das Gegenteil der Fall ist. Hier im Tessin, wo neue in Ausbreitung begriffene Vegetationselemente sind, zeigen sie wärmeres Klima und heutige Veränderungen des Klimas. In Sizilien sind *Ilex aquifolium* wie z.B. auch *Taxus baccata* bloss noch Reste von einer ehemaligen laurophyllen Vegetation, welche heute nicht mehr da ist. Sie haben ihre Refugien innerhalb der *Fagus* - Stufe. So bevorzugt beispielsweise *Taxus baccata* die feuchteren und schattigeren Standorte innerhalb der *Fagus* - Wälder, wie dies in Sizilien auf der Nordseite der Nebroden Gebirgen und gegen das tyrrhenische Meer der Fall ist.

Answer: G. CARRARO

Concerning the *Castanea* forests: We know that *Castanea* is originally from SW-Asia and the two relevés were taken from sites which hadn't been utilized, so they can be looked upon as virgin forests, so we think that these *Castanea* forests are nearly natural. But I agree that now there are *Castanea* forests in the Ticino with similar constitution or association like *Quercus*, or *Quercus-Tilia* forests.

In the ecotonal area in the Nebrodi mountains, Parco delle Madonie, in Sicily, we have really a residual vegetation of the Tertiary. It is very important to know the similarities and differences compared to the situation we have here. In this way we must study these limitrophous regions for a better understanding of the conditions in different other regions.

Recent shifts in vegetation boundaries of deciduous forests,
especially due to general global warming, pp. 53–60
edited by F. KLÖTZLI and G.-R. WALTHER

Causes of shifts in vegetation in the past and present

Peter VOSER
FORNAT, CH-8708 Männedorf, Switzerland

Abstract

Apart from the greenhouse effect several other factors are responsible for Laurophyllisation. It might even be stated, that the other factors are more decisive than anthropogenic climate warming. Most of the area in which Laurophyllisation presently occurs had been cultivated land such as fields or vineyards until thirty years ago. Due to this intensive management, no exotic species succeeded to establish in the vicinity of the settlements. Only after these areas were abandoned, the spread of exotic evergreen species begun. It occurred very sparsely in woodland and chestnut-Selvas and was mostly restricted to the former areas of vineyards and cultivated fields.

Personal observations lead to the hypothesis, that within the next two to three decades more of the evergreen broad-leaved individuals might grow up to the tree layer. However, the area with laurophyllous species will probabely not expand much further and these new established forests with evergreen exotic species are likely to remain a mixed forest type with evergreen broad-leaved and indigenous deciduous shrub and tree species.

Introduction

In the Atlanticum, from about 7000 BP onwards, the temperature- and precipitationregime as well as the soil conditions have probabely been suitable for evergreen broad-leaved species. A lot of Mediterranean species immigrated and remained later as relict stands at the southern foot of the Alps. However, the famous tertiary flora was restricted to the Macaronesian Islands from where there was no possibility to recolonize the European Mainland.

The speed for the recolonization must not be overvalued. Beech, for example, needed almost 6000 years to reach the border of the Alps.
Man was also an important factor in this process. Woodland was cleared and kept open with fire. Agriculture and domesticated animals contributed to suppress the recolonization of woody species. There was no interest in exotic ornamental species and the exchange of people and goods between the continents used not to be as intense as it is nowadays. Land for cultivation was rare and the slopes were formed in terraces. *Castanea sativa* was cultivated in so called "Selva" and "Palina" (coppice forest).
Until 1850 the "little ice age" limited cultivation experiments. Later, palms, laurel, camphor trees and the Japanese Honeysuckle (*Lonicera japonica*) were introduced in the newly established gardens and parks. Garden deposits from these were removed and disposed in natural habitats such as gorges.
With the industrialisation new cultivation techniques were introduced. The least suitable areas for agriculture on the steep slopes were abandoned and became available for the colonization of all sorts of woody species, indigenous and exotic ones.

Verification of Laurophyllisation on a local scale

In a forest area of 20ha in the vicinity of Locarno-Solduno the cover densitiy of laurophyllous species was mapped in detail (Figure 1). The evergreens are concentrated on the lower belt of the slope, which is terraced. An old Selva suppresses the spread of neophytes due to shadow and competition with the dense grass-layer. In such places only few tiny palms and small *Elaeagnus* individuals persist. In gorges with thicker and moister soils the conditions for the evergreen exotics are more favourable. In regularly managed forests, so called coppice forest, there is almost no spread of Neophytes. On the one hand, the aspect changes from Querco-Fraxinetum to Querco-Betuletum on rocky sites, on the other the competition increases due to an already occupied ecological niche.

Legend to figure 1: Percentage of laurophyllous species

0	no occurrence		
+	singles or up to 1 %	● (grey)	11 up to 50 %
◌ (dotted)	2 up to 10 %	● (black)	more than 50 %

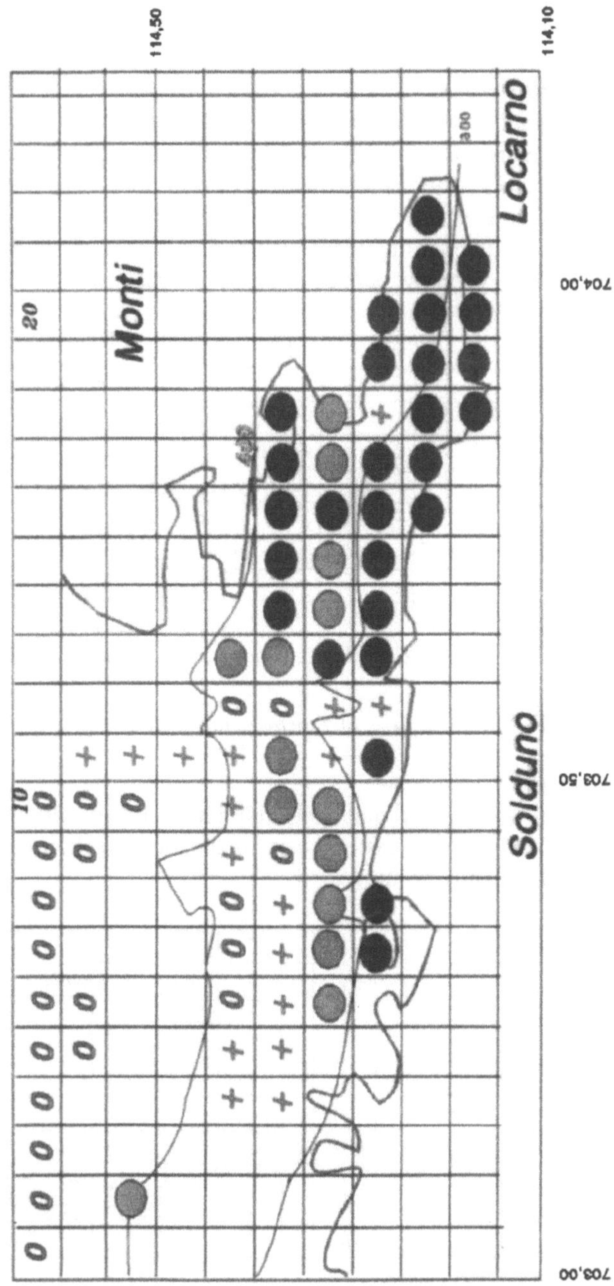

Figure 1: Map with coverage of laurophyllous species (winter 1995/96)

A more detailed view on particular species is given with the following figures:

Trachycarpus fortunei

Trachycarpus fortunei is widely distributed, partly also on 400m a.s.l. growing very vigorously (Figure 2). It establishes fruits also in gardens located higher up the slopes. In places where forest used to occur, only tiny individuals of *Trachycarpus* grow, in gorges a little taller ones can be found. The highest site with *Trachycarpus* saplings in the investigation area is located on 480m a.s.l. in a gorge. There are only very few palms with inflorescens in the forest.

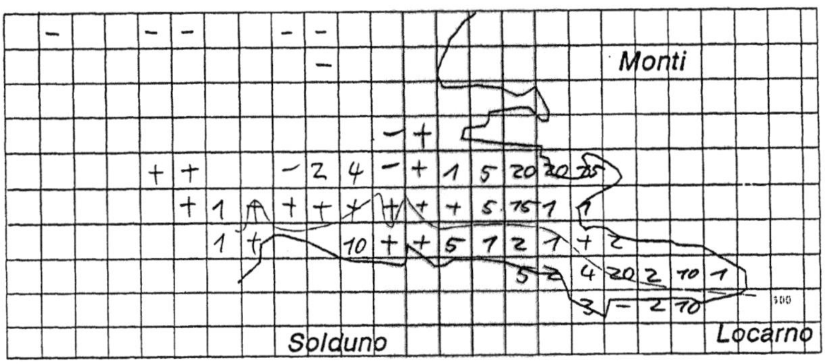

Figure 2: Mapping of the occurrence of *Trachycarpus fortunei* (density in %)

Elaeagnus pungens

Elaeagnus pungens establishes quite tall individuals up to 430m a.s.l. Below, *Elaeagnus* occurs only in the eastern part of the investigation area, in more dense stands and frequently fruiting.

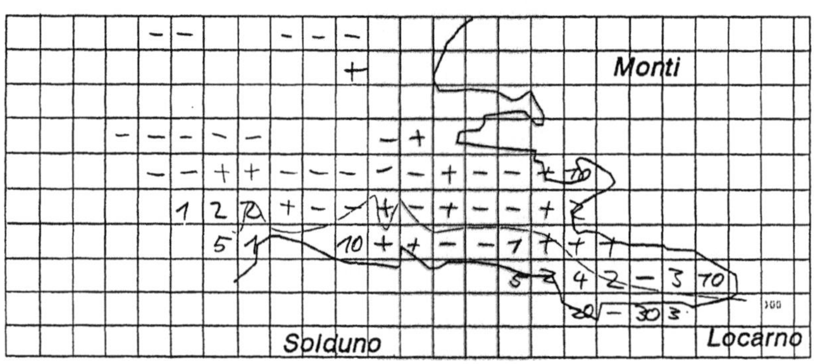

Figure 3: Mapping of the occurrence of *Elaeagnus pungens* (density in %)

Prunus laurocerasus

Prunus laurocerasus is widely distributed, but only in parts where forest has been very sparse. It occurs also on higher altitude, e.g. on 490m a.s.l. in gorges. Individuals with fruits can be seen frequently.

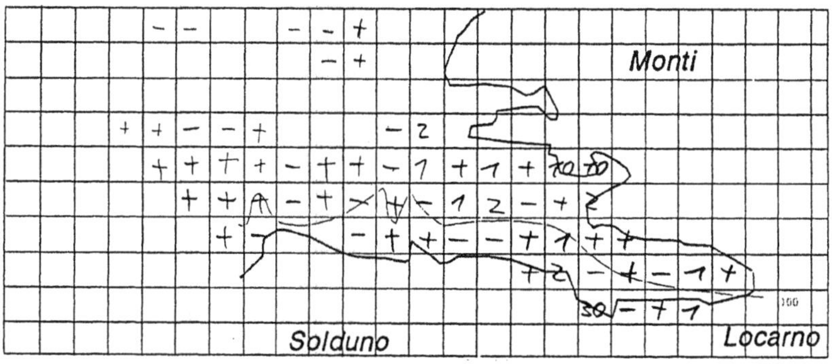

Figure 4: Mapping of the occurrence of *Prunus laurocerasus* (density in %)

Laurus nobilis

Laurus nobilis prefers the lower belt of the investigation area. There are individuals establishing inflorescens and fruits.

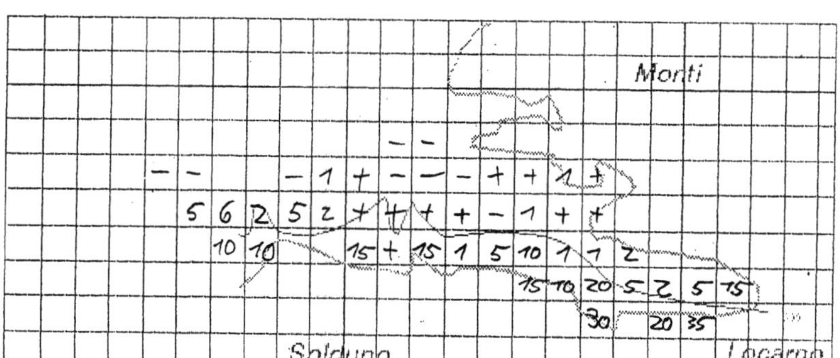

Figure 5: Mapping of the occurrence of *Laurus nobilis* (density in %)

Cinnamomum glanduliferum

In the eastern part of the investigation area, there are some individuals grown up to the higher tree layer and with DBH of more than 20cm. It is uncertain if they develop inflorescens. The highest site with occurrence of *Cinnamomum glanduliferum* is located in a gorge on 380m a.s.l. Although there are some fruiting camphor trees close to the forest above Solduno in the central part of the investigation area, no occurrence of seedlings or saplings of *Cinnamomum glanduliferum* could be observed higher than 300m a.s.l. In the western part there are no seed producing trees.

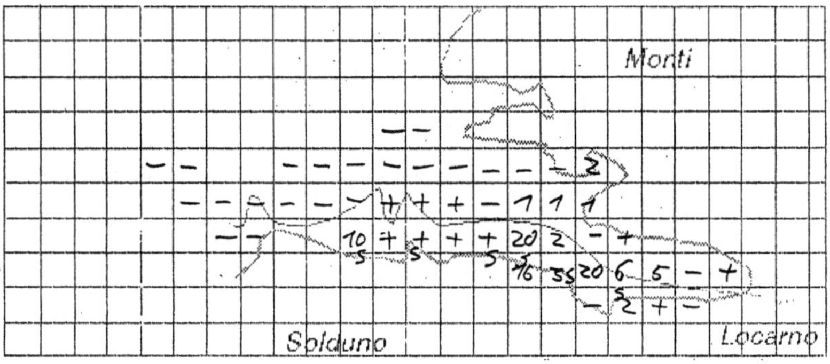

Figure 6: Mapping of the occurrence of *Cinnamomum glanduliferum* (density in %) (SS = tall seed producing trees)

Lonicera japonica

Lonicera japonica is a typical species for abandoned terraced slopes/vineyards up to about 400m a.s.l. It occurs very sparsely in permanent forest sites and Selvas.

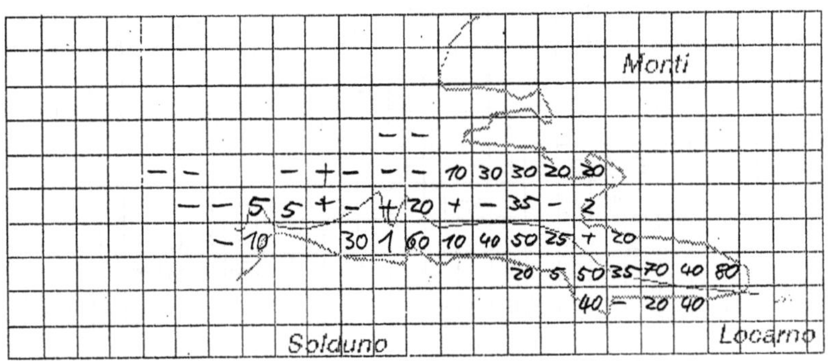

Figure 7: Mapping of the occurrence of *Lonicera japonica* (density in %)

There are traces of forest fires in the whole investigation area. But the laurophyllous species did not profit from the opened niches. With the natural recolonisation exotic species invade the forest slowly.

Conclusions

Two different situations in the process of colonisation of laurophyllous species must be distinguished. On the one hand, there are the abandoned fields and vineyards. With the presence of seeds and seedlings of exotic species at the early beginning of the recolonisation of such areas with forest species, it was easy for laurophyllous species to grow up simultaneously and establish a dense stand with a mixed forest of exotic and indigenous species. The situation is different in areas, where forest occurred or where the management of chestnut Selvas was carried out. The seeds of laurophyllous species have been brought in by birds but the species need much more time to become established. The process of vigorous Laurophyllisation that we could observe in the past was due to the vast amount of abandoned land. Therefore, once the colonization of abandoned areas of former cultivated fields and vineyards has been realized, the spatial expansion of laurophyllous species in established forests proceeds very much slower. There is no more abandoned land available for these species. Areas with laurophyllous mixed forests will remain restricted to a rather small area in the vicinity of settlements and on the lower part of southern exposed slopes in the area influenced by the lakes.

Plenary discussion

Question: H. BRUELHEIDE

What's the size of the squares in your mapping? Is your estimation of cover based on the hole area of these squares?

Answer: P. VOSER

The squares are fifty by fifty meters, but they are not fixedly marked. I just wanted to show where there are places with a lot of evergreen species. I assessed the area covered with evergreen broad-leaved species and put it in relation to the whole area of the square. Often it is like a continuum and therefore it is quite difficult for an exact estimate but, in my opinion, for a succession study it is still better than the Braun-Blanquet scale.

Question: V. PILLAR

Is there any place in your investigation area that hasn't been cultivated and which is below 400m a.s.l.?

Answer: P. VOSER

Yes there is an uncultivated area on a steep ridge between two ravines. There you can't find any laurophyllous species because of the dryness of this site. Another place with very few evergreens is an old Selva with large chestnut trees, and in that region I guess it won't be easy for laurophyllous species to invade. There are some palms and one or two *Laurus* or *Elaeagnus* but only very few. The mature vegetation resists in that place.

Statement: F. KLÖTZLI

I have to add two or three things: First of all, I am also convinced that anthropogenus influence is playing a role and that in old vineyards or in old Selvas you can have an invasion of laurophyllous species under the circumstances. But I know the Ticino since I was about five years old, and when I came here I was always interested in flowers etc. and certain areas I know therefore quite well. Not as well as my friends from the Ticino but still not too bad. So I would say, knowing areas which have been neglected after having had a cultivation period and knowing areas where there has been no cultivation (or maybe just a little bit sylvicultural management), both areas have now laurineous or laurophyllous species since a period which dates back to the seventies. So non-cultivated areas before the seventies had no laurophyllous species neither, except where they were planted. So this is against the assumption that it is just the management that has provoked the appearence of laurophyllous species. Again I say I am convinced that under such circumstances the hole development is accelerated.

Another point I have to take up concerns northern Switzerland: Northern Swiss forests of course are cultivated rather intensively by the foresters, but if I compare their activities with certain areas in a "Selva" this is neglectable. So, why are there invasions of laurophyllous species in northern Switzerland and why is the invasion in the Ticino only dating from the seventies, because in the old relevés even in very quite areas where there has been no management we had no laurophyllous species. So your theory should be modified; it is true what you said but it is not true that it is only in places where management has changed, that is absolutely clear, even if we compare the procedures with the northern Swiss parts. And, by the way, all of us will go to such places on Wednesday afternoon. It is a small walk and you can also have a look at the spreading and the diversity of these eastern asian and other species.

General discussion

Chairman: Prof. Peter J. EDWARDS (Geobotanical Institute, ETH Zürich)

Statement: P. EDWARDS

We had three presentations today: One was from Frank Klötzli talking about northern Swiss forests, in which he showed that at the level of the individual relevé there has been enormous change in plant species composition. If you make an ordination of the data you see that the changes are very clearly directional. Overall there appears to be a loss of species. The changes partly reflect increasing abundance of more thermophilous species , but there are many other kinds of dynamics are at work. Then we had a presentation from Gabriele Carraro mainly on the southern Swiss forests which showed that there have been all kinds of environmental changes during this century: changes in fire frequency, changes in litter collection, changes in wood utilisation. In particular the climate data show clearly that patterns of temperature over the first fifty years of this century were very different from those in the second. There was a much higher frost frequency, minimum temperatures were lower and so on in the first fifty years. Then he showed us some of the data from the transect through Switzerland, reiterating some of the points that Frank made. Not all of the species changes since the 1960's can be interpreted in terms of climate. For example some changes clearly reflect more nitrophilous conditions; also there is a loss of many heliophilous species. However these explanations alone are insufficient. His argument was that we are right on the edge of where these laurophyllous species naturally could occur and the changes in species composition are partly due to climatic warming. And I must say that at that point I was quite convinced by the argument.

But then we had a bit of demolition work from Peter Voser. His detailed mapping in the area of Locarno showed that there has clearly been a spread of these laurophyllous species from the original seed plants in gardens. However most of the species haven't actually gone very far. Furthermore, it seems that the areas into which they have spread are those which have seen major land use change. At that point I found myself quite confused. And so we come to the crucial question: how good is the evidence that the species changes we have seen and heard about are due to climatic change. Ladies and gentlemen - you heard the evidence; what is your verdict?

Statement: G. CARRARO

I must say one thing; I have the same idea of Peter Voser because in the region of the house of Peter Voser it is a situation that thirty years ago and now it is sufficient warm for a Laurophyllisation. So the increasing temperature of the last twenty years in this particular place is not decisive. But if you compare with other places with different mesoclimatic conditions but with all the other factors being the same: distance from the seed plants, the association of the *Castanea* forest (Carpinion) and soil conditions. Then we are sure that it is only a difference in temperature. Therefore if we compare the temperature of these stations in the climatic ecotonal situation, we see that the difference in average January temperature is 2°C, average of the absolute minimum temperature is 3°C and we have a difference of 50% less frost days. Under these conditions there is no Laurophyllisation or only very few. You can see this change in site conditions even within 20m as we showed you this morning on Mte Verità. On the one side, in the southern exposed slope an evident Laurophyllisation, but just of the opposite side of the hill, within the same distance of the seed plants but with different mesoclimatic conditions, you could see a completely different degree of Laurophyllisation.

Why do we think that the warming of the last thirty years has played a very important role? Because the difference in the absolute minimum temperatures of 2°C is exactly the same as I showed before. The forst days diminished by 50% and finally the difference of the average January temperature is also about 2°C. So we have an evident similarity in differences in climatic conditions between the different regions with different stages of Laurophyllisation compared with the differences caused by climate warming within the last thirty years. Therefore the climatic change of the last thirty years enlarged the potential area for Laurophyllisation. But of course there are some particular places as the one mentioned by Peter Voser where climatic conditions ever have been sufficient for Laurophyllisation. And there the recent change in temperature parameters just sponsored the quantity and vitality of laurophyllous species but hasn't been the responsible factor to initiate the process of Laurophyllisation itself.

Statement: H. BRUELHEIDE

As I see it, the only way to combine the statements of Peter Voser and of Gabriele Carraro is to gather exact climatic data for areas that have been mapped exactly. In one study the climatic evidence is missing, in the other the area was not mapped with sufficient detail.

Another point refers to the climate diagrams of Walter & Lieth. I share Peter Voser's view that the Ticino climate was already appropriate for broad-leaved evergreen species in the first half of this century. Walter & Lieth's atlas was first published in 1960 and it lists cities that had been observed for 40 years or more (e.g. Locarno, Bellinzona). Therefore, it can be concluded that the climatic data in the atlas refers to the first half of this century. Some cities in the atlas show an annual temperature of about 13°C or above (Chiavenna, Dongo) and most stations show precipitation of more than 1500 mm. These are just the right conditions to support broad-leaved evergreen forests (Klötzli, F. 1988, Veröff. Geobot. Inst. ETH Stiftung Rübel 98: 169-196 and Satoo 1983: Temperate broad-leaved evergreen forests in Japan. In: Ovington, J.D.: Ecosystems of the world. 10. Temperate broad-leaved evergreen forests. Elsevier, Amsterdam: 169-189).

Consequently, we do not have to claim that a climatic change in the second half of this century was a necessary precondition for the growth of evergreen shrubs and trees in the Ticino area.

Statement: F. KLÖTZLI

I have to point out distinctly that this is a wrong assumption. Because I showed clearly that Insubria was outside of the evergreen broad-leaved area, if you take the average temperature and precipitation values, but there is one thing which is much more important and you have not considered that and these are the threshold values given by the traumatic events. And the traumatic events we had before, they just killed the saplings and the young plants. And then we came over a certain threshold value and that's when they came into the game, all these laurophyllous species. They came to show themselves after we had left the traumatic low values. If you look at the sensitivity of laurophyllous species (and we may discriminate them in different groups) then we have laurophyllous species which die when you come down to -5°C. There is a man among us, Mr. Walder, who has done a lot of measurements in his large garden and he was able to show, at temperatures of -9 to -10°C, how many species of the different discriminate groups had suffered or died. And it was absolutely clear we have threshold values for certain species which e.g. can subsist below -5°C but mostly die down to -10°C, or then they get into troubles at -4°C and that's *Acacia dealbata*, *Fuchsia*, *Phoenix*, *Musa*. But then you have species which are hardier, much hardier and they start to have troubles at -15°C and this is for instance *Aucuba*, *Berberis* div spec., *Camellia japonica*, *Eucalyptus*, especially the snow Eucalypts, *Magnolia grandiflora*, *Rhododendron* species, *Skimmia* etc. And then there is still another group where we have let's say less troubles at -15°C and only when it gets lower than -15°C they freeze back; e.g. *Prunus laurocerasus*, *Lonicera pileata*, *Rhododendron ponticum* and some *Cotoneaster* species they lose all the supraterranian parts, in their subterrainian parts they remain alive and you get sprouts again. But when it is a young plant then the hole young plant is destroyed. So please don't forget first traumatic moments and don't forget there are discriminant subgroups between these different laurophyllous species. For all those who would like to have a more detailed view it is published and the paper is included in the documentation folder.

And I just want to show some other aspects which point out the fact that it is not just an affair of laurineous species. If you look at this list of species with *Hedera*, *Ilex* and *Taxus baccata*, these three species they have increased their presence in the whole of this transect from Schaffhouse to Chiasso. They're everywhere more plentyful. This fact is endorsed by statistical tests which we haven't explained so far. Well, this is nothing new for you. But this other list of species is new for you and that's also very important. All these species, they are all more or less montane species (*Abies alba*, *Sambucus racemosa*, *Lonicera alpigena*, *Rosa pendulina* and then the herbs *Galium rotundifolium*, *Knautia sylvatica*, *Veronica latifolia*, *Aruncus sylvestris*, *Polystichum lobatum*) they have decreased because of the changes in temperature. So we have a discrimination of species on both sides: We have some more thermophilous species which have been able to increase, and we see distinctly that there are species which have decreased and that's a clear fact on the corridor through the whole of Switzerland. This just as an addition.

Statement: A. Borhidi

I don't want to discuss the Laurophyllisation because I am convinced that this process exists. My impression is that we have to find the causes, why? Climate is changing, that's true. Why, especially here, the original vegetation is so vulnerable? And I think it is an important point, that in the native vegetation we have a lot of species which have their southern limit of distribution or the lower one. This is *Fraxinus excelsior*, wihich is a central european species and southern of the Alps it is in an isolated situation. *Tilia cordata* as well. *Vaccinium* is at the southern edge of its existence. *Anemone nemorosa* also. I think we have here a very vulnerable community. Even those species you mentioned as montane species are shifting upwards. So I think it is a very good situation really to understand what is on. The other thing is, I'm from south-east Europe, in this situation in south-east Europe, there is a very broad forest belt with *Quercus cerris*, *Quercus pubescens*, *Quercus virgiliana*, *Quercus dalechampii* and *polycarpa*, closely related to *Quercus petraea*. Here all these species are in a very restricted situation, where they couldn't cover their real ecologicalrole and instead of this there are several hybridasation and hybridpopulations in the ecologicalgaps showing that all these species are at the edge of their distrubution and they are really not well situated ecologically. I think this is something which is worth to think about.

Statement: N. Zazanashvili

In some areas of the Caucasus there is an unusual situation. In different places the relation between the change of natural vegetation and some climate data is not so evident, which is in coherence with microclimatic diversity. In the Southern Colkheti there is a tendency for lowering of some *Picea orientalis* communities down to 700m a.s.l. and on the other hand border of evergreen species ssuch as *Rhododendronponticum*, *Rh. caucasicum* and other species in different places are decreasing. This fact needs checking and confirmation. I mean that all the diversity of changing must not always be in the context of global change of the climate.

Answer: F. Klötzli

Laurophyllous is not synonymous with thermophilous. That's why I also spoke about the threshold values. They have a certain sensitivity. And I think sometimes *Rhododendron* is misleading because they need snow cover. We have the ±subalpine species *Rhododendron ferrugineum* in the whole area here in the Ticino, coming down to 250m a.s.l., but in many parts of their distribution during the critical periods these *Rhododendron* are either in a sheltered position or higher up, where they are under snow. So their laurophyllous leaves need snow cover, they need protection. That's why we have certain species which can stand very low temperatures but are laurophyllous and on the other hand we have certain more or less subtropical species such as *Cinnamomum* which are rather sensitive to some of these lower temperatures. So temperature influence is quite clear, and *Rhododendron* is a very good genus to depict the influence on the laurophyllous types. But we have to distinguish these laurophyllous types needing cover. You could also take *Buxus*. *Buxus sempervirens* is a funny species, it is actually quite hydrostable but it is also growing under more or less similar conditions as here in

Ticino, where on the other hand it is growing with *Quercus pubescens* in the Jurassic mountains on southern slopes. So we have certain types which are not absolutely clear in their grouping or significance.

Question: K. FUJIWARA

For me one point is not clear yet. Therefore I have a question to Prof. Klötzli. You mentioned that montane species have decreased. Is this in the mountain area or in the surroundings of the Locarno area?

Answer: F. KLÖTZLI

No, I have shown you only examples of the submontane belt. And in the submontane belt you get shadowy slopes where we get montane forest associations coming further down to about 300m a.s.l. They have these montane species. But if we compare the relevés of such montane or pseudomontane sites in the sixties and in the nineties we see that clearly *Aruncus*, *Polystichum*, *Galium rotundifolium*, etc. have decreased. So these montane or pseudomontane shadow slopes became warmer and therefore montane species have decreased.

Question: K. FUJIWARA

And the laurophyllous species have increased not only on the surroundings of the lake. So species like *Aucuba* or *Elaeagnus pungens* just spread on the mesic sites, but what has happend on the drier sites?

Answer: F. KLÖTZLI

Well, the dry sites have not much difference. These montane species I have shown, except *Knautia sylvatica*, they all grow on mesic moist sites. So I haven't shown a montane species growing on dry sites. There is an other equlibrium. I haven't spoken about that. The dry sites have also changed but not this way. On the dry sites we have a distinct decrease in heliophilous and acidophilous species and that's a bit queer because we have acid rain, but we also have nutrient rich rain. So nutrient indicators have increased to the detriment of acidity indicators. Acidity indicators, also in Bazzanio-Abietetum on Stagnogley or Pseudogley, they have retreated and other species especially nitrophilous species they have confirmed their positions. Again I refer to the general changes on this corridor from Schaffhouse to Chiasso.

Statement: P. VOSER

I noticed this study of WSL. But for me I'm sorry he made a mistake. Because he didn't notice enough that light factor also changed. At the beginnig of this century also in northern Switzerland the forests were cleared and you had a lot of light on the floor. And now they grew up, you have the input of NO_3 and all this resulted in a denser tree layer and in a more fertile soil, you have more humus in the soil, and all this made that the acidophilous group which needs more light than the other groups diminished. For me personally the light factor is enough to explain why this acidophilous group decreased.

Question: V. PILLAR

I am trying to put the pieces of this puzzle together: reduced wood harvest, reduced litter harvest, reduced gracing within the forest areas and less bush fires. About the species change: there was a decrease in heliophilous species and it may be related to increased light competition as a result of a decrease in the removal of biomass. Then comes the problem, how can you distinguish these vegetation changes from the ones caused by climate change? I am not saying that climate has not changed. For instance, Prof. Klötzli showed an ordination diagram and there is a clear vegetation shift. Can you find out whether this shift has been a result of change in the pattern of use or change in temperature? My question is also if these 300 relevés are all from protected areas or have these areas also been managed?

Answer: G. CARRARO

To the second question: This areas were not managed or if managed only with light impact. They cut only single stems but very rarely, according to the findings in the field, because the same stems were still there and we could see which stems are missing.

About the changes we must clear something which is very important: I did not come here to say all these changes are due to global warming, assolutamente no! I explained a lot of additional changes of ecologically important factors like bush fire frequency, litter harvest intensity and so on, are of much more importance for these changes in these forests. This is very, very important. But now the situation here in Insubria is so interesting and we asked that it is also possible that the climate has sponsored some changes that now are naturally going on in the forests. And so what we can do and this is the only possibility that we can control is to go in the forests which have the same history and the same conditions but only a difference in the mesoclimate, exposition and so on. And we control there whether we have more or less laurophyllous species. And these species are very sensitive to these changes of one to two degrees. So we are sure that the change in climate we had in last forty years has sponsored the spread of these species. But it is still a complex relationship. Among us we understand this complexity of this phenomenon quite easily, but if you talk with other people, they link the spread of the palm or *Prunus laurocerasus* very easily with global warming, although it is not as easy. We are sure that the warming helps pushing in this direction, we can control this on ecotonal sites, where the conditions are similar, but in natural ecosystems you never have only one factor, there are always a lot of different factors to take into account.

Statement: V. PILLAR

There is another piece of the puzzle that I would like to add: I see that you are noticing change much faster in the areas which were used by agriculture. In these areas there is less inertia, in natural areas the change would be much slower. So it is explainable: there is no agriculture any more, the changes are faster, these species are entering in a time-scale that we can notice.

Answer: G. CARRARO

For understanding the changes of the different species you must have an idea of the climax of the region. You must visit some forest which are near to the climax, not virgin forest, we don't have one hectar of virgin forest in Insubria. But they are near to the climax. Look in there, which species do you find. These species are probabely the species which live in the climax stadium or at least live better than the others. And this is very important to structure the different criteria that we have utilized for understanding the different changes in the different species. That's first. Secondly you must look in a region with different mesoclimatic conditions which species do you find in thermophilous sites or in ombrophilous colder sites. And so comparing the two criteria you can evaluate or estimate, but it is not a calculation, if you have a shift in the direction of global warming. And the direction is given by an analogues estimation. Go into the forest and look at the structure. It is closer to the structure of a subtropical forest than forty years ago. So the direction is based only on a qualitative evaluation, we can't calculate. With the works of Peter Voser we can also calculate anything in particular areas. But you must have regional climatic models you must know the fire history and all the other factors, and it is probabely that at the end you can say yes, we see something that only this 10-20% of the changes can be sponsored by the changes in temperatures.

Statement: F. KLÖTZLI

Just a quick addition: I think that Gabriele CARRARO put one point quite clear and this should be reminded: It is not the shift of one group. It is the increase and the decrease of different, contradictory groups and that is important. We take the problem from both sides. And the second point I just wanted to put into consideration is the fact that also in gardens in villages (not in townships) cultivated plants have changed. In the Zurich area no *Trachycarpus* was seen in my youth. And now, let's say, courageous people they have already *Trachycarpus* in Zurich in the gardens and *Araucaria* and *Cryptomeria* etc. Rather sensitive garden species have increased now. And not only in Zurich. I think this is also another sign which should be considered.

Statement: P. EDWARDS

I feel we made a very good beginning by going straight to the heart of the problem. One of the big criticisms of vegetation science is that its ability to establish causality is weak - that it relies too much upon correlation as evidence. As we have seen today, the problem with correlation is that there are often confounding variables. Later this week we have a discussion about what are the research needs in this area. People want to know about the likely effects of global change and I think the vegetation science community has to come up with credible methodology. My guess is it will be a combination of correlative methods (in other words, our traditional field survey methods), plus experimental work, plus modelling.

Recent shifts in vegetation boundaries of deciduous forests,
especially due to general global warming, pp. 69–84
edited by F. KLÖTZLI and G.-R. WALTHER

Vegetation shift and laurophyllisation: the possible role of forest fires

Marco CONEDERA[1], Claire HOFMANN[1], and Willy TINNER[2]
[1]FNP Sottostazione Sud delle Alpi, CH-6501 Bellinzona
[2]Geobotanisches Institut der Universität Bern, Altenbergrain 21, CH 3013 Bern

Abstract

Southern Switzerland is a fire prone area where fire has to be considered as a natural environmental factor. In the past decades, fire frequency has tended to increase due to changes in landscape management. The most common type of fire is surface fire which normally breaks out during the vegetation resting period. Usually this type of fire shows short residence time (rapid spread), low to medium fire intensity and limited size. South-facing slopes are particularly fire-prone, so that very high fire frequency is possible: under these conditions passive resistant species and postfire resprouting species are favoured, usually leading to a reduction in the number of surviving species to a few fire adapted sprouters.

Evergreen broadleaves are extremely sensitive to repeated fires. A simulation of the potential vegetation of southern Switzerland under climatic changed conditions evidenced the coincidence of the potential area of spreading forests rich in evergreen broad-leaved species with the most fire-prone area of the region.

Therefore, in future, wildfires could play an important regulating role: most probably they will not stop the large-scale laurophyllisation of the thermophilous forests of southern Switzerland, but at sites with high fire frequency the vegetation shift could be slowed or even prevented by fire-disturbances.

Introduction

Next to human activities, fire is the most ubiquitous terrestrial disturbance (Bond & van Wilgen 1996), changing ecosystem dynamics, removing

organisms and providing space for new colonisations. Thus, in fire prone areas ecological studies cannot be carried out without reference to fire.
The goal of this contribution is to present and discuss the possible role of forest fires in the vegetation dynamics of the deciduous forest belt of the southern slopes of the Alps.

Methods

The long-term fire history is described on the base of palynological studies (Tinner & Conedera 1995). For the reconstruction of the present fire situation, the forest fire data base of the FNP Sottostazione Sud delle Alpi was used (Conedera et al. 1996). The relationship between forest fire and postfire vegetation pattern is discussed on the base of both, the palynological approach (Tinner & Conedera 1995) and the phytosociological approach (Delarze et al. 1992, Hofmann et al. 1996). The computer model of Kienast et al. (1994) was used to simulate the impact of a changing climate on the potential vegetation cover and on fire regime in southern Switzerland (see also Brzeziecki et al. 1994 and Conedera et al. 1996).

Forest fires in Southern Switzerland

Fire history

Long-term fire history helps in understanding the ecological role of fire in a given region. Questions such as how natural are wildfires to forest ecosystems or how high is the natural level of forest fires are of central importance in the study of the role of fire in forest dynamics.
Tinner & Conedera (1995), studying the vegetation and fire history at Lago di Origlio during the Holocene, show that the fire history of Southern Switzerland is usually closely related to forest dynamics and that it can be divided into four main periods (Figure 1):
(1) Palaeo- and Mesolithic, with a moderate concentration of charcoal particles, probably representing the natural fire level ascribed to the corresponding climatic and vegetational conditions; (2) Neolithic, with an increase in concentration of charcoal particles, probably due to early human activities, (3) Bronze and Iron Age, with a considerable increase in the charcoal concentration, probably due to the slash and burn practice used by men to create open land; (4) Roman period to the present, with a dramatic decrease in charcoal concentration, probably due to early fire suppression practice. In fact, the introduction of chestnut in that period represented a real revolution in landscape management in southern Switzerland, changing from the slash and burn agricultural approach to chestnut forest grove and coppice management. From the chestnut forests all the biomass was used (fruits from the trees, litter

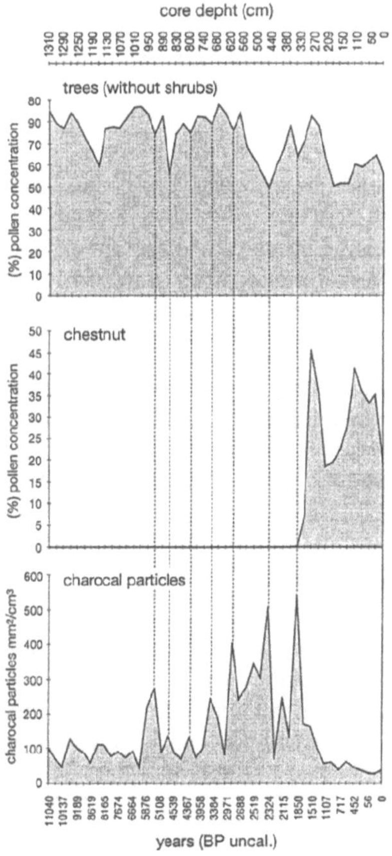

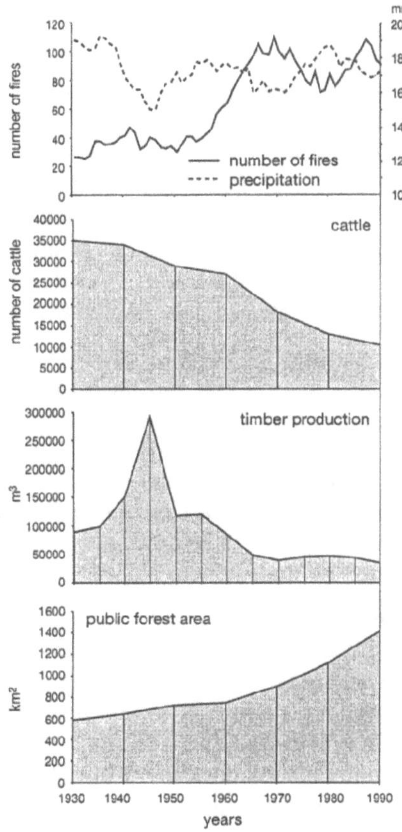

Figure 1: Long-term fire history in Southern Switzerland (Source: Tinner & Conedera 1995 modified on the basis of additional radiocarbon data)

Figure 2: Landscape management and fire history of the past in Ticino (Southern Switzerland)

for the animals, wood for the poles and for the fireplace, pasture, ...) so that nearly no necromass was left for wildfires.
Conedera et al. (1996), studying the forest fire history of southern Switzerland of this century, show that the fire level has increased by a factor 3 since the fifties. Figure 2 shows that the relationship between fire level and precipitations is overlayed by the effects of the changes in landscape management and to the decrease of the agricultural and sylvicultural activities in particular.
From this fire history the following can be concluded with regard to forest dynamics: southern Switzerland and probably the whole of the south slopes of the Alps is a fire prone area; fire has to be considered a natural environmental factor. Fire frequency is influenced by both climatic conditions and human activity. Due to the changes in landscape management, for the future an increase in fire frequency is possible.

Fire regime

The effects of forest fires differ widely, depending on when fire occurs (fire season), how fiercely it burns (fire intensity) and how often fire occurs (fire frequency). The combination of these three elements is termed the fire regime (Bond & van Wilgen 1996). The response of the vegetation to fire is fire regime dependent. We can describe the fire regime in Southern Switzerland as follows:

Fire season

Figure 3 shows that the main fire season in southern Switzerland is between December and April, during the vegetation resting period. In this period also the greatest area is burnt. The summer peak (July-August) is mainly due to lightning fires in the coniferous forests and in the upper deciduous forest belt. Usually these fires are not so large in terms of burned area.
The main fire season during the vegetation rest is common to all the regions located on the southern slopes of the Alpine Bow from France to Italy.

Fire intensity

No precise information is available about fire intensity in southern Switzerland. However, the intensity class can be estimated from the type of fire. Figure 4 shows that during the winter time the most common type is rapidly spreading surface fire. Therefore we can estimate the fire intensity as being from low to medium, compared with the high fire intensity reached in the crown fires during the summer time in the mediterranean regions.

Fire frequency

Figure 5 shows that south-facing slopes are more often burnt. This fact, due to the stronger irradiation, leads in certain cases to very high fire frequencies (up to 5 fires within 30 years) on south-facing slopes. There, only a fire adapted vegetation can survive.

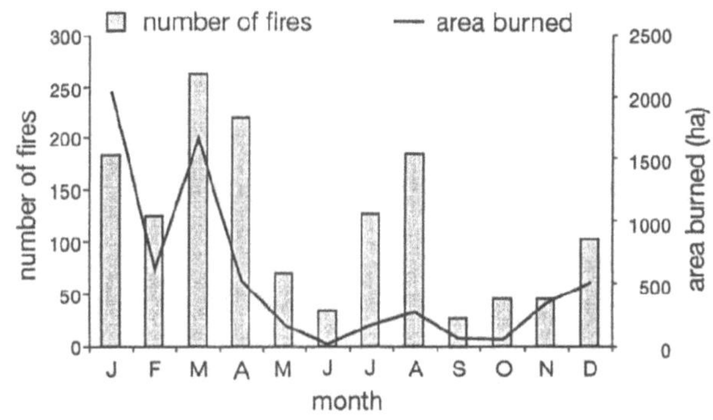

Figure 3: Monthly distribution of forest fires and burned area in Southern Switzerland (period 1981–1994)

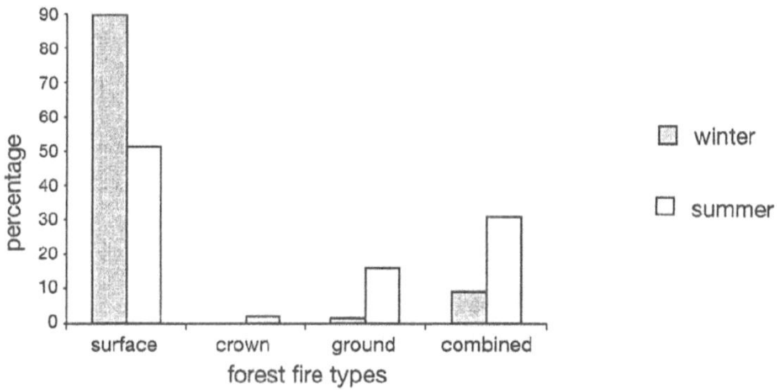

Figure 4: Distribution of forest fire types in Southern Switzerland (period 1981–1994)

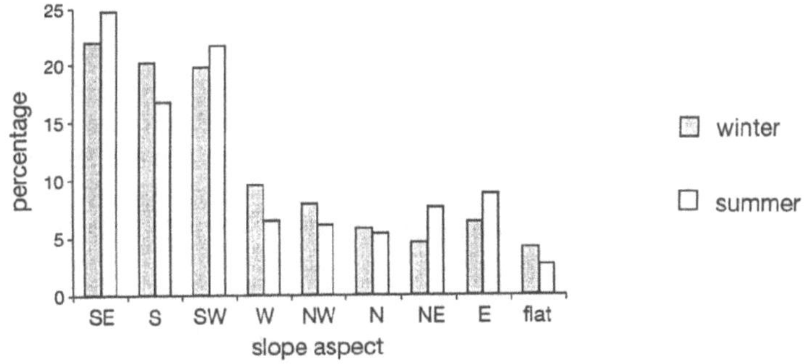

Figure 5: Fire frequency by slope aspect in Southern Switzerland (period 1981–1994)

From the fire regime the following can be concluded with regard to forest dynamics: in southern Switzerland the main fire type is surface fires during the vegetation resting period. Usually this type of fire shows short residence time (rapid spread), low to medium fire intensity and limited size. On south-facing slopes very high fire frequency is possible.

Fire ecology in southern Switzerland

Fire surviving strategies

The response of plants to fire in terms of survival and/or reproduction is important in understanding the effects of fire on vegetation dynamics. The ecological effects of fire depend on both the magnitude of fire-induced mortality, which frees space, and the importance of fire stimulated recruitment to utilize it (Bond & van Wilgen 1996). Figure 6 shows the most common fire surviving strategies.

vegetative survival
- active resistance (bark thickness)
- passive resistance (fire-stimulated sprouting)
- *above-ground sprouters*
- *below-ground sprouters*

reproductive survival
- fire-stimulated seed release
- fire-stimulated seed germination
- fire-stimulated flowering

Figure 6: Fire survival strategies (from Bond & van Wilgen 1996, modified)

Due to the particular fire regime of the deciduous forest belt of southern Switzerland (fires with limited intensity during the vegetation resting period) vegetative survival plays a very important role in postfire forest dynamics. We can find well documented cases of both active resistance (*Quercus petraea*) and passive resistance (*Castanea sativa*, *Pteridium aquilinum*, *Molinia coerulea*, ...). The efficiency of the postfire recruitment of the sprouting species reduces the reproductive survival to play a secondary role, at least as regards the long- term effects of repeated fires (Delarze et al. 1992, Hofmann et al. 1996).

Fire effects on Forest vegetation

Plant and tree species show very different fire life histories. From palaeoecological studies on fire and vegetation-history in southern Switzerland (Tinner & Conedera 1995), we know that some species react very sensitively to fire (Figure 7), while others are stimulated by fire (Figure 8). Studying the present vegetation Delarze et al. (1992) and Hofmann et al. (1996) found similar patterns. Resprouting capacity is confirmed as playing a basic role in the fire survival of species.

The selective effect of fire increases with high fire frequencies, because of the fire-interval dependent effects (Bond & van Wilgen 1996, Delarze et al. 1992, Hofmann et al. 1996) and because of the long-term effects of fire on site conditions (Delarze et al. 1992, Hofmann et al. 1996).

In the long term the selective effect of repeated fires reduces the number of plant species to a few fire adapted plants which usually produce an increased amount of flammable fuel, which leads in certain cases to an increased fire hazard (Delarze et al. 1992).

In conclusion, long-term forest fire effects on vegetation consist in general of a reduction of the plant species. In the mixed deciduous forest on limestone this corresponds to an impoverishment of tree species (Hofmann et al. 1996). In abandoned chestnut forest on acid soils this corresponds to a blocking of the natural evolution toward mixed deciduous forests (Stanga 1995).

Possible fire effects on Laurophyllisation

Evergreen broadleaves are thought to be in general very prone to fire (Walther 1995). This assumption was confirmed by Hofmann et al. (1996) at Monte Brè near Lugano, studying the effects of different fire frequencies on mixed deciduous forest on limestone in the foothill region of southern Switzerland.

As shown in Table 1, evergreen broadleaves are not resistant to fire and tend to disappear after the second forest fire within 25 years.

According to Gianoni et al. (1988), in the peroceanic foothills of the southern Swiss Alps (characterized by hygric-submediterranean vegetation) there are ecological niches for thermophilous forests rich in laurophyllous species.

Walther (1995) already registered a clear laurophyllisation gradient **in unburned areas** between the warmer point of the hyperinsubric lake region (Ronco s./Ascona) and the region of Bellinzona in Canton Ticino. The border situation of Insubria between deciduous forest and evergreen broadleaved forests, as demonstrated from the climatic parameters (Klötzli 1988), lead Klötzli et al. (1995) to explain this laurophyllisation process as a vegetation shift sponsored by a climatic change (specially winter temperature above the normal values) as well as a more extensive type of forest management.

plot number	35	34	27	33	25	26	24	18	32	19	23	20	1	12	7	3	4	2	5	9	14	22	21	8	30	31	29	28	6	13
number of fires since 1969	0	0	0	0	0	0	0	1	1	1	1	1	1	1	1	1	1	1	1	2	2	2	2	2	2	2	2	2	2	>2
shrub layer																														
Buglossoides purpurocaer																+		+												
Trachycarpus fortunei	+	+	+	+				+	+	+																				
Eleagnus x submacrophyl	+			+							+																			
Ilex aquifolium								1	2	+																				
Laurus nobilis	3	2	1	+	1		+	+	+	+																				
Lonicera japonica	1			+																										
Prunus laurocerasus		1					1																							
Pyracantha coccinea			+					+		+																				
Taxus baccata				+			+				+																			
herb layer																														
Ilex aquifolium									+	+																				
Laurus nobilis	1	1	+	+	+			1	+	+																				
Lonicera japonica	+	+							+																					
Prunus laurocerasus	+							+																						

Table 1: Presence of evergreen broadleaves in relation to fire frequencies at Monte Brè (Hofmann et al. 1996)

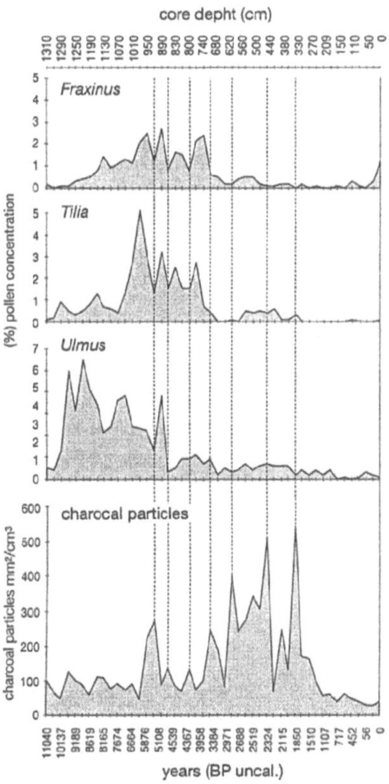

Figure 7: Fire prone species in Southern Switzerland: palaeohistory of *Ulmus*, *Tilia* and *Fraxinus* spp. (Source: Tinner & Conedera 1995).

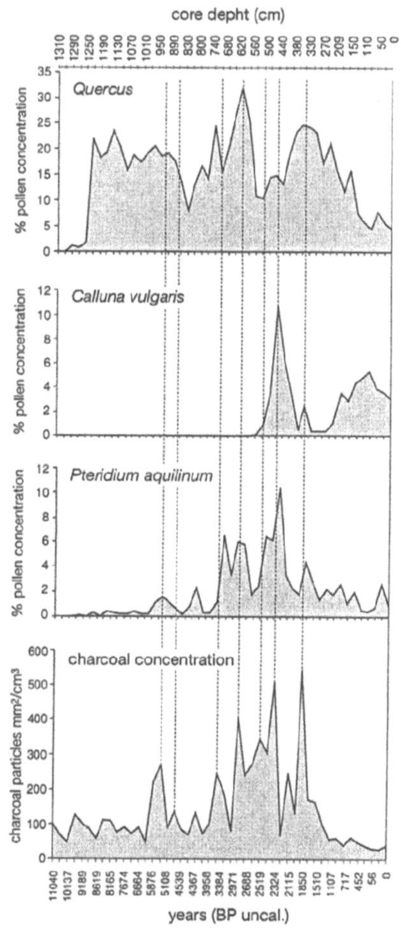

Figure 8: Fire stimulated species in Southern Switzerland: palaeohistory of *Pteridium*, *Calluna* and *Quercus* spp. (Source: Tinner & Conedera 1995).

Conedera et al. (1996) used a vegetation model of Switzerland to simulate the impact of an increase of 2.4°C in the annual average temperature on the potential vegetation cover (Kienast et al. 1994, Brzeziecki et al. 1994) and referred the vegetation shift to the present fire distribution. Figure 9 shows that the area of the thermophilous vegetation of southern Switzerland (which could easily correspond to the forests rich in evergreen broad-leaved species) thus obtained corresponds to the most fire-prone area, where more than 50 percent of the wildfires between 1969 and 1993 took place.

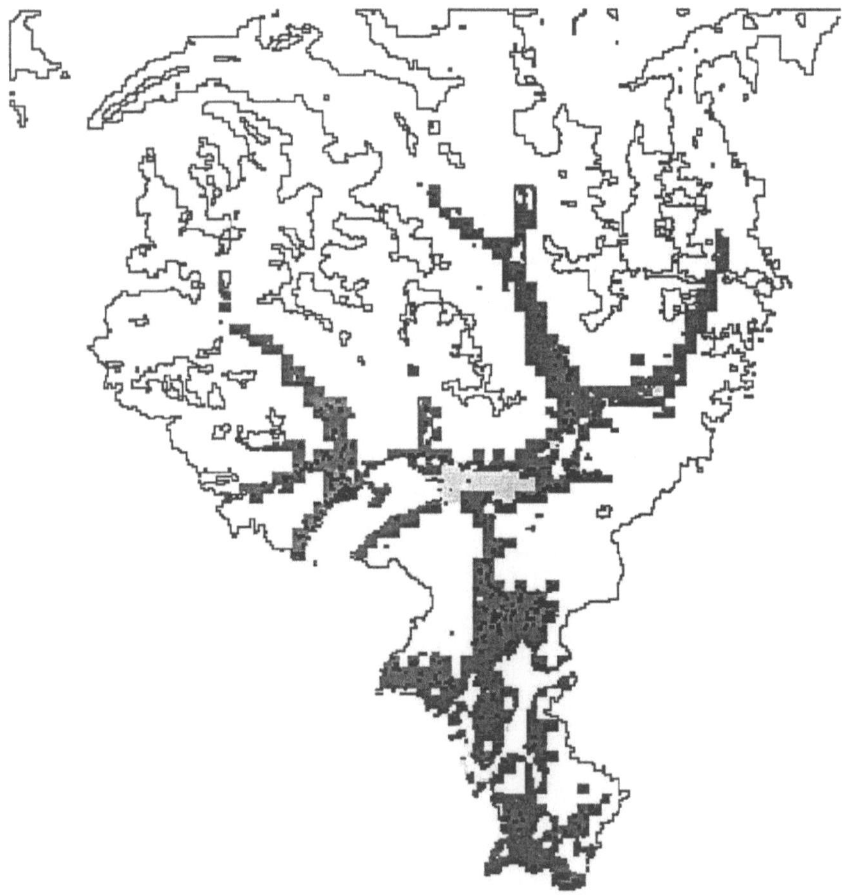

Figure 9: Simulation of the enlargement of the thermophile area of Southern Switzerland as a result of an increase in the average annual temperature of 2.4°C (Source: Conedera et al. 1996)
present distribution simulated future area
starting points of fires (1969–1991)

Thus wildfires in future are expected to play an important regulatory role in the laurophyllisation processes in southern Switzerland.

Conclusions

The southern foothills of the Alps are fire-prone. Fires take place mainly in wintertime in the colline and thermophilous region of the area, where the favourable climatic conditions allow the spread of an evergreen broadleaved vegetation. The particular sensivity of evergreen broadleaves to repeated fires endows wildfires with a regulatory role in the laurophyllisation process.
In the future this role could increase in importance with the enlargement of the potential laurophyllous area due to climatic change, because fire frequency in the colline region is not expected to be reduced in the coming years. The selective effect of fire will depend on the postfire recruiting capacity of the individual evergreen species and of the whole stand, once that a mature development stage is reached.
Fire will not stop large-scale laurophyllisation, but in special conditions (sites with high fire frequency) the vegetation change may be delayed or even prevented by its disturbance-effect. Where laurophyllisation takes place, fire could have a regulatory effect among evergreen broadleaved species.

References

Bond W.J. & van Wilgen B.W. (1996): Fire and Plants. Chapman & Hall, London.

Brzeziecki B., Kienast F. & Wildi O. (1994): Potential impacts of a changing climate on the vegetation cover in Switzerland: a simulation experiment using GIS technology. Price M.F. and Heywood D.I. (eds): Mountain Environnements and Geographic Information Systems, pp 263-279.

Conedera M., Marcozzi M., Jud B., Mandallaz D., Chatelain F., Frank C., Kienast F., Ambrosetti P. & Corti G. (1996): Incendi boschivi al Sud delle Alpi: passato, presente e possibili sviluppi futuri. Rapporto di lavoro PNR 31. Zürich, Hochschulverlag an der ETH.

Delarze R., Caldelari D. & Hainard P. (1992): Effects of fire on forest dynamics in southern Switzerland. Journal of Vegetation Science 3: 55-60.

Gianoni G., Carraro G. & Klötzli F. (1988): Thermophile, an laurophyllen Pflanzenarten reiche Waldgesellschaften im hyperinsubrischen Seenbereich des Tessins. Ber. Geob. Inst. ETHZ, Stiftung Rübel, Zürich, 54:164-180.

Hofmann C., Conedera M., Delarze R., Carraro G. & Giorgetti P. (1996): Effects des incendies de forêt sur la végétation au Sud des Alpes Suisses. Bellinzona.

Kienast F., Brzeziecki B. & Wildi O. (1994): Computergestützte Simulation der räumlichen Verbreitung naturnaher Waldgesellschaften der Schweiz. Schweiz. Z. Forstwes. 145(4): 293-309.

Klötzli F. (1988): On the global position of the evergreen broad-leaved (non ombrophilus) forest of the subtropical and temperate zones. Veröff. Geobot. Inst. ETHZ, Stiftung Rübel, 98: 169-196.

Klötzli F., Walther G.-R., Carraro G. & Grundmann A. (1995): Anlaufender Biomwandel in Insubrien. Verhandlungen der Gesellschaft für Ökologie 26: 537-550.

Stanga P. (1995): Evoluzione naturale dell'areale castanile. Forestaviva 13: 59-62.

Tinner W. & Conedera M. (1995): Indagini paleobotaniche sulla storia della vegetazione e degli incendi forestali durante l'Olocene al Lago di Origlio (Ticino Meridionale). Bollettino della Società Ticinese di Scienze Naturali 83(1/2):91-106.

Walther G.-R. (1995): Ausbreitung und Grenzen laurophyller Arten im Südtessin. Diplomarbeit am Geobotanischen Institut der ETH Zürich.

Plenary discussion

Question: V. PILLAR

Do you have any information about the history of use of these areas. You noticed an increase in fire frequency in the last thirty years. Why was this? Was it related to a change in grazing regime?

Answer: M. CONEDERA

We have a couple of indicators. We know that the number of cattle decreased since the sixties as well as the timber harvest. So more biomass is left in the forest. In addition the forest area is increasing. Therefore there are more contact points between human activity and fuel. In other words there is now a greater chance that a fire starts.

Question: G. Doyle

Thank you very much for your presentation. You clearly indicated that the fire regime has changed since about 1950. Following our discussion yesterday, we were trying to unravel the influence of climate from other ecological influences that impact on the forests. It appears from your presentation that the changing fire regime since 1950 may have promoted laurophyllisation, and that fire management with constant burning might reverse that process.

Answer: M. CONEDERA

I don't believe that within the next ten years we will be able to link this evolutionary process. We are not able to decide, to have any influence, but I have to say that I don't see any reason to stop this evolution.

Question: F. KLÖTZLI

If I look at the fire tendencies, I guess, it is first of all triggered by the occurrence of litter of *Pteridium* and of *Castanea*. And how would it be if there was less *Castanea* but more *Quercus*? There is also a lot of litter, but this is less dangerous, I guess.

Answer: M. CONEDERA

It is possible. In addition in oak stands you would have a crown cover already since the first summer after the fire, because they don't die and so maybe there is less chance for *Pteridium* to develop and produce a lot of "fuel". But this is a hypothesis, maybe *Quercus* is not so natural here and it was also increased by fire. We should work more on such hypothesis using the palynological approach. But I do believe that maybe the high oak frequency that we got before the Romans is the result of human caused fire. We should try to study this point.

Statement: F. KLÖTZLI

That's where I just wanted you to get in. In case you look at *Castanea* as a non indigenous species, introduced by the Romans, so nature is trying to wipe it out to the benefit of *Quercus*, which then prevents further fires. So actually I see a natural regulation happening in all this. It is the "fight of nature" if you want to use this term, a "fight of nature" in a wider sense against the foreign species. So in case this carries on and in case *Endothia parasitica* is still active, in the long run there would be a certain change which is against fire and then laurophyllisation might be spreading, theoretically spoken.

Answer: M. CONEDERA

It could be, but we are going to have low frequency of fire and not no fire! But I agree, this is quite a good scenario.

Statement: F. KLÖTZLI

A word to the effects of bracken: If I look at the evergreen broad-leaved forest I know, especially in the tropical mountains, they are also fire prone, now and then they get a fire, and why do they get a fire? Very often they have a lot of bracken, also in the tropical mountains. And that's why they also burn readily, but they don't burn in the crowns. So this is a similar effect. It wouldn't, let's say, eliminate laurophyllous species, it would just give another herb layer in a way and another regeneration capacity for certain species.

Answer: M. CONEDERA

About the chestnut evolution: We have studied it with aerial photographs from the sixties and from present time and we saw that chestnut groves are decreasing very rapidly. There is no chance to keep it without management. I can give you the concentration of the pollen of *Fraxinus* and *Betula* as an indicator of it. As you can see, the pioneer species *Betula* is increasing since the sixties very rapidly, and *Fraxinus* which follows *Betula* in the forest is increasing dramatically since the eighties now. And we know if you look around, there are no pure chestnut forest stands anymore. If you don't manage it, chestnut is not able to regenerate. And this is one aspect which is very important for the place here, because as you know the pollen of *Fraxinus* and *Betula* are very allergenic. So in southern Switzerland the percentage of allergic people is getting higher and is now about twice what you have e.g. in Zurich. And this will be a major problem for the future for us. It is like we switched off the management thirty years ago and now we start to feel the effects: fires, laurophyllisation, pollen and so on. And we don't know exactly what is coming next. These are just first spots, but I do believe that there are more effects that we still don't know.

Statement: F. KLÖTZLI

Maybe we have no sensitivity to new species, we have to develop it first. So it would be in favour for us if we would have more *Cinnamomum*.

Statement: P. EDWARDS

We saw yesterday some nice data showing that the climate has changed in the last forty years or so, with many fewer frost days and higher minimum temperatures. I wondered if there is a correlation between these temperature changes and e.g. the frequency of burn days or days in which fire would be a real danger in this area. The real point behind this question is that we have regarded fire as a management change, but it could indirectly be due to a climatic change. In contrast when we were talking about the spread of some evergreens, it was suggested that this was a response to climatic change and we didn't need to think about management.

Answer: M. CONEDERA

In case of fires the management does overlay everything. It is much more important, if you look at it in the long run. Of course the climatic shift plays a role to fires. But, even in the short run, social factors play a role: we constructed a model to predict the fire probability each day and we saw that the human activities play a significant role. For example in Saturday and Sunday the fire frequency is higher than on weekdays. But if you have a closer look you even can see that in the sixties and seventies the frequency of fire was higher on Sundays because people still did work on Saturday. And then since we have a five days week, the situation has changed and now the fire frequency on Saturdays is higher than on Sundays. Another good example for the human effects is the introduction of the summer time. Since 1981 we have the summer time in Switzerland and if you look at the starting hours of fires you can see an one hour shift in the distribution in the five years following 1981. So the human behaviour is able to mask meteorological effects. Maybe only traumatic climatic changes or meteorological situations would overlay these anthropogenic factors.

Statement: A. BORHIDI

I would like to ask you if according to your results fire is restricting laurophyllous species. But among them, there are some which are really fire protected plants like palms, bamboos, ... And I think that maybe - if not directly - but indirectly by germination of the seeds of the palms for instance and the underground shoots of the bamboos they can well apply and even be supported by fires.

Answer: M. CONEDERA

It is possible. At the present stage of the laurophyllisation it is maybe to early to see this effect in Southern Switzerland. In addition the vegetative reaction of the present species – *Castanea* in particular – is very strong, so that laurophyllous species still have little resprouting chance.

Statement: F. KLÖTZLI

Just a word to *Trachycarpus*. If you measure the temperature during a fire in the heart of a palm (of course I haven't done it with *Trachycarpus* because I can't put fire here but I have done it several times in the Savannah areas) there is no change in temperature compared to the outside temperature. So the meristematic tissues survive if a fire is going through which is mostly rather quick because there is a lot of litter around the palms. So palms have a certain advantage, that's true. And if I look at a palm which has been killed by fire, well by accident or so, then there are root suckers soon and everywhere in the shape of a ring of root suckers all around the old stem. So palms have a lot of defender strategies.

Statement: A. BORHIDI

Palms are believed as Savannah trees although they are not, low palms are virginally scrub elements. Their survival is supported by fire and therefore after the bush fires they remain as elements of the secondary Savannahs.

Statement: F. KLÖTZLI

If I look at the Japanese palms we name *Trachycarpus* it is actually in a way not endangered by fire, yes when it is a sapling, but the moment it has grown up a little bit and fire frequency is not very high, then we have a small stem and this stem can't be killed very easily. So there is a chance for *Trachycarpus* to survive fire prone formations.

Question: K. FUJIWARA

Do you have any kind of fire resistance in the laurophyllous species? We saw many patches in the forest especially on mesic sites. Many laurophyllous species occur and make small patches. In the northeastern United States, in South- and North-Carolina they have this phenomenon also. Of course this is a special type of vegetation. This kind of vegetation has just appeared around the laurophyllous forest, but always under fire management. Then it changes to evergreen shrubs in wetlands like *Persea* or evergreen *Magnolia* and *Lyonia*, everything is evergreen. Then often, in summer they are burnt but they can sprout very well. It is well known that it becomes tall forest without fire. Also in Japan, where we had the earthquake in Kobe City, some part of the burnt area had very small shrub areas with *Aucuba* or *Fatsia* and sometimes with *Cinnamomum camphora* plantations. These resisted the fire. They just shut out the fire. They were burnt but afterwards sprout and regenerate again. So even when this area had a large fire, evergreen shrubs sometimes stopped the fire. Do you have any similar example?

Answer: M. CONEDERA

It is still to early at the moment, but it could happen later as Prof. Klötzli said. Maybe if we have a progression of this shift, then we will have this situation. It could be the future for us.

Statement: K. FUJIWARA

Yes, I think so. In southern Japan we have many burnt forests, especially red pine (*Pinus densiflora*) secondary forests. They always burnt, then the people thought to stop the fire and they planted evergreen shrubs. Because, evidently, evergreen trees were just at the boundary of the former fire and they could clearly stop the fire.

Recent shifts in vegetation boundaries of deciduous forests,
especially due to general global warming, pp. 85–89
edited by F. KLÖTZLI and G.-R. WALTHER

Potential impacts of climate change on species richness in mountain forests – an ecological risk assessment

Norbert KRÄUCHI and Felix KIENAST
Dept. of Landscape Ecology
Swiss Federal Institute of Forest, Snow and Landscape Research (WSL/FNP)
CH-8903 Birmensdorf, Switzerland

In this ecological risk assessment we evaluated potential climate-induced vegetation changes in mountain forests of Central Europe and possible impacts on species richness. The analysis was performed on all 1km grid points of the Swiss forest inventory (c. 11500 points) as well as on two subsets representing the forested points within the geographical limits of two inventories of legally protected reserve areas. The core of the approach is (a) a conceptual model of the movement of climatic ranges along altitudinal gradients as a result of global warming and (b) a spatially explicit forest community simulator that predicts a potential natural vegetation unit for any given 1 km grid point of the Swiss forest inventory for both current climate and for altered climate regimes. The model is derived from empirical data featuring the relationships between quasi-natural vegetation types and measured site variables. Species richness for any modelled forest community is estimated on the basis of phytosociological relevés.

The conceptual model showed that out of 109 legally protected landscapes, sites and natural monuments of national importance (minimum area > 1km^2) about 40 - 50% have an altitudinal or degree-day range that is able to host migrating species within the reserve limits if climate is warming in a moderate way (increase of mean annual temperature between +1.0 and +1.4°C). In the case of strong warming (increase of mean annual temperature between +2.0 and +2.8°C) only 20 - 30% fullfil this criterion. The spatially explicit forest simulator showed that under a temperature increase without simultaneous increase in precipitation (warmer and more xeric due to increased evapotranspiration), Fagus-dominated communities in the colline-submontane belt might eventually be replaced by oak-hornbeam (Carpinion) communities. In the montane belt, the dominance of conifers will be seriously threatened by an invasion of deciduous species from the low montane and submontane belt. Under warmer and wetter conditions the vegetation shifts might not be as drastic as under warmer and more xeric conditions and the shift towards oak and oak-hornbeam communities on the Plateau is not supported. Concerning species richness, the models showed that in the case of warmer temperatures and constant precipitation (warmer and more xeric due to increased evapotranspiration) overall species richness is increasing on all 1 km points of the Swiss Forest Inventory as well as on all selected subsets representing the forested points within legally protected reserve areas. In the case of warmer and wetter conditions the risk assessment does not show any drastic changes in the long-term species richness.

(The original paper was published in: *Biological Conservation Vol. 83, No.3, pp. 291-305, 1998; Elsevier Science Ltd.*)

Plenary discussion

Question: P. EDWARDS

How can you take account of the inertia in a system.? In developing your spatial model you took the Ellenberg & Klötzli classification as your baseline, with the implication that this represented an equilibrium between environmental conditions and vegetation. But clearly that's not true -we heard yesterday how the climate changed so much during this century and the vegetation cannot have adjusted fully to these changes.

Answer: F. KIENAST

The paradigms of the spatial model are "equilibrium". In this model, you can for example say that we have the potential for a certain species assemblage but within this community we allow this and that dynamic features. This is the way that we go now with the treeline model. We allow a certain species assemblage and within this frame we allow a dynamic evolution of the site. It is definitely a first step to guess what the potential of a site is.

Question: A. BORHIDI

I would like to ask how it is calculated that synchronously with temperature change there are two other processes on the way: acidification and eutrophication. And the other problem is that in the model of changes we are thinking of the actual phytosociological units but it is not sure that these units will migrate as units. They perhaps will be desorganized and reorganized at other places. These are the difficulties to be considered.

Answer: F. KIENAST

Again, I completely agree. We have to consider this under different paradigms and one paradigm is the community paradigm, another paradigm is the equilibrium. Under these two paradigms, the model is probably the best guess of a potential of a site. It is a pragmatic way of looking at heterogeneous aggregation of sites in space. But I completely agree, we have to understand these models under certain paradigms. The communities might brake up completely under changing climates. But if we change to individual species models we do in principal accept the same paradigms because you can never define life history strategies of species independent of competition. In both cases - single species or communities - we simulate realized niches and not physiological or potential niches.

Statement: F. KLÖTZLI

I remember your first models many years ago and we have discussed these models many times and I have to say I am deeply impressed by the progress you have reached in assuring us a let's say sensible model. But there are maybe two factors which in due time should be included in these models: One of these factors I mentioned already many years ago that's the pathogenic moment. If I look at beech and its sensitivity to fungus attack on the bark then I think 11/2°C more will give more difference than you depict without thinking of the pathogenic moment. That's one thing and the other thing are the vectors. These vectors are many times organismic, especially birds. If a bird population changes, now just for instance if there are pathogenic moments in the black bird population, then we have a completely different view on the seed dispersal. So in future you have to opperate with more external factors to show the possibilities we have. And the third thing I want to say are possible open "windows". Don't forget laurophyllisation is maybe a new process, but the species are here since 1670. They have all been as we say in german "Gewehr bei Fuss". They always had been ready on a point of let's say invading but they had no chance to invade the areas, they rested and had their peace in the parks and the gardens. And all of a sudden you have this invasion, so the "window" is larger than you think. We have a quiet period where the species were present but not in a way giving viable seeds to the forests. And now this picture has changed. So I would suggest that maybe you try to use the larger window and try to simulate a future development with all these different factors which we have already discussed. Just three suggestions, thank you.

Question: V. PILLAR

The map of Switzerland with the relevés of the sixties has been shown. Can you estimate actual temperature changes in these different locations based on climate data and the digital terrain model?

Answer: F. KIENAST

We could calculate the temperature increase as Norbert Kräuchi has shown for approximately 50-80 stations in the last sixty years and we could couple them with the digital terrain model.

Statement: K. FUJIWARA

I understand you compared your climatic potential vegetation, it is very good. Also I agree with Prof. Klötzli, I wanted to say something about the seed dispersal by birds. When a bird population changes, then how does the forest change? That is also a potential difference. But when you are discussing actual vegetation, that is a different matter because vegetation shift is not always due to temperature. When temperature increases then maybe bird dispersal of plants and seed distribution and also fire can occur more frequently. So if that is the case, then actual vegetation shifts are quite different from the potential ones. Also after fire, soil may erode very much. So if you put other factors into the model other interesting phenomena may be found as well.

Statement: H. BRUELHEIDE

Thank you very much for your very stimulating presentation. I appreciated the predictive vegetation maps of Switzerland very much and would like to have such a map for Germany, too. In my opinion, one difficulty is obtaining appropriate climatic data. Did you include absolute minimum temperatures in your climatic model or did you only use average temperatures?

Answer: F. KIENAST

It is always a question of data availability. We started with very crude estimates of spatial data. And then we ended up with more and more sophisticated input parameters. At the moment we have 50 longterm stations with temperature minima and here I will try to do a spatial extrapolation with these fifty stations to get the number of days for each month where minimum temperature drops below zero. Here I would like to do a frequency analysis and then do a spatial extrapolation, because we also have the impression that at the upper timberline the shifts that might take place there are due to these minimum temperatures. For example the distribution of *Pinus cembra*.

Statement: F. KLÖTZLI

I would very much like to endorse that. We should consider the frequency of extremes, because many parts or in nearly all the parts of the world the vegetation or the distribution of certain types of the biomes is due to extrema. If I look at Steppe - forest boundaries it's the extrema of drought periods; this could also be valid for our Valais. And in other parts it is the extrema in the deep temperatures, the low temperatures especially in the deep frost periods. So I think it is absolutely necessary, also now coming back to laurophyllisation, to judge the amount of laurophyllisation you have to expect.

Statement: G. CARRARO

A question to the increasing number of plots where the vegetation is not so adapted. We saw yesterday that it is very important to know the natural climax of the vegetation of a site. In Ticino, in southern Switzerland, we know that a lot of places with *Castanea* and *Quercus* are a product of fire and it isn't directly related to climate. So we should have in these places beech forests. We have some beech forests on the foothills, at 200-300m a.s.l. And so for me the first important thing is to make better use of the regional knowledge about the potential climax vegetation and not only the old relevés of the sixties.

The second point in the case of the Ticino: an increase in temperature would decrease the number of the plots which are not adapted, because we have oak and *Castanea* forest now on places which are to high and so with climate warming we get more adapted forests than today.

Answer: F. KIENAST

In the analysis together with Marco Conedera we saw that comparing actual species composition in the Ticino with the model we also saw many plots that become better adapted with climate change. I remember that because that was a surprise for us. I mean you see that also in other parts of the country, but here in Ticino you see that more pronounced. I mean in principal if you have the static model you have always the realized niche and the model is always very week in the border regions of these niches. And the Ticino is one of the border regions not only politically but also from the relevés. It would be maybe interesting to have Italian data here, but I'm not so sure because of this biome shift. I'm not so sure if you can simply say, oh we go 200km south take all the relevés and then we have solved this problem here, I'm not so sure, because maybe here we are at the biome border and the statistical model principal has a tendency to react very week on biome borders and this is the case here in Ticino.

Statement: A. BORHIDI

Just a remark about the Holdridge model, you showed. It is a very attractive model but doesn't work in the reality. It is very dangerous that several latin American countries are mapped upon this base, but practically these maps are useless. I made the vegetation map of Cuba upon field work and upon the model of Holdridge with the result of two completely different maps.

Statement: F. KLÖTZLI

Don't forget Holdridge had been working on the other side of the sea, in Costa Rica.

Answer: F. KIENAST

I mean we wanted to be very honest here in showing all the advantages and disadvantages of modelling and I think that's the way to go. We shouldn't consider models as being more intelligent that we are, we should consider models as a technical tool and for enhancing our knowledge of systems.

Recent shifts in vegetation boundaries of deciduous forests,
especially due to general global warming, pp. 91–120
edited by F. Klötzli and G.-R. Walther

Changing climate and the effects on flora and vegetation in Central European Cities

Herbert Sukopp[1)] & Angelika Wurzel[2)]
[1)] Institut für Ökologie der Technischen Universität Berlin,
D-12165 Berlin, Germany
[2)] Deutscher Rat für Landespflege,
D-53179 Bonn, Germany

Urban climate in Central Europe

As a result of climatological and environmental research some scientists expect a global warming within the next decades. The causes seem to be man-made. The question is whether a warmer climate has an influence on flora and vegetation.

Since the 1850s the effects of a global climatic warming have been anticipated by the rise of temperature in many big cities. Cities have particular climatic characteristics (Landsberg 1981, Kuttler 1993):

- higher air pollution: gaseous pollution is 5-15 times more, condensation nuclei about 10 times higher;
- altered radiation: 5-15% less hours of sunshine, 20-25% less direct solar radiation, ca. 10% less surface albedo, 12% more atmosphere back radiation lead to an increased net radiation of 11% at noon or 47% in the evening;
- the relative humidity is between 2% (winter) and 10% (summer) lower; on clear days this difference can reach 30%;
- wind speeds are reduced by 10-20% due to the roughness of the city surface, calm days are 5-20% more often;
- the annual mean precipitation is increased by up to 20%.

The most important ecological result of these effects is a higher temperature. Cities are „heat islands“ or „hot spots“ on the surface of the earth. The difference depends on the size of the city, it can reach 12°K on clear days, or 1 to 2°K in the yearly mean temperature.

Climatic conditions within a city can vary considerably, depending on the type of construction, paving, location in the city and, especially the distance to large

vegetated areas. According to these changes, different climatic zones usually more or less concentric, can be distinguished.
The city heat island usually covers the same area as the built-up area, but changes in wind direction can mean other areas are also temporarily heated.
Air pollution is caused by traffic, heating of buildings, power stations and industry. Trace gases, excepting low-level ozone, occur at concentrations five to fifty times higher than usual.
The warmer climate in cities causes reductions

- of the frost period and the frost intensity and
- of the number of frost days (to nearly half of the normal amount) and of snow-days.

The effects on the vegetation in cities are

- a prolongation of the vegetation period (e.g. in Vienna about 10 to 20 days yearly) and
- a shifting of phenological phases.

As a result of the above mentioned facts, higher temperatures and aridity are specific conditions for the existence of flora and vegetation in cities.

Changes of flora[1] and vegetation in Central European cities

Changes of flora and vegetation in Central Europe have well been documented (e. g. Haeupler & Schönfelder 1989). Special investigations concerning the interrelationships between boundaries of plant areas and climate are available too (Jäger 1968). Flora and vegetation of cities have been described and mapped for a long time (Sukopp & Wittig 1993). It is possible to draw conclusions about environmental changes, e. g. also climatic changes, using historical data in comparison with present conditions (Sukopp 1973).
Lists of urban plants have been published by the "Arbeitsgruppe Methodik der Biotopkartierung im besiedelten Bereich" (1993) and by Frank & Klotz (1990). Some examples of non-native urban plants are presented in Table 1 for spontaneously growing plants.

Remark by the editors: For details in Zürich compare Landolt (1994, 1995, 1996, 1997), Bot. Helv. 104, 105, 106, 107.

[1] Nomenclature according to Rothmaler (1988) or to Stace (1991)

	Tempe-rature value (Ellenberg 1991)[1]	Moisture value (Ellenberg 1991)[2]	Remarks (Haeupler & Schönfelder 1989)	Examples	References
Ailanthus altissima (Miller) Swingle	8	5		Berlin Münster, Essen, Düsseldorf	Kunick 1982 Kowarik & Böcker 1984 Wittig et al. 1985
Amaranthus retroflexus L.	7	4	Neophyte, possibly of North-american origin, cultivated since 1794 in Germany, mapped in spontaneously stade in 1815. Today found fully integrated in many regions.	Braunschweig	Brandes 1977
Artemisia absinthium L.	7	4		Braunschweig	Brandes 1977
Buddleja davidii L.	7	4		Berlin Münster, Essen, Düsseldorf Berlin Halle, Leipzig Ruhr area	Kunick 1982 Wittig et al. 1985 Kunick 1982 Frank & Klotz 1990 Dettmar & Sukopp 1991
Chenopodium botrys L.	7	4	In Berlin fully naturalized, building special communities; at other places inconstant. Mainly in cities naturalized.	Leipzig Berlin Halle, Leipzig	Gutte 1971 Sukopp 1971 Kunick 1982 Frank & Klotz 1990
Corydalis lutea (L.) DC.	7	6	Since middle ages at many places fully naturalized, especially at old walls.	Halle, Leipzig	Frank & Klotz 1990
Cynodon dactylon (L.) Pers.	7	4		Leipzig	Gutte 1971
Eragrostis minor Host	7	3		Braunschweig Leipzig Halle, Leipzig Münster, Essen, Düsseldorf	Brandes 1977 Gutte 1971 Frank & Klotz 1990 Wittig et al. 1985
Ficus carica L.			Fig-trees appear occasionally, especially in wine-regions, where they were planted. They are inconstant and synanthropic. At some places, for example at the port of Neuss, they are able to fruit.	Berlin	Kunick 1982
Hordeum murinum L. (Critesion Raf.)	7	4		Essen, Düsseldorf Münster Köln Bremerhaven	Wittig et al. 1985 Wittig et. al. 1985 Kunick 1984 Kunick 1979
Lepidium latifolium L.	6	5		Leipzig Poznan	Gutte 1971 Zukowski 1971
Rhus typhina L.			This species is not completely mapped; it occurs partly in a spontaneous and partly in a naturalized manner.	Halle, Leipzig Rhinelands	Frank & Klotz 1990 Adolphi 1995

Table 1: The distribution of some non-native plant species in Central European cities (Sukopp & Wurzel 1995, shortened).

[1] Meaning of the temperature values (scale from 1-9):

Altitude	Air-temperature (annual-mean temperature)
5 = montane to planar	6,0°C
6 = submontane to planar	7,5°C
7 = collin, planar (warm low-lands)	9,0°C
8 = only in extremely favoured valleys in the south-west	10,0°C
9 = at the warmest places	> 10,5°C

[2] Meaning of the moisture values (scale from 1-12):

2= situated between 1 and 3 (indicates strong desiccation and desiccation).
3= desiccation indicator, is more often found on dry soils than on fresh and humid soils.
4= situated between 3 and 5.
5= indicates freshness, is mainly found on medium wet soils, missed on wet and on often dry soils.
6= situated between 5 and 7.
7= moisture indicator, is mainly found on humid, but not wet soils.
8= situated between 7 and 9 (indicates humidity and wetness).
x= indifferent.

In cities the indicator values of a specific ruderal flora can be used. Many of these are short-living plants reacting quickly to climatic changes. The cosmopolitan character of many urban plants is potentially a tribute to the ubiquity of man's modification of environmental conditions and to his efficiency as an agent of dispersal (Salisbury 1961). The percentage of heat-resistent plants among ruderals is high according to their origin from warmer regions. Only in favourable conditions like in cities they are able to build up stable populations in Central Europe, e.g. C_4-plants, plants which need warmth to germinate and plants sensitive to frost and cold. The prevailing influence of climate on ruderal plants and their ability to react quickly to climatic changes make them a suitable indicator species. Climatic changes will thus be seen in the distribution of ruderals in cities as well as in changes of their northern borderline and their altitude in mountain areas (Hügin 1992, 1995).
A summary of the ecological demands of the urban spontaneous flora can be attained by comparing the indicator indices like those of Ellenberg (1979,1991). Kunick (1982b) made such an analysis and compared many cities with each other; the results for certain indices are shown in Figure 1.

Establishing of plants sensitive to frost and cold

Distribution and dispersal of non-native plants in Central Europe have been changed during the last centuries resulting in a higher percentage of plants sensitive to frost and cold today.

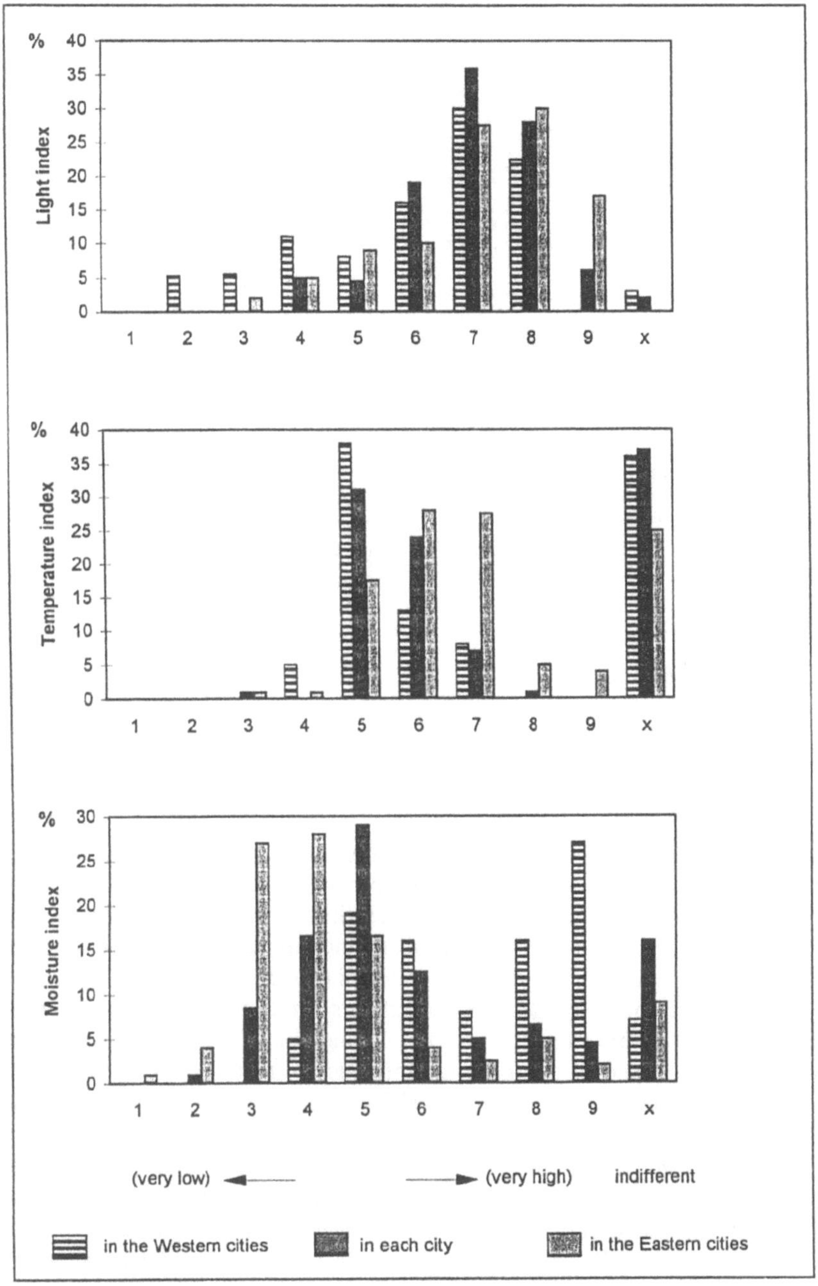

Figure 1: Ecological demands of the urban flora for certain indices in Central Europe (Kunick 1982).

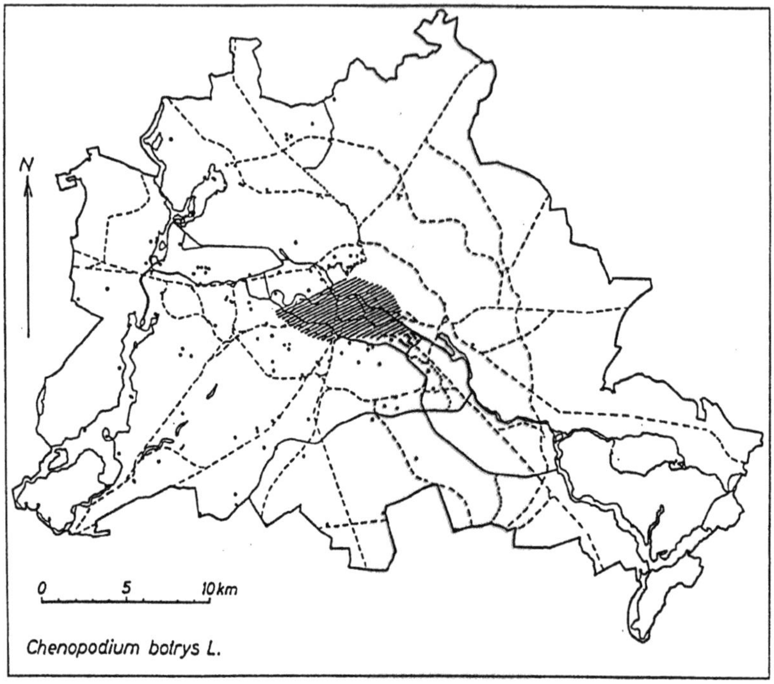

Figure 2: Distribution of *Chenopodium botrys* L. in Berlin 1947 to 1971. The area with the highest density is hatched (Sukopp 1971).

Chenopodium botrys of south-eurasian-mediterranean origin has spread over large areas of Central and Western Europe in consequence of increasing transport, trade and land use changes. It was first introduced in Berlin in 1889 (Sukopp 1971). The original habitats of this species are sandy and stony places along riversides and debris of rocks, altogether special habitats with low competition. Corresponding secondary habitats in Central Europe are road verges, fields and rubble sites. Under natural conditions the extent of such sites is low in Central Europe. Only under the influence of man the establishment of open calcareous sites with sand and gravel has been increased. Concerning the characteristics of ideal weeds *Chenopodium botrys* is distinguished only by high seed production under favourable environmental conditions. For germination on the other hand very specialized conditions in quality and intensity of light, photo-period and storage of seeds must be realized.
In Central European cities *Hordeum murinum* is an indicator of dry and warm sites on permeable soils. Since Roman times in Central Europe, Wittig (1991) describes it for the northern part of Central Europe as extremely urbanophilous. By mapping this species in Osnabrück 1978 and 1989/90 Hard & Kruckemeyer (1990) showed clear expansion without contemporarian climatic changes. They

indicated that this expansion did not depend on climatic change but on the extensification of land use and modified ideas of clean and well-kept cities.
Robinia pseudoacacia has been planted in Europe since 1623 and has been spreading sponteanously on the rubble of bombed cities but only in areas with relatively continental and summer warm climate (Kohler & Sukopp 1964). In Central Europe the invasion of *Robinia pseudoacacia* (Figure 3) has been more successful in areas with a subcontinental or submediterranean climate than in areas under oceanic influence.

Robinia pseudoacacia on areas with ruins or rubble:
O missing or rare *● often or common*

Figure 3: *Robinia pseudoacacia* L. comes from North-America. It is one of the non-native species which now have established in parts of Middle and South Europe (Kohler & Sukopp 1964).

The same is true for *Ailanthus altissima*: in the Mediterranean or Pannonian regions, *Ailanthus* has spread abundantly through a broad range of sites. In central Europe, however, it is virtually confined to warmer regions or to urban-industrial sites with a more favorable local climate. Considering that the continental climate is characterized by a longer growth period with an increased heat sum, it can be assumed that the spreading of *Robinia pseudoacacia* had been promoted by the change to a more favorable climate in the 19th century. The invasion was noticed in 1924, but only in the second part of the last century did this species become frequent. The initial spread of *Ailanthus* was only noticed at the beginning of this century. Actually, it is more frequent in the warmed zones of Berlin (Figure 4, Table 3) than on the urban fringe, and in Brandenburg it has been reported mainly in cities (Kowarik & Böcker 1984, Kowarik 1992b).

Figure 4: Distribution of *Ailanthus altissima* (Miller) Swingle in West-Berlin.
The map shows 1. the points of location of the trees and 2. the different temperature-zones; the warmest zone is cross-hatched (Kowarik & Böcker 1984).

Apart from these species, *Prunus laurocerasus* exemplifies the possible limitations to invasions by low winter temperatures (Adolphi 1995). *Prunus laurocerasus* has been cultivated in Berlin since 1663, but the first seedlings were not observed until 1982. In regions with milder winter temperatures this species obviously does better: it is common in London, in cities in the West of Germany (Table 2), and has become a permanent member of natural forests in the Insubrian region (Gianoni et al. 1988). In Zurich, it has only begun to invade forests in recent years.

Name	Origin
Berberis julianae L.	Nemoral zone, humid mountain regions of West-China with a monsun-climate; summer green deciduous forest
Cytisus multiflorus (L'Her. ex Aiton) Sweet	Meridional zone, winter-mild nemoral zone of semihumid regions; mediterranean hardleaf forest, dry and summer-mild
Cytisus striatus (Hill) Rothm.	Like *C. multiflorus*; introduced with seeds from Spain and Portugal for the plantation of highway banks
Lonicera nitida E. Wilson	Humid, nemoral mountain regions of the meridional zone of the southwest of China; summer green deciduous forest, laurophyllous forest
Lonicera pileata Oliver	Nemoral zone of North-America; summer green deciduous forest
Mahonia aquifolium (Pursh.) Nutt.	Nemoral zone of North-America; summer green deciduous forest
Prunus laurocerasus L.	Nemoral zone, humid, winter mild regions from southeast Europe to the Caucasus; summer green deciduous forest

Table 2: Evergreen broad-leaved plants which recently appear more often spontaneously in the Rhinelands (Adolphi 1995)

Buddleja has been invasive since 1952, but it is actually less frequent in Berlin than in regions with mild winters.

Other species which are considered as thermophilous (indicator value for temperature > 7 in Ellenberg et al. 1991; time lags between the introduction and the beginning of spreading in parenthesis, and Table 3) are *Laburnum anagyroides* (198 years), *Quercus cerris* (161 years) and the submediterranean *Colutea arborescens*, which is native to the warmest sites in Southwest Germany. In Brandenburg, 265 years elapsed before the spread of *Colutea* began in 1859. There is an even larger time lag in *Vitis vinifera* and in *Juglans regia*, both of which have been cultivated in Brandenburg since 1200 A.D., but were first reported as invading in 1860 and 1968. The warmer climate may also have encouraged the spread of the frequently occurring *Prunus persica* and of the rarer *Prunus armeniaca*, both of which spread 300-400 years after their first cultivation (Kowarik 1995).

Species of trees	Year of first intro-duction	Year of first observation of spontane-ous dispersal	Time lag	Species of shrubs	Year of first intro-duction	Year of first observation of spontane-ous dispersal	Time lag
Acer negundo L.	1736	1919	183	*Buddleja davidii* L.	1796	1852	56
Aesculus hippo-castanum L.	1663	1787	124	*Colutea arborescens* L.	1594	1859	265
Ailanthus altissima (Miller) Swingle	1780	1902	122	*Cornus alba* L.	1773	1857	84
Juglans regia L.	before 1200	1860	> 660	*Cornus stolonifera* Michx.	1785	1861	76
Laburnum anagyroides Medikus	1663	1861	198	*Ligustrum vulgare* L.	1594	1787	193
Populus x canadensis Moench.	1787	1952	165	*Lonicera tatarica* L.	1770	1864	94
Prunus armeniaca L.	1657	1965	308	*Lycium barbarum* L.	1769	1839	70
Prunus mahaleb L.	1785	1839	54	*Mahonia aquifolium* (Pursh) Nutt.	1822	1860	38
Prunus serotina Ehrh.	1796	1825	29	*Philadelphus coronarius* L.	1656	1839	183
Pyrus communis L.	before 1594	1787	> 193	*Prunus persica* (L.) *Batsch*	before 1594	1965	> 371
Quercus cerris L.	1796	1957	161	*Ribes aureum* Pursh	1822	1883	61
Quercus rubra L.	1773	1887	114	*Symphoricarpos albus* (L.) S.F. Blake	1822	1887	65
Robinia pseudoacacia L.	1623	1824	201	*Syringa vulgaris* L.	1663	1787	124
Sorbus intermedia agg.	1796	1908	112	*Vitis vinifera* L.	before 1200	1860	> 660

Table 3: Time lag between the introduction and the spontaneous dispersal for a number of non-native woody species of Brandenburg (Kowarik 1992b).

The example of *Syringa vulgaris* shows that the mode of spreading also has to be considered. This species was reported as being invasive early on, but it usually spreads by clonal growth enlarging previous plantations. Even on rocky outcrops in the Rhine valley with a favourable local climate, fruits of *Syringa* do not ripen regularly (Lohmeyer & Sukopp 1992). In Brandenburg, *Syringa* spreads mainly by root suckers, but recently some established shrubs were discovered fruiting along abandoned railway areas in Berlin. It is assumed that the fruit ripening on these sites is promoted by the warmer urban climate. In this case, climatic changes would have enabled a species to enlarge its repertoire of spreading strategies (Kowarik 1995).

Whereas many woody species are still in their critical stage, one species has become fully established as a characteristic companion of *Robinia*-stands: *Mahonia aquifolium* thrives very well at the older *Robina*-stands and characterizes the *Sambucus*-group, since it is present in nearly every older stand. *Ribes aureum* may be suitable as well which, while growing up to the shrub-level, cannot withstand the shadow pressure as well as *Mahonia* (Lohmeyer & Sukopp 1992, Kowarik & Langer 1994).

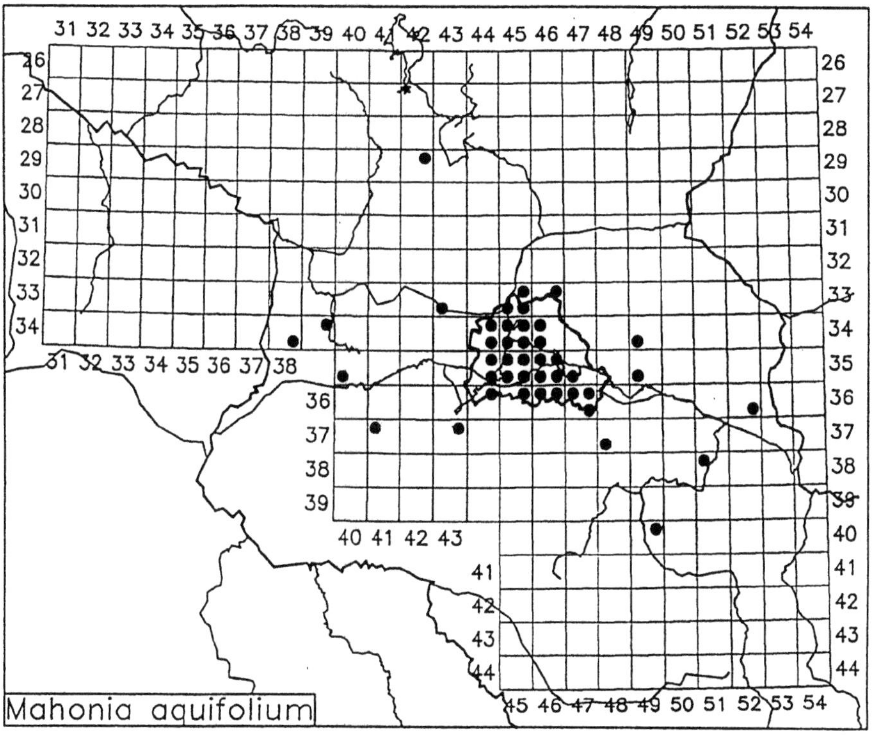

Figure 5: Spontaneous distribution of *Mahonia aquifolium* (Pursh) Nutt. in Brandenburg and Berlin (the border of Berlin is marked with a thick line) (Kowarik 1992b).

During research in the Leipzig-Bitterfeld region Auge (1997) found several reasons why *Mahonia aquifolium* is such a successful invader: it easily can adapt to disturbed and even contaminated soils, it profits from clonal growth and furthermore it is quite fertile.
Erkamo (1956) analyzed the capability of woody species to regenerate by seeds. He found that from the beginning to the middle of this century, a number of species were extending for example *Quercus robur*, which is native to the South of Finland and alien species such as the North American *Amelanchier spicata* (Figure 6).

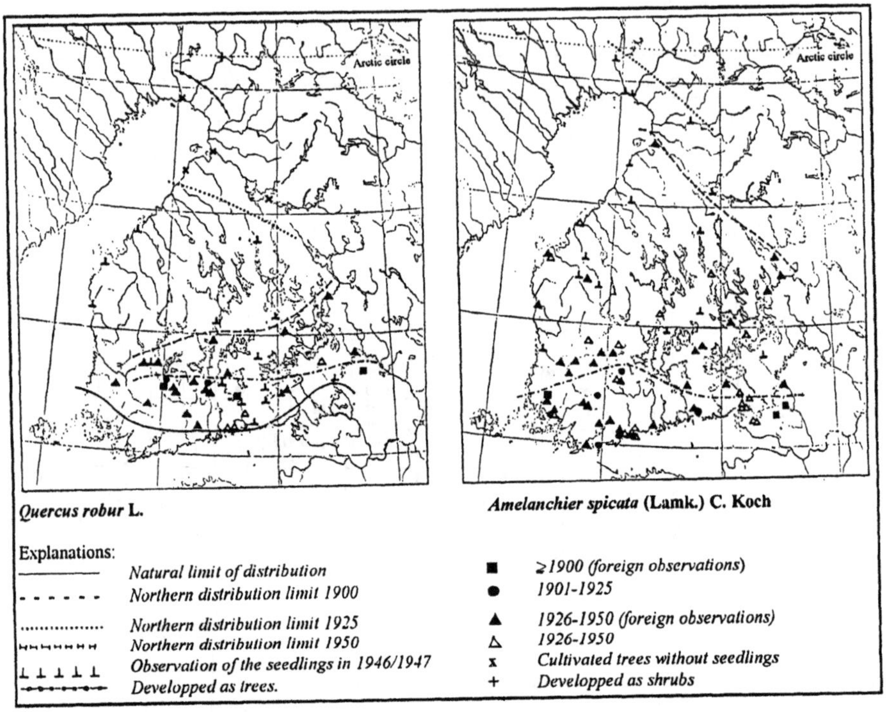

Figure 6: Changing distribution limits of native and spontaneous growing shrubs (1900, 1925 and 1950) and observations (1946, 1947) of the seedlings near the mother plants of *Quercus robur* L. and *Amelanchier spicata* (Lamk.) C. Koch in Finland (Erkamo 1956).

To estimate the effect of local or global warming it is necessary to develop models which take into consideration the history of populations expressed in terms of the duration of time lags. Davis (1986) illustrated big differences in both the magnitude and the timing of species' responses to the same climatic trend. She referred mainly to changes in abundance, demography, and geographical distribution of native species. Different response patterns to climatic changes however, may also be a key factor in determining the variation in lag phases preceding the beginning of biological invasions.

The examples mentioned above show that the recent spatial distribution of spontaneously growing flora in cities has different causes: Land use, care and substrate play an important role as well as the city climate. Thus the distribution patterns mirror building and economic structure of the city (Sukopp 1969, Kunick 1982) and its social structure (Hard & Otto 1985). The meso and macro climate of the city overlies the distribution patterns by central-peripheral gradients.

Phenological investigations in cities

The special temperature situation in its spatial differences is represented also by systematic phenological investigations. Maps can be drawn on which points are shown where a certain species enters at the same time into a particular phase of its life cycle: budding, leafing, flowering, fruiting. The different starting points of the different phases indicates temperature relations in the city. It is possible to map heat islands and cooler areas with relatively small technical difficulties.

In many cases such phenological phases start several days earlier in the center of the city than at its edge or in large parks. As an example the investigation of the flowering phase of *Tilia euchlora* in Berlin is shown (Zacharias 1972, Figure 7). *Tilia euchlora* is chosen because of the genetic homogeneity of the cultivated plants being off-springs of one individuum.

In the inner city the first flowers are seen eight days earlier than at the edge of the city. Comparing with the measurement of temperature one day earlier equals one degree C. This is in good accordance with the distribution of phenological phases in Europe in general. In cool valleys and bogs the flowering could start even two days later than at the city basis station. The steepest gradient is found at the boundary between forests and built up areas.

The relationship between mean air temperature and leafing of *Aesculus hippocastanum* at Geneva is shown in Figure 8 (Kuttler 1987). For Geneva the air temperature has been measured since 1808. From the same year on the date of emerging of young leaves has been taken on records (in days after the beginning of the year). Between 1808 and 1980 the date of leafing came earlier from April 5th to March 5th. Because this phenomenon is depending on temperature it could be concluded that in this time also the air temperature has been changing. This proved to be really true. After some fluctuations the air temperature has been increasing from 9.6° and 9.3° C (1831/60) to 10.4° C in 1964. Both curves show a high degree of correspondence. This is astonishing because the air temperatures are mean annual values whereas the starting of leafing should depend on the temperature of last spring (Keil & Schnelle 1981). This example shows that plants are suited to prove climatic changes in cities provided a long period of measure has been used.

Figure 7: Differences in the flowering phases of *Tilia x euchlora* K. Koch in Berlin (mapped during spring 1967 at the edge of the city and at stations in the city; the numbers indicate days before (or after) flowering at the basic station (Zacharias 1972).

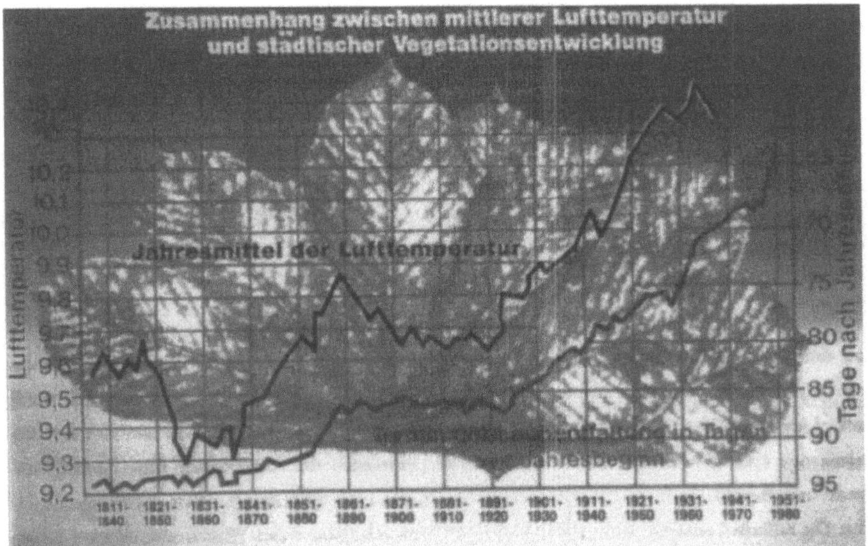

Figure 8: Relationship between mean annual air temperature and leafing of *Aesculus hippocastanum* L. (Kuttler 1987; slide: Landesbildstelle Nordrhein-Westfalen).

Specificy of urban flora and vegetation

The biological specificy of urban areas compared with the surrounding area has been recognized by Gutte (1972) when studying the ruderal vegetation of Saxony. The urban regions of Leipzig, Chemnitz and Dresden were separated on the bases of occurrences of thermophilous vegetation units. In his map of the naturalness of vegetation Schlüter (1992) revealed the big cities as specific too.
The urban flora is remarkably different from the flora of the rural hinterland caused by different human influence. A comparison of the flora of big agglomerations in Central Europe shows, however, only little difference among them. Such differences are caused solely by large scale climatic differences. They grow with the distance of urban agglomerations from each other. However, the differences are still astonishingly minor even between cities as distant as Warshaw and Brussels (Kunick 1982b).
"Urban" in a narrow sense is the vegetation of ruderal places, e. g. rubble, railway and port areas, ruins and walls, waste areas. The occurrence of new non-natives on these sites under the influence of man (Thellung 1915) early stimulated many investigations concerning adventive flora. The newcomers were recorded and categorized according to the time of introduction, way of introduction and degree of naturalization. As one interesting result the close coincidence between the human population growth and the number of newcomers should be shown in Figure 9.
Highlights of these studies were the fast colonization of bombed-out areas of cities (London, Stuttgart) and the climatic changes of the last decade.
The spatial and chronological aspects of the urbanization of flora and vegetation are shown in the following figure.

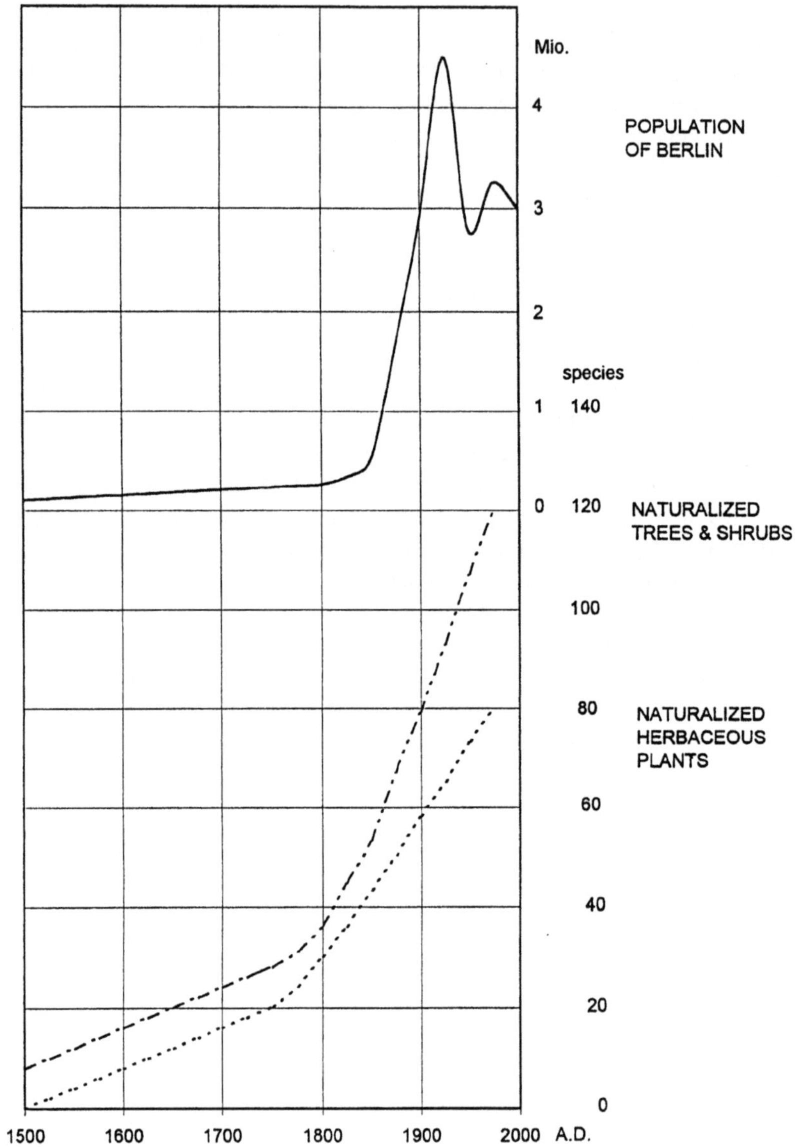

Figure 9: Coincidence between the human population growth and the number of two groups of non-native plants (naturalized neophytic ruderal plants excl. trees and shrubs according to Scholz 1960, and non-native trees and shrubs according to Kowarik 1985).

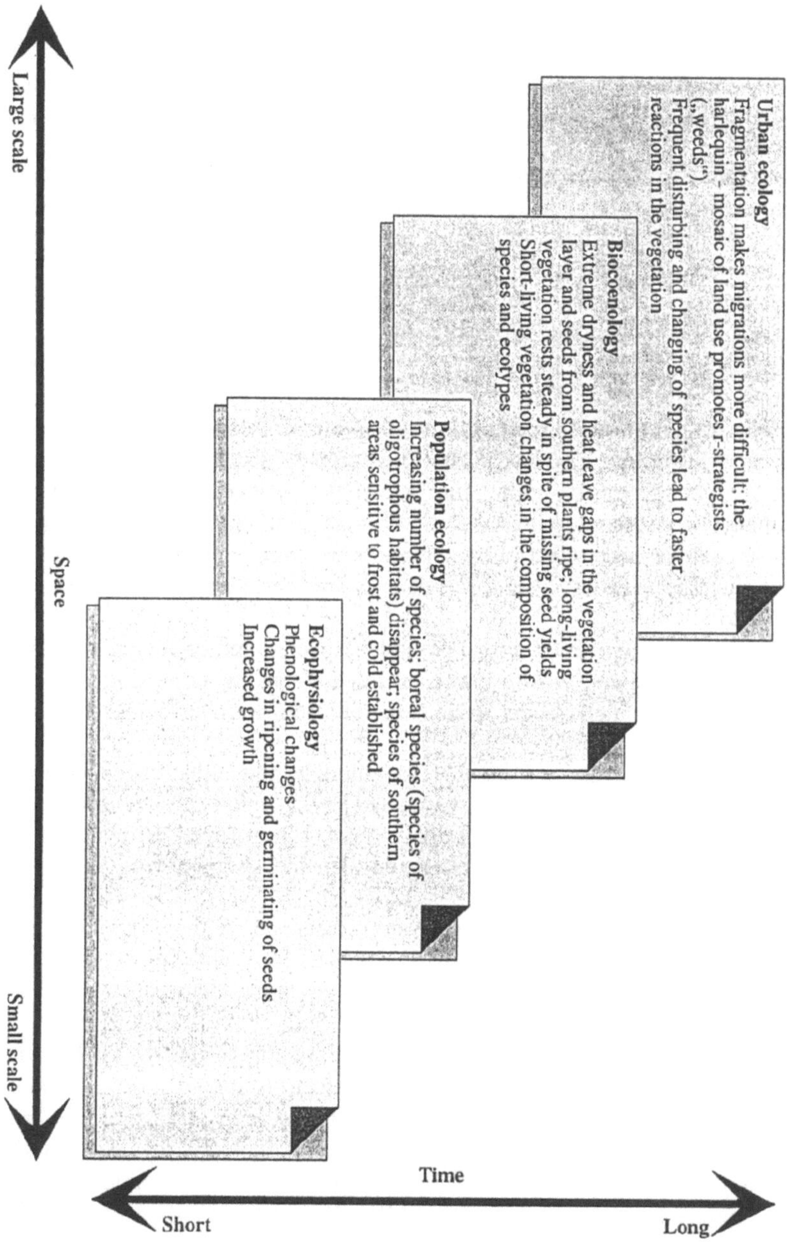

Figure 10: Spatial and temporal aspects of the urbanization of flora and vegetation.

Conclusions

The question was put forward whether a warmer city climate alone shows effects on flora and vegetation. As mentioned above there are not only climatic deviations but also other ecological and social-economic factors affecting the vegetation.

The introduction of non-native species by transport and trade or for reasons of horticulture was a prerequisite for the dispersal of thermophilous plants. In many cases this meant the abolition of insurmountable barriers between former areas. In cities native plants come together with those which would have never reached the new area without human interference. The high degree of non-native species is a characteristic feature of urban areas.

Erkamo (1956) has demonstrated - as mentioned above - that the rise in temperature during the first half of this century caused responses in the occurrence of single plant species as well as at the community level. Over the last decade it became evident in the international discussion that ecological theories on changes at the population or community level are inadequate when the climate is considered to be constant. Climatic parameters change over time, and these changes may affect plant communities, even if only ecologically short periods such as decades or a few centuries are considered as the time scale (Davis 1986).

Urban areas not only show a retreat of native species and archeophytes; at the same time cities are and will be in the future the starting point and centers of abundance of newcomers. Their origin is mainly from warmer regions of Europe, Asia and America.

Also many native plants are capable to colonize new urban sites. They are called apophytes (Rikli 1903/1904). Kowarik (1992a) found 32% of Berlin native plants established on urban sites. Falinski (pers. comm.) even estimated that all native plants of the Bialowieza forest could also live as apophytes in one form or another.

Urban plant species show a relatively slow dispersal in the surroundings of cities. This process can last several decades to centuries (Davis 1986, Kowarik 1992b). The introduced plants have quite uneven chances for dispersal.

10% of all introduced species in Central Europe are able to spread, 2% become a permanent member of the flora and only one percent is able to survive in natural vegetation. *Robinia pseudoacacia*, *Acer negundo* and *Ailanthus altissima* were introduced in the 18th and 19th century. It took 100 to 180 years for spreading these adaptable and very fertile plants for which suitable sites were available before the devastation of World War II (Figure 11).

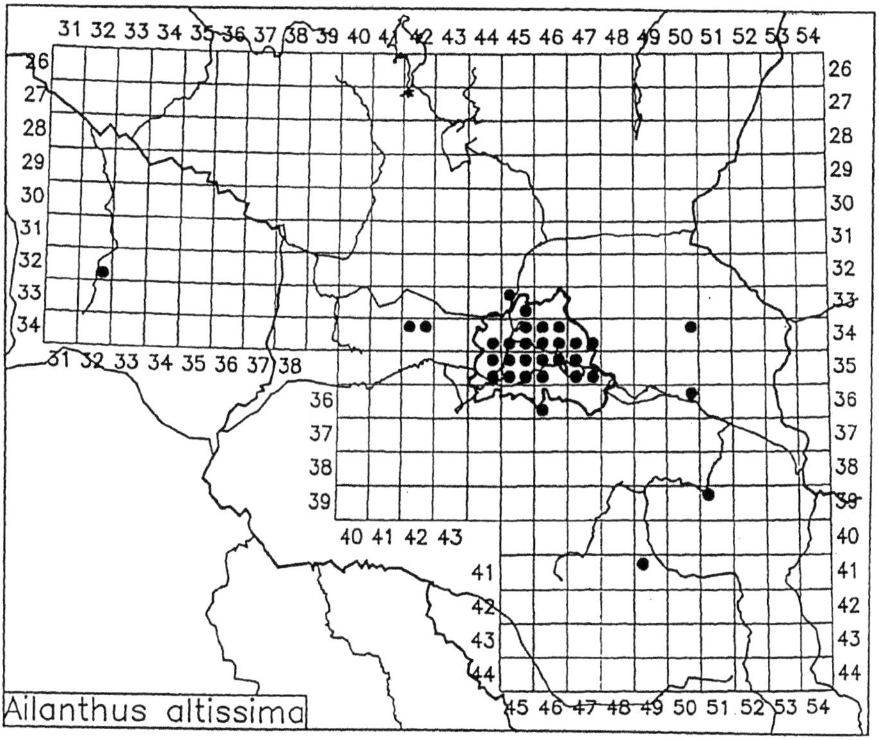

Figure 11: Spontaneous distribution of *Ailanthus altissima* (Miller) Swingle in Brandenburg and Berlin (the border of Berlin is marked with a thick line) (Kowarik 1992b).

For big cities, the effects of the climatic warming since the 1850s have been exacerbated by the rise in temperature of the urban climate. In the case of Berlin this involved a shift from a city with about 170.000 inhabitants in 1800 to a metropolitan area with 3.7 million inhabitants in 1910. Calculations of the increased warming effect of the urban climate are: 0.2°C for 1798-1804, 0.7°C for 1831-1837 and 1.4°C for 1886-1898. For the period 1961-1980, there was a difference in the annual mean air-temperature of more than 2°C between the centre and the surroundings of Berlin. This warming is correlated with a significant reduction of frost days (<64 days in the centre, >102 days in the surroundings (von Stülpnagel et al. 1990). Consequently, an accelerated invasion of non-native species with higher temperature tolerance can be expected. This hypothesis is supported by Figure 12 showing an obvious increase in woody species which began to invade by the middle of the last century, coinciding with the changes in temperature: In 1756-1847, the winters were colder by -0.7°C compared to 1848-1907, and there was a significantly higher frequency in extremely cold winters in the period before 1846.

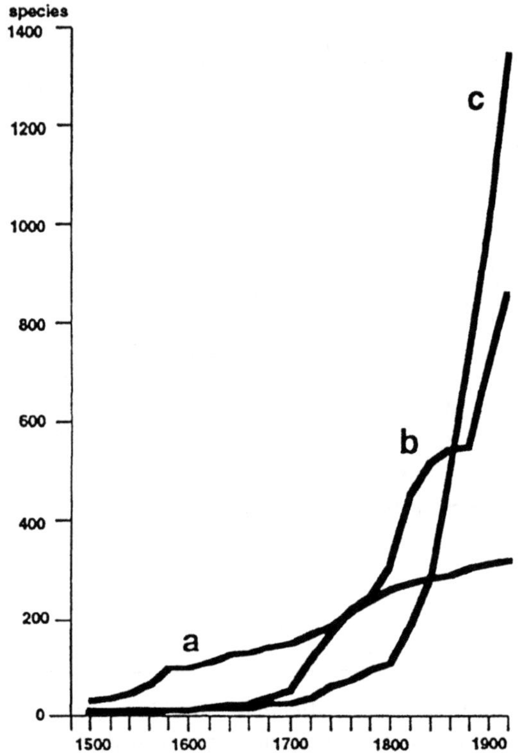

Figure 12: Introductions of woody species to nemoral Europe during the period between 1500 and 1916 (Kowarik 1995).
The species are grouped according to their origin from
a) other parts of Europe including the Mediterranean (n=309),
b) America (n=857), and
c) Central and Eastern Asia (n=1351)
(cumulative curves; additional 128 species introduced from West Asia and of cultural or unknown origin are not shown)

Similar trends have been observed for annual and perennial herbs in Berlin and for all established non-native plant species in Germany. At first they had been explained only by a huge increase in incidental introductions and subsequent dispersal, promoted both by new transport systems and the increasing commercial exchange of goods, which coincided with a diversification of sites in the urban environment. But many of the non-native invaders derive from warmer areas and they are considered to benefit also from a more favourable local climate, even on a small spatial scale (Kowarik 1995).

References

Adolphi K. (1995): Neophytische Kultur- und Anbaupflanzen als Kulturflüchtlinge des Rheinlandes. Nardus 2. Wiehl.

Arbeitsgruppe "Methodik der Biotopkartierung im besiedelten Bereich " (Hg.: Schulte W.; Sukopp H.; Werner P.) (1993): Flächendeckende Biotopkartierung im besiedelten Bereich als Grundlage einer am Naturschutz orientierten Planung. Natur und Landschaft 68/10: 491-526.

Auge H. (1997): Biologische Invasionen: Das Beispiel *Mahonia aquifolium*. In: Feldmann H., Henle K., Auge H., Flachowsky J., Klotz S. & Krönert R. (Hg.): Regeneration und nachhaltige Nutzung. Konzepte für belastete Regionen. Springer: Berlin, Heidelberg, New York. 124-129.

Brandes D. (1977): Über die Flora der Stadt Braunschweig. Mitt. Techn. Univ. Carolo-Wilhelmina Braunschweig 12(1/2): 49-59.

Davis M. B. (1986): Climatic instability, time lags, and community disequilibrium. In: Diamond J. & Case T.J. (Ed.): Community Ecology. Harper and Row, New York. 269-284.

Dettmar J. & Sukopp H. (1991): Vorkommen und Gesellschaftsanschluß von *Chenopodium botrys* L. und *Inula graveolens* (L.) DESF. im Ruhrgebiet (Westdeutschland) sowie im regionalen Vergleich. Tuexenia 11: 49-65.

Ellenberg H. et al.(1979): Zeigerwerte der Gefäßpflanzen Mitteleuropas. Scripta Geobotanica 9, 2. Auflage.

Ellenberg H. et al.(1991): Zeigerwerte von Pflanzen in Mitteleuropa. Scripta Geobotanica, Vol. 18.

Erkamo V. (1956): Untersuchungen über die pflanzenbiologischen und einige andere Folgeerscheinungen der neuzeitlichen Klimaschwankung in Finnland. Annales Botanici Societatis Zoologicæ Botanicæ Fennicæ 'Vanamo' (Ann. Bot. Soc. 'Vanamo'), 28, (3): 1-290.

Frank D. & Klotz S. (1990): Biologisch-ökologische Daten der Flora der DDR. Unter Mitarbeit von W. Westhus. Beiträge Martin-Luther-Universität Halle-Wittenberg. 2. Auflage.

Gianoni G., Carraro G. & Klötzli F. (1988): Thermophile, an laurophyllen Pflanzenarten reiche Waldgesellschaften im hyperinsubrischen Seenbereich des Tessins. Ber. Geobot. Inst. E.T.H. Stiftung Rübel 54: 164-180.

Gutte P. (1971): Zur Verbreitung einiger Neophyten in der Flora von Leipzig. Mitt. d. Sektion Spezielle Botanik 2: 5-24.

Gutte P. (1972): Ruderalpflanzengesellschaften West- und Mittelsachsens. Feddes Repert. 83: 11-122.

Haeupler H., Schönfelder P. (Hg.) (1989): Atlas der Farn- und Blütenpflanzen der Bundesrepublik Deutschland. Stuttgart.

Hard G. & Kruckemeyer F. (1990): Die Mäusegerste und ihre Gesellschaft in Osnabrück 1978-1990. Über den Zusammenhang von Stadt- und Vegetationsentwicklung. Osnabrücker Naturwiss. Mitt. 16: 133-156.

Hard G. & Otto G. (1985): Die vegetationsgeographische Gliederung einer Stadt. Erdkunde 39(4).

Hügin G. (1992): Höhengrenzen von Ruderal- und Segetalpflanzen im Schwarzwald. Natur und Landschaft 67/10: 465-472.

Hügin G. (1995): Höhengrenzen von Ruderal- und Segetalpflanzen in den Alpen. Flora 190(2): 169-188.

Jäger E.J. (1968): Die pflanzengeographische Ozeanitätsgliederung der Holarktis und die Ozeanitätsbindung der Pflanzenareale. Feddes Repert. 79(3-5): 157-335.

Keil K. & Schnelle F. (1981): Phänologische Beobachtungen und Klimaschwankungen. Meteorologische Rundschau 34, H. 6: 180-181.

Kohler A. & Sukopp H. (1964): Über die Gehölzentwicklung auf Berliner Trümmerstandorten. Zugleich ein Beitrag zum Studium neophytischer Holzarten. Ber. Dtsch. Bot. Ges. 76(10): 389-407.

Kowarik I. (1985): Die Zerreiche (*Quercus cerris* L.) und andere wärmeliebende Gehölze auf Berliner Bahnanlagen. Berliner Naturschutzblätter 29(3): 71-75.

Kowarik I. (1992a): Das Besondere der städtischen Flora und Vegetation. Sch.-R. des Deutschen Rates für Landespflege, H. 61 "Natur in der Stadt", 33-47.

Kowarik, I. (1992b): Einführung und Ausbreitung nichteinheimischer Gehölzarten in Berlin und Brandenburg und ihre Folgen für Flora und Vegetation. Verh. Bot. Ver. Berlin Brandenburg, Beiheft 3.

Kowarik, I. (1995): Time lags in biological invasions with regard to the success and failure of alien species. In: Pysek P., Prach K., Rejmànek M. & Wade M. (Ed.): Plant invasions - General aspects and special problems. SPB Academic Publishing, Amsterdam, The Netherlands. 15-38.

Kowarik I. & Böcker R. (1984): Zur Verbreitung, Vergesellschaftung und Einbürgerung des Götterbaums (*Ailanthus altissima* (Mill.) Swingle in Mitteleuropa. Tuexenia 4: 9-29.

Kowarik I. & Langer A. (1994): Vegetation einer Berliner Eisenbahnfläche (Schöneberger Südgelände) im vierten Jahrzehnt der Sukzession. Verh. Bot. Ver. Berlin Brandenburg 127: 5-43.

Kunick W. (1982): Zonierung des Stadtgebietes von Berlin West - Ergebnisse floristischer Untersuchungen. Landschaftsentwicklung u. Umweltforschung, 14.

Kunick W. (1982b): Comparison of the flora of some cities of the Central European lowlands. In: Bornkamm R., Lee J. A., Seaward M.R.D. (Ed.): Urban ecology, 13-22.

Kunick W. (1984): Verbreitungskarten von Wildpflanzen als Bestandteil der Stadtbiotopkartierung, dargestellt am Beispiel Köln. Verh. d. Gesellschaft für Ökologie 12: 269-275 (Bern 1982).

Kuttler W. (1987): Stadtklimatologie; Westfalen im Bild - eine Bildmediensammlung zur Westfälischen Landeskunde. Reihe Grundlagen und Probleme der Ökologie 4. Landesbildstelle, Landschaftsverband Westfalen-Lippe (Hg. W. Linke). Münster, 62 S., 24 Dias.

Kuttler W. (1993): Stadtklima. In: Sukopp H. & Wittig R. (Ed.): Stadtökologie. Stuttgart. 113-153.

Landsberg H. (1981): The urban climate. International Geophysics Series 28. New York.

Lohmeyer W. & Sukopp H. (1992): Agriophyten in der Vegetation Mitteleuropas. Schr.-R. f. Vegetationskunde, 25.

Rothmaler W. (Begr.) (1988): Exkursionsflora von Deutschland. 2. Bd. 16. Auflage. Jena.

Rikli M. (1903/1904): Die Anthropochoren und der Formenkreis des *Nasturtium palustre* DC. Ber. Zürch. Bot. Ges. 8: 71-82. In: Ber. Schweiz. Bot. Ges. 13.

Salisbury E. (1961): Weeds and aliens. London.

Schlüter H. (1992): Vegetationskundliche Analyse der Flächenutzungsmosaike Nordostdeutschlands. Natürlichkeitsgrad der Vegetation in den neuen Bundesländern. Naturschutz und Landschaftsplanung 24/5: 173-180.

Scholz H. (1960): Die Veränderungen in der Berliner Ruderalflora. Ein Beitrag zur jüngsten Florengeschichte. Willdenowia 2(3): 379-397.

Stace C. A. (1991): New flora of the British Isles. Cambridge.

Sukopp H. & Wittig R. (Hg.) (1993): Stadtökologie. Jena.

Sukopp H. (1969): Der Einfluß des Menschen auf die Vegetation. Vegetatio 17: 360-371.

Sukopp H. (1971): Beiträge zur Ökologie von *Chenopodium botrys* L. I. Verbreitung und Geschichte. Verh. d. Botanischen Vereins der Provinz Brandenburg 108: 3-25.

Sukopp H. (1973): Die Großstadt als Gegenstand ökologischer Forschung. Schriften des Vereins zur Verbreitung naturwissenschaftlicher Kenntnisse in Wien. Selbstverlag.

Sukopp H. & Wurzel A. (1995): Klima- und Florenveränderungen in Stadtgebieten. Bundesamt für Naturschutz (Hg.), Reihe Angewandte Landschaftsökologie, H. 4, Bonn-Bad Godesberg, 103-130.

Thellung A. (1915): Pflanzenwanderungen unter dem Einfluß des Menschen. Engl. Bot. Jb. 53 (3/5): 37-66.

von Stülpnagel A., Horbert M. & Sukopp H. (1990): The importance of vegetation for the urban climate. In: Urban Ecology (Ed. Sukopp H. et al), The Hague, 175-193.

Wittig R. (1991):Ökologie der Großstadtflora. Flora und Vegetation der Städte des nordwestlichen Mitteleuropas. Stuttgart.

Wittig R., Diesing D. & Gödde M. (1985): Urbanophob - Urbanoneutral - Urbanophil. Das Verhalten der Arten gegenüber dem Lebensraum Stadt. Flora177:265-282.

Zacharias F. (1972): Blühphaseneintritt an Straßenbäumen (insbesondere *Tilia* x *euchlora* Koch) und Temperaturverteilung in Westberlin. Diss. Freie Universität Berlin.

Zukowski W. (1971): Zmiany we florze synantropijnej miasta Poznania w latach 1950 - 1970. Mat. Zakl. Fitosoc. Stosov.U.W. 27: 115-132.

Plenary discussion (The paper was presented by F. KLÖTZLI)

Question: A. BORHIDI

I think that in the process of the nicheal distribution of the climate there is also the far spreading of plants, which are colonizing islands and isolated so called foreposts. So I think in the warming of the climate, practically in the cities, they will act as foreposts of the different laurophyllous or sclerophyllous plants. I think perhaps in cities sclerophyllisation will be an interesting process because the warming of the cities is linked with an increase in drier conditions.

Answer: F. KLÖTZLI

Well, this can be endorsed in the surroundings here. The more thermophilous but dry sites, they are able to carry *Olea europaea* and of course *Quercus ilex* and other species of the mediterranean area. It is quite clear, since we have introduced *Quercus ilex* on Mte Caslano in the nineties of last century, now *Quercus ilex* has trees on Mte Caslano and natural regeneration. So it is possible what you said. And in case you get the opportunity to pass Bellinzona there is a huge, beech-like *Celtis australis* on a warm hill-side above Bellinzona and *Olea* trees of stem diameters of about 30–40cm, so they are very old, and of course also *Quercus ilex* is growing there. So it is absolutely true what you mentioned, also sclerophyllous species can spread or can get installed under more dry conditions, locally dry conditions.

Statement: G.-R. WALTHER

One remark to laurophyllisation in the context of invasion: I personally never talk of invasion when I'm talking about the laurophyllisation here in this area, because talking of invasion you normally consider introduced species spreading in an area with very similar conditions to the one, where they are coming from. E.g. an introduced deciduous species can easily spread in deciduous vegetation of another region. But the difference in this area here is, firstly: special climatic conditions as we saw in the lecture of Prof. Klötzli. The Insubrian climate is just at the boundary to the climatic conditions of the evergreen broad-leaved biome. The other thing is that it is not an ordinary fact, that every species now spreading in this area has the same characteristics: it is dispersed by birds and it is evergreen broad-leaved - although they are spreading in deciduous forests - and there is no exception in all these species. So I don't look at this process here in this area as an ordinary invasion problem, in which we just have to wait until the lag time is over and the population of introduced specimens of the species is big enough, and then they start spreading. But in this area there must have happened other, additional adaptations until the process of laurophyllisation could start. Therefore I would like to invite you to make a difference between the "ordinary" invasion processes we know up to now and this new phenomena of laurophyllisation. Whereas in "ordinary" invasion processes the invasive species have been of the same type as the invaded vegetation, we see in the process of laurophyllisation evergreen broad-leaved species spreading in deciduous forest. So there is that crossing of a biome boundary, that makes the difference.

Question: K. Fujiwara

How about weed species in this area, are there any subtropical weed species coming to this area? For example in Japan in all areas with temperate forests we have subtropical weeds like *Digitaria*, *Portulaca*, etc. Or if we go to the tropics then we can see them frequently. These weeds will come to the warm-temperate forest region. In Okinawa, which is subtropical, tropical weeds occur. So in this area here, already laurophyll species have invaded the forest. How about weeds?

Answer: F. Klötzli

Actually we have Soya bean cultivated in the Magadino plain (the easterly continuation of Lake Major touching Ascona) and there we have also the companions of Soya bean. There are Malvaceae species (*Abutilon theophrastii*) from the warmer parts of the United States and they spread in the Magadino plain. And this is still absolutely impossible in the Zurich area, but here in the neighbourhood of Mte Verità it came over the climatic threshold and it is installed.

Statement: A. Borhidi

Among the invading Gramineae for instance, there are also subtropical species. *Cynodon dactylon* and *Eragrostis minor* are originally subtropical species. I think that invasion and invading plants distinguish not mainly by the origin of the plant. Perhaps it is better to distinguish them by the manner of migration.

Question: N. Zazanashvili

I have a question to Mr. Walther: You pointed out a difference between laurophyllisation and invasion, a very good atempt. But if there were non-native species spreading, introduced from a far away area, e.g. in the Caucasus from North America or from Japan (as *Robinia pseudoacacia* , etc.) the process I think is not invasion, or is it? But we should recognize invasion of species from neighbourhood regions (e.g. *Castanea sativa* from the Caucasus to the Mediterranean area) and the movement of exotics, which are more evident indicators of global climatic changes among them - deciduous exotics. Maybe we may widen this term to laurophyllisation.

Answer: G.-R. WALTHER

That's why I distinguish laurophyllisation from "ordinary" invasion. Of course laurophyllisation is also some kind of invasion, that's for sure, but a special type of invasion. As we have seen, although these laurophyllous species have already been here for a long time, they didn't invade. Therefore, there must have been a change in environmental conditions first to make it possible for these laurophyllous species to spread. So we have now an invasion process, but to make this invasion process possible, the environmental conditions had to change first. That's why our point of interest is so much focused on climatic factors. Of course it is an invasion process, but different from the "ordinary" invasion processes, where the spread of introduced species is not depending on a change in local environmental conditions.

Statement: G. DOYLE

But surely that is a feature of invasion, there must be change. The crux in this discussion is whether or not such invasion is the result of climatic change or related to alterations in grazing patterns and forest management practices.

Statement: V. PILLAR

I see here a problem of distinguishing the causes of the absence of a species. Is it absent because it cannot be at the site or because it did not reach that site yet? So the distinction is between invasion and laurophyllisation. Sometimes laurophyllous species may not be at a site because they could not reach that point yet or because the conditions are not right. I may mention the problem we have with *Eragrostis plana*, an invader species introduced from Africa, in which we are trying to find out, based on surveys, the conditions in that this species would establish. We have not yet succeeded due to the same problem of confounding causes of absences.

Statement: H. BRUELHEIDE

I would like to return to the point we discussed yesterday in the context of the invasion of laurel-like broad-leaved evergreen species and the climatic requirements for their growth. I had a further look at the climatic diagrams in Walter & Lieth for the period before 1960. Locarno, for example, has an absolute minimum of -8.3°C, a mean annual temperature of about 12°C and about 1900 mm annual precipitation. Apart from the quite low annual temperature of 12°C, which differs by 1°C from regions with broad-leaved evergreen forests in Japan, the climate is suitable for sustaining evergreens. All broad-leaved evergreen trees, including *Cinnamomum japonicum* and *C. camphora*, resist freezing down to just -8°C (Sakai, A. (1971): Freezing resistance of relicts from the arcto-tertiary flora. New Phytol. 70: 1199-1205). My question: Which parameters did you use for these species in your model? How did you define the climatic envelope for evergreen broad-leaved species?

Answer: F. KIENAST

That was simply mean annual temperature. But as I told you today we no longer use mean annual temperature.

Statement: H. BRUELHEIDE

I suppose we had suitable climatic conditions for the growth of evergreen broad-leaved species even in former times. Other factors besides climate could have contributed to the expansion of these species, such as a change in land use or adaptation to local dispersal agents like birds etc.. I am still not convinced that we can attribute it to a change in climate alone.

Statement: F. KLÖTZLI

I have given my arguments yesterday. If you look at the area of the laurophyllous biome, I have not denied it is not near but it was not in the area of the laurophyllous biome, it was on the edge which is the same as in certain Colchic areas, where you have not a laurophyllous tree layer. Now we get laurophyllous trees in the lower and sometimes in the upper tree layer and this is new! As I have pointed already this is absolutely new; and they are fast growing. And as I have shown in my transparency, this region of Insubria is now overlapping with the laurophyllous biome a fact since about twenty years. That's a clear difference and I haven't denied that all these species were able to grow in the gardens, but that's not the same thing as to grow in the wild. And to grow in the wild this is new, again absolutely new. And there are some new events which I should also show. If you look at the heat days, in the way of days with temperatures higher than 30°C, you see the heat days up to 1990 for the area where we are. And all of a sudden you get heat days which are doubtlessly more frequent than in all former years. This is quite clear. It is a new situation for our decade. And the other thing I wanted to mention is field work of Gabriele Carraro and his coworkers, where he has shown a distinct difference between the warmest spots in winter and the spots which are less warm and have no laurophyllous species. There is a distinct difference in the micro- and mesoclimate between the areas with the laurophyllous species and areas without laurophyllous species. So we have two factors which are favourable for the spreading: that's the lower winter temperatures which have risen and it is the dampness which is more obvious in our summers (and, as you know, dampness can be calculated on the basis of air humidity and temperature). This dampness, of course, is in favour for the establishing seedlings because they have less stress in the hotter days than in the case with lower air humidity. With the high air humidity and higher winter temperatures the saplings were favoured and that's why we have so many saplings now. There are much more saplings around as you would have been able to see in all the years before. We always have had saplings but as you have seen now in the surroundings this is unusual.

Statement: K. FUJIWARA

I would like to go back to Prof. Sukopps paper. He mentioned many invading species in Berlin or other urban areas. But vegetation shift due to climate change and invasion are different. For example *Robinia* invades in deciduous forests. In the United States *Robinia pseudoacacia* is also a pioneer species in eroded areas. So it occurs on very unstable areas. *Ailanthus* also occurs on unstable areas. They can also occur in urban areas. But *Cynodon dactylon*, *Commelina communis*, etc. this kind of warmer or subtropical weed species should be checked. We must check if they are weeds or invader species, thermophilous species or the normal situation from deciduous areas, or then they are pioneer species. Then we can check if there is a shift related to climate change. That's what we should check.

Statement: P. EDWARDS

To some extent I'm repeating something which was said yesterday. What kind of proof do you need in order to demonstrate that vegetation change is occurring due to climate warming? This is something which I think that vegetation scientists need to be very concerned about. There has been a massive attack on the methods of phytosociology and vegetation science in the last few years. I can point to a number of articles by distinguished ecologists such as Crawley and May who argue that vegetation science has lost its way in ever more complex multivariate correlations which produce no kind of evidence of causality. You may profoundly disagree with these statements, but it is the case that vegetation science has a bad image in some circles.

At this meeting we have heard very convincing correlative evidence that the vegetation is changing in response to climate change. But how do we prove it? In the last fifteen to twenty years ecology has seen a growing emphasis on the role of experiments. However, it is increasingly realized that there are some things you just can't test by experiment. Effects of global environmental change are an example. Modelling has become increasingly recognized as powerful alternative. Now coming back to this particular case, I feel that to establish the credibility of vegetation science we have to use a range of different approaches. The presentations at this meeting have pushed the correlative evidence as far as possible. As we have seen there are so many confounding variables; so much apart from climate have changed since 1950. A promising approach, if Felix Kienast can get all the parameters he needs for his model, is modelling.

I feel strongly that there is also an important role for manipulative experiments. One of the great thing about experiments is that it forces you to examine very closely what your assumptions are. We've been talking about invasion and why invasion hasn't happened in the past. I haven't heard anyone explicitly say why they think this is so. Is it because viable seed is not produced, is it because the dispersal agents were not there, is it because there is something about the vegetation canopy which doesn't make it invasible? I think the main hypothesis is that climate has changed so that vulnerable seedlings are now able to survive. That is something which could be tested. I know Gian-Reto Walther is going on to work in this area. For example, he could take *Trachycarpus* seedlings in flowerpots and plant them on a transect up to the top of the mountains; he could spread seeds out in plots at different altitudes and see to what extent the establishment occurs, and so on. Now, this kind of approach has not been a normal part of the repertoire of vegetation science up till now. I think this is a great opportunity to perform a few elegant experiments which help to clarify what are the main processes at work. At the same time such studies will do a lot of good for the credibility of the subject.

Statement: G. CARRARO

About this distribution of seeds and invasion: We have only five relevés of the sixties with some small laurophyllous species in the herb layer. Only five relevés. And these five relevés are not situated in the favourable climatic zone, where we can see laurophyllisation today. And what can you observe thirty years later? The species are still there but they haven't grown up. We found *Trachycarpus* and *Prunus laurocerasus* in the forest, but they haven't grown. This means that the species could escape from the garden into the forest but, if they don't find sufficiently warm conditions, they just stay there without very much prospect for further development. So these five relevés in the Locarnese are situated about 400 to 600m a.s.l and in northern exposition. There you can't see one place with laurophyllisation but the seeds are still coming.

Statement: A. BORHIDI

What about the seed? Is it still fertile or not? It is an important question as Pigott showed in England that *Tilia cordata* has two timberlines, one of the vegetative propagation and the other of the generative propagation, the pollen can not fertilize the ovarium because of the low temperature which doesn't in the latter case permit the germination of the seed. Here may happen a similar situation and perhaps for this reason the temperature of March or April may be important or of special interest.

Statement: H. BRUELHEIDE

Experiments like Pigott and Huntley's are probably what Prof. Edwards had in mind when he said that we needed another kind of evidence to prove our assumptions. The authors showed that the failure of *Tilia cordata* to propagate at its northern distribution border is caused by insufficient pollen-tube growth due to low temperatures at flowering time as well as by incomplete seed development due to low summer temperatures (Pigott and Huntley 1981).

I very much appreciate Prof. Edwards comment, because this was also the starting point for our working group in Göttingen. There have always been assumptions about the reasons why plants show distribution boundaries but very few ecological experiments to support them. Probably the reason for the lack of clear evidence is that it is extremely difficult to design the ultimate experiment that would prove the distribution of a species to be limited by climate.

I believe it would be very useful to conduct transplantation experiments with *Trachycarpus fortunei*. Seedlings or even plants which are some years old and already have stems could be put in pots and transplanted beyond the actual distribution boundary up the slopes. At the same time, the temperature and observations about growth, fructification, germination etc. should be recorded. This kind of experiment does not need a lot of equipment. Of course, it could take some years to collect the evidence, but it would be worth doing.

Answer: G. CARRARO

A short answer: We did control *Trachycarpus*. It is cultivated in gardens up to 700m a.s.l. but you don't find a regeneration in the forests around these gardens. It is almost impossible. We found regeneration in the forests only below 500m,

600m a.s.l. is the maximum altitude. So the species can grow in the garden, produce fruits, but in the forests the conditions are not sufficient for regeneration or for the competition with native species.

Statement: F. KLÖTZLI

Actually I'm very much in favour as well as my good colleague Peter Edwards, namely, that we should go on for experiments. I'm not denying that. It's only a matter of time and organization of course, you just can't pick somebody on the road and put him to watch the pots. But such experiments are in preparation. Then another thing: What I can't understand now is that you still have these doubts regarding the situation. First of all I have shown you that the shift is new, the climate stations in Insubria were not in the biome of laurophyllous species. And that they shifted into the new biome in the last thirty years, this is new. And another thing which can't be denied is: for the installation of saplings, seedlings we need a concoction, a new concoction of environmental parameters. And here is the new situation, which is not just simply higher winter temperatures, it is not just simply more damp days, it is not just simply the shift in the climate diagram, no, it is the three changes together which enable these plants to build up a new population: The milder winters, the more damp summers and the general shift. So of course - and I am not denying that - we need experimental prove, but the facts that also have been mentioned by Mr. Carraro, that we have already a certain amount of garden experiments, is also clear: only up to about 600m a.s.l on southern slopes and 400m a.s.l. on northern slopes we have success. So this is clearly a distinct correlation which has to be seen by the actual meteorological development of winters and summers. And if we go higher on the slope one of these factors is not suitable any more, or two: it is not damp and warm enough in summer or it is too cold in winter. So please keep in mind we keep the traumatic events in our eyes but we keep also the gradient in our eyes, which is a gradual shift from the non-laurophyllous biome to the laurophyllous biome.

Statement: F. KIENAST

The question here is: what is causal and what is a simple correlative effect. In experiments as well you have correlations and then the physiologists assume that this is a causal effect. Because it is a correlation on a very low spatial level we call it a causal effect. If you look at landscapes in an experimental sense, we should use different sites, arrange them in a succession (space for time). If we do not accept the paradigm "space for time" we would have to perform warming and cooling experiments in the landscape.

Recent shifts in vegetation boundaries of deciduous forests,
especially due to general global warming, pp. 121–150
edited by F. Klötzli and G.-R. Walther

An experimental study on the impact of winter temperature on the distribution of *Euphorbia amygdaloides* L. in Central Germany

Christian Schulz & Helge Bruelheide
Albrecht-von-Haller-Institut für Pflanzenwissenschaften,
Abteilung Ökologie und Ökosystemforschung, Universität Göttingen
D-37073 Göttingen, Germany

Abstract

The general range of *Euphorbia amygdaloides* L. in Europe and Asia is characteristic of a subatlantic-submediterranean distribution type. The main factor limiting its eastern distribution may be the minimum winter temperature. Field experiments were conducted to show the direct effects of decreasing temperature in a west-east gradient on individual plants. *Euphorbia amygdaloides* plants were transplanted along a transect beyond the distribution boundary in NW-Thuringia, Germany. Air temperature and soil temperature were recorded with data loggers at eight experimental plots. Frost damage was analysed at the end of the winter of 1995/96 and during the winter of 1996/97. Frost tolerance experiments were conducted in a freeze chamber during the same periods. Furthermore, frost damage in plots with and without snow cover was analysed. Frost damage was quantified by determining rates of electrolyte leakage (Murray et al. 1989).
Additional information on the species' population biology and growth patterns near the distribution boundary is given. Photosynthesis measurements in response to changes in temperature provide data on the autecology of *Euphorbia amygdaloides*.

Introduction

Climate can be considered as the most important factor limiting plant ranges (e.g. Woodward 1987, Grace 1987). Consequently, there are numerous approaches in plant geography to determine the relationship between the geographic range limits of species and climatic constraints. To show coincidences, researchers have mainly employed a broad geographical scale with a low resolution of 50 km or more (Dahl 1951, Jäger 1968, 1972, 1990, 1992, 1995a, Meusel & Jäger 1989, Box et al. 1993, Huntley et al. 1995). Investigations on a local or regional scale are scarce although range boundaries on this scale can clearly be encountered in nature. Ultimate boundaries of plant ranges are often well-known to plant geographers and therefore offer good opportunities for research. Such local work has been conducted in northern England (Grime 1990) and in Norway (Holten 1993). We are not aware of any Central European studies.

Among all climatic parameters, winter temperature can be considered the most significant in explaining the northern and eastern range limit of plant species (Sakai & Larcher 1987: 1, Begon et al. 1996: 61). By causing frost damage and killing whole individuals or important reproductive organs, low temperatures directly influence plant distribution. This was recognised quite early by vegetation scientists. Grisebach (1880) for instance tried to compare range limits of plants in Northwest Germany with isotherms of the coldest month in1847. Up to the present, climatic maps with isotherms of the mean temperature of January as the coldest month of the year have proven useful to reveal coincidences with plant ranges (Hjelmqvist 1940, Woodward & Williams 1987). Closer relationships between plant ranges and temperatures can be determined by directly using temperatures measured by official weather stations and noticing whether the species under consideration is present or absent near the station (Iversen 1944, Hintikka 1963, Huntley et al. 1989, Box et al. 1993). Another way to utilise station data directly is to interpolate them to grid cells of distribution maps, as was done by Huntley et al. (1995) with Atlas Flora Europaea maps.
Recently, plant range climate correlations have gained much importance by offering the opportunity to predict range shifts as a response to anthropogeneous climatic changes (Holten 1993, Hendry 1993, Jäger 1995a,b, Huntley 1991, Huntley et al. 1995). Climatic warming may cause an expansion of plant ranges of ornamental species that have been introduced from other climatic regions, were kept in gardens and now find suitable conditions to regenerate and escape into the surrounding area (Klötzli et al. 1996). To some extent, it is even possible to predict the range extension of anthropochorous weeds (Jäger 1988). A problem with predicting the range of adventitious species is that they are most probably not in equilibrium with range limiting factors other than climate, e.g. dispersal agents, herbivores or parasites. In contrast, this equilibrium can be assumed for most indigenous species. Therefore, indigenous species may be better indicators for climatic shifts.

Despite the general consensus that frost is a key limiting factor, there is disagreement as to whether frost should be characterised by mean temperatures or by extremes. Jäger (1995a) argues that range boundaries reflect averages because they are the result of a long-term confrontation with the plants' abiotic environment. This may be true for heat or radiation requirements but not for frost damage. As far back as 80 years ago Eichler et al. (1927:305) argued that damage is not caused by low monthly or annual averages but by single frost events. They were able to relate the absence of suboceanic-submediterranean species such as *Ilex aquifolium*, *Tamus communis* and *Buxus sempervirens* in some areas in Southwest Germany to the occurrence of absolute minima below -23°C measured by weather stations. Even when using extremes, however, the plant may experience temperatures that differ considerably from those measured in the air two meters above ground, which is the standard for weather stations (Grace 1987). We do not know of any field studies that directly relate frost damage observed close to a range boundary to minimum temperatures measured in the field.

Since different plant species show distinctly different responses to frost, autecological knowledge about the species is essential. This is especially true if climatic data are used to form a hypothesis on changes in the distribution pattern of a species (Holten 1993). Unfortunately, knowledge about the climatic factors limiting plant distribution is scarce and mainly available for species of commercial value (Holten 1993). Whereas in most cases detecting the autecological, climate dependent factor that limits a plant's range is a difficult and complex undertaking, (Pigott & Huntley 1981, Jäger 1989), the mechanisms of frost damage and frost hardiness are well understood (Sakai & Larcher 1987). Correlations between range limits and frost hardiness under controlled conditions in the laboratory were found by Till (1956), Kappen (1964), Polwart (1970 in Bannister 1976) and Callauch (1986). However, there is still no evidence (1) to show that minimum temperatures which cause damage in laboratory experiments also cause damage under field conditions and (2) that these temperatures differ inside and outside of the range of a given plant on a regional scale. The aim of this study is to fill this gap.

As a test species we chose *Euphorbia amygdaloides* because we considered it a suitable indicator of changing winter temperatures. The main question was whether the eastern range boundary of this species is determined by a minimum winter temperature.

We approached this question in various ways:

- by conducting a transplantation experiment along a transect beyond the distribution boundary
- by recording winter temperatures at sites with native and transplanted occurrences
- by quantifying frost damage in a controlled environment and in the field
- by performing biometrical investigations of native and transplanted individuals
- by determining the population pattern at the distribution boundary
- by measuring photosynthesis
- and by generally characterising site conditions (pH, light, soil moisture).

Study Object and Study Area

Euphorbia amygdaloides is characteristic of beech and oak-hornbeam forests with favourable nutrient supplies. The species mainly occurs on moist rendzinas on limestone with high pH and high base saturation. The plant is an evergreen perennial with a thick stock and tufted stems. During the winter it has shoots of about 30 cm in length which bear bunches of leaves at the top. The survival of shoots over the winter is essential for generative propagation. Only the previous year's shoots blossom at the end of April or beginning of May. Fruits ripen at the end of July. The generative shoots then die.

The general range of *Euphorbia amygdaloides* in Europe and Asia is characteristic of a subatlantic-submediterranean distribution type. According to Jäger (1970) *Euphorbia amygdaloides* belongs to the *Globularia*-type as a characteristic element of submeridional thermophilous deciduous forests. The species does not occur in the continental and northern parts of Europe but colonises the east up to the pannonical region and even into the colchic and hyrcanic provinces (Figure 1). The north-eastern distribution limit may be due to avoidance of low winter temperatures. The north-western limit indicates a distinct heat requirement (Meusel et al. 1978), since hard winter frosts can not be considered as a limiting distribution factor in Great Britain.

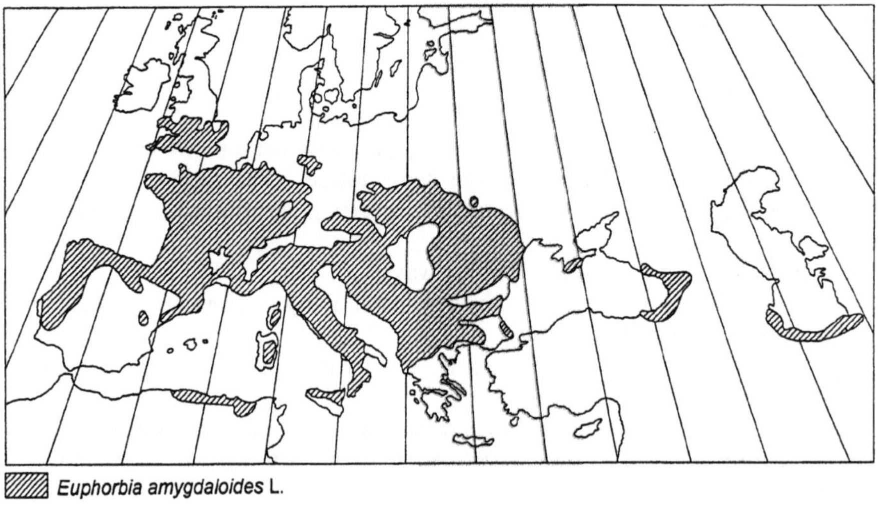

Figure 1: Distribution map of *Euphorbia amygdaloides* (modified after Meusel et al. 1978).

In southern Lower Saxony and Thuringia the range features an outpost with Göttingen in its centre (Figure 2).

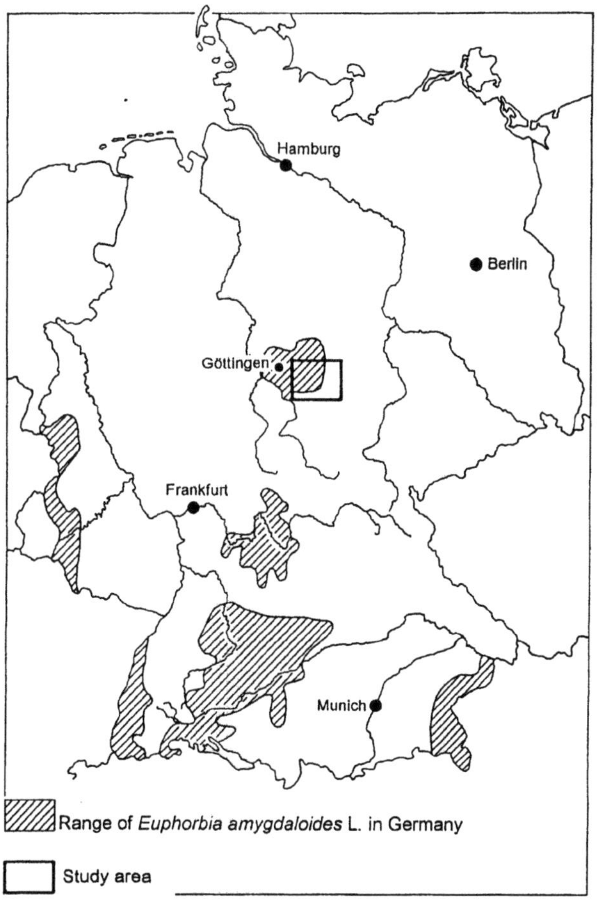

Figure 2: Distribution map of *Euphorbia amygdaloides* in Germany (modified after Haeupler & Schönfelder 1988 and Benkert et al. 1996).

There an exceptionally well-suited study area was found with a limestone mountain ridge called the Hainleite. A floristic inventory of this area was first carried out by Weinitschke (1959). He already pointed out the occurrence of distribution boundaries for several plant species along the mountain ridge. One of the plants he mentioned was *Euphorbia amygdaloides*, which occurs in the western Hainleite but not in the eastern part despite the existence of suitable soil conditions and vegetation types (Figure 3).

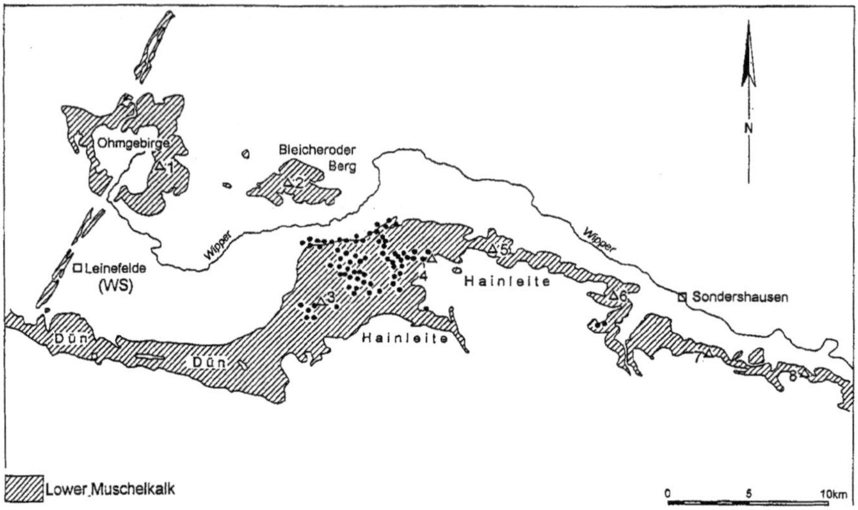

Figure 3: Study area (according to the geological map 1:200,000):
WS: Weather station at Leinefelde
Δ: experimental plots. Plots 1 to 4 are within the range of *Euphorbia amygdaloides*, plots 5 to 8 are outside.
The black spots (•) mark populations mapped by Weinitschke (1959) in the Hainleite area at its distribution limit.

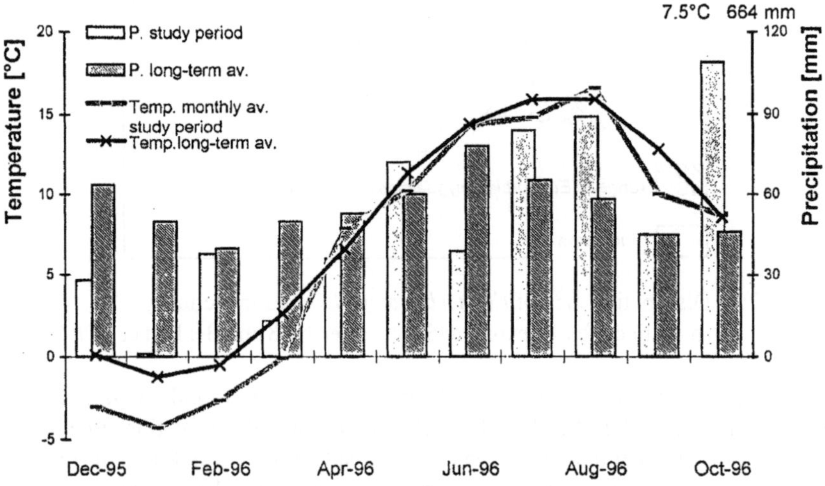

Figure 4: Monthly mean temperatures and precipitation for Leinefelde, 365m above sea level, for the investigation period from December 1995 to October 1996 and for the long-term average (1961-1990).

Figure 4 shows the climatic conditions for Leinefelde, the weather station closest to the study area, both for the investigation period and for the long-term average. The annual mean temperature is 7.5°C and average sum of annual precipitation is about 660 mm (Deutscher Wetterdienst 1995/1996). The precipitation in 1996 differed greatly from the long-term average due to a drier winter and a wetter summer. The winter of 1996 was exceptionally cold and therefore offered excellent conditions for investigating frost damage. The mean temperature recorded for January 1996 at the weather station was -4.3°C, which is 3.1°C colder than the long-term average (1961-1990). The mean temperature in February 1996 was -2.6°C, compared with a long-term average of -0.5°C. The average in March 1996 was still below the freezing point (the March average is normally about 2.6°C). The lowest temperature recorded by the weather station at the standard height of 2 m was -15.5 °C on the 7th of February 1996. The lowest temperature recorded near soil level was -19.1°C on February the 9th.

Methods

Transect

A transect was set up from WNW to ESE across the distribution boundary along the Hainleite mountain ridge (Figure 3). In December 1995 experimental plots were established about every six km along this transect. Plots 1 to 4 were placed within the *Euphorbia amygdaloides* range, and plots 5 to 8 were placed outside.

All plot sites chosen were comparable and fulfilled the requirements for *Euphorbia amygdaloides* habitats. The criteria were:

- *Fagus sylvatica* forests
- Plateau locations
- Limestone sites
- The plots within the *Euphorbia amygdaloides* range had to contain native *Euphorbia amygdaloides* populations
- Easily reached by car

Since the mountain dips gently to the east, the plots' elevation could not be kept completely constant (Table 1). The highest was plot 3 at 474 m above sea level. This plot was located within the *Euphorbia amygdaloides* population, second nearest to the distribution boundary. The plot with the lowest elevation was plot 8, located furthest towards the east, at 365 m above sea level.

The light conditions ranged between 2.8% and 4.6% with the exception of plot 2, which showed a significantly higher (8.8%) relative light intensity than the others (Table 1).

The pH at all sites was higher than 5.5 except for plot 7, which had a pH of only 4.6 due to loess influence. This plot was also the driest at the end of summer (Table 1).

All plots were equally installed in December 1995 (Figure 5).

	Inside range of *Euphorbia amygdaloides*				Outside range of *Euphorbia amygdaloides*			
	Plot 1	Plot 2	Plot 3	Plot 4	Plot 5	Plot 6	Plot 7	Plot 8
Elevation [m a.s.l.]	465	457	474	422	415	410	390	365
Rel. intensity of light [%]	3,1	8,8	4,3	3,4	2,8	4,6	3,1	3,0
pH (H_2O)	6,1	5,5	7,2	6,5	6,3	6,1	4,6	5,4
Soil humidity [vol. %]								
21-May-96	37.5	35.0	34.4	36.7	38.3	36.3	33.2	36.9
09-Aug-96	22.9	22.8	25.3	22.9	23.0	24.3	13.5	21.4

Table 1: General characterisation of the eight experimental field plots.

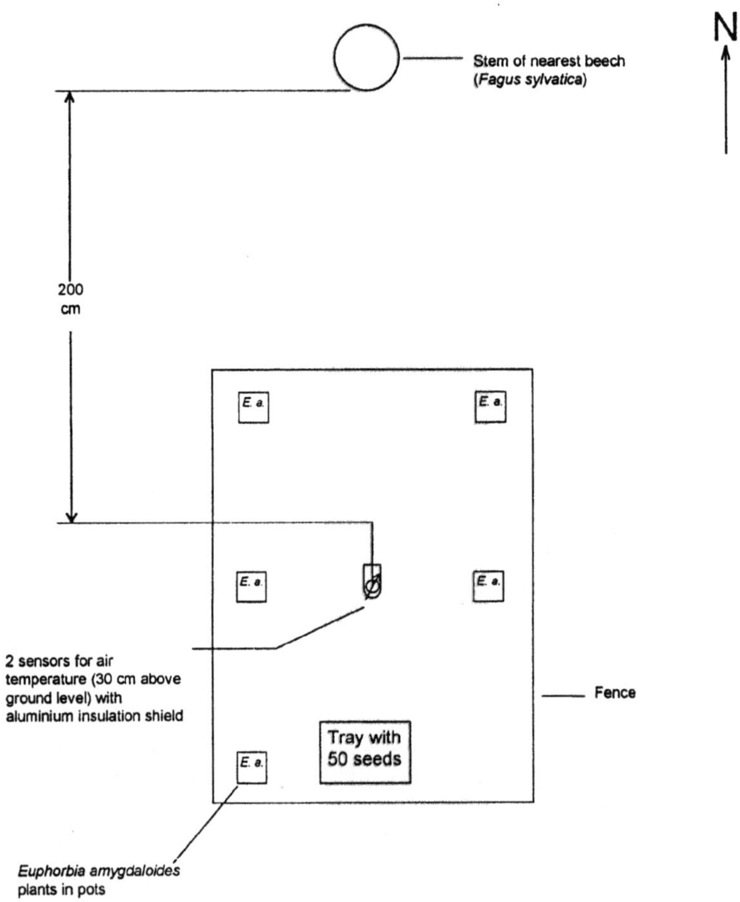

Figure 5: Schematic plan of the experimental plots.

Euphorbia amygdaloides plants which had previously been taken from a large population close to plot 1 and planted into pots were randomly transplanted throughout the eight plots along the transect. Each plot received five plants. Furthermore, each plot was provided with a tray filled with soil and 50 *Euphorbia amygdaloides* seeds in order to observe the climatic influence on germination.
Air temperature was recorded at 30 cm above ground level, which corresponds to the average height of *Euphorbia amygdaloides*, instead of at the standard height of 2 m, which probably would have yielded significantly higher temperatures than closer to the ground. The temperature was recorded every 1.2 hours using data loggers (Tinytalk-Temp, Miniature Temperature Loggers, Orion Components Ltd.) during the winter and during the following vegetation period. The sensors were insulated by an aluminium shield. Soil humidity was measured regularly using time domain reflectrometry (TDR, Soilmoisture Equipment Corporation). Relative light intensity was measured on three different dates in the period from 08/20/96 to 09/02/96 at each plot. We used two Li-Cor data loggers (LI-1000) with PAR quantum-sensors. One of them, which was used as a reference for 100% light, was placed outside of the wood as close to the plot as possible. The other measured light intensity at the height of 30 cm close to the transplanted *Euphorbia amygdaloides* plants. The light conditions had to be diffuse.

Frost tolerance and frost damage

1. Frost tolerance experiments in freeze chambers were carried out at the beginning of April and in December 1996 on individuals taken from the same population that had been the source for the transplantation experiment. The plants were randomly exposed to a non-freezing control temperature (+4°C), and to -6°, -12°, - 18°, -24° or -30°C (Figure 6). Additionally, some plants were treated with liquid nitrogen (-196°C).
2. All transplanted individuals were examined for frost damage in March 1996 and in December 1996.
3. Figure 7 shows the snow cover along the transect in March and November 1996. The snow cover followed more or less the west-east gradient.
   The snow cover at plot 4 close to the distribution boundary was 8 cm in March 1996. This provided the opportunity to perform a further field experiment because the snow covered only parts of the native *Euphorbia amygdaloides* population. Plants were collected which were either completely covered by snow or which stood out from the snow.

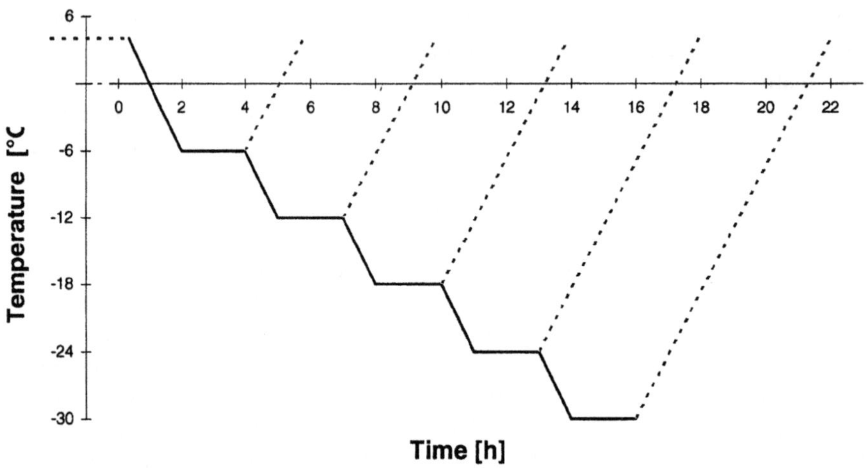

Figure 6: Freezing treatment in the frost chamber experiment: test plants were taken out of the freeze chamber after being exposed to a frost level for two hours. Afterwards the chamber with all remaining plants was cooled down to the next level.

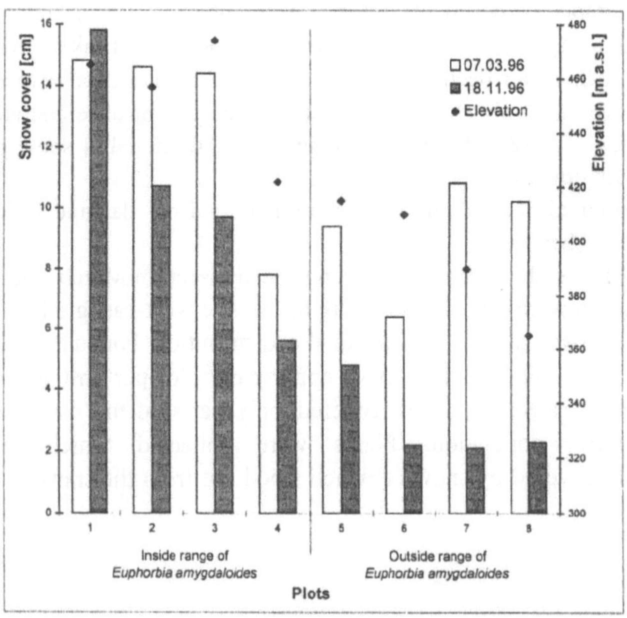

Figure 7: Snow cover distribution along the transect in March and November 1996.

In all experiments the test for frost damage was based on electrolyte leakage (Murray et al. 1989). Plant tissue was put in test tubes with demineralized water and 3 % propanol. The electrical conductivity of the solution was measured (LF 90, LF 2000, WTW) at the beginning (C_0) and then regularly for about 4 days (C_t). The more the tissue was affected by frost damage, the more rapid an increase in conductivity could be detected. The solution's conductivity after being boiled for 20 minutes (C_{boiled}) was used for reference.
Each time series of relative conductivity (RC) was fitted (SAS 6.02: proc regr) using an exponential function (1). The resulting parameter k increases with the rate of electrolyte leakage and can therefore be evaluated as a measure of frost damage:

$$RC = \frac{C_t - C_0}{C_{boiled} - C_0} = 1 - e^{-k * t} \qquad (1)$$

Biometrical investigations of transplanted individuals

The growth rate of the transplanted individuals was recorded during the course of the vegetation period as was the germination rate of the seeds, the seedling survival rate, fructification, seed production and seed weight.

Population pattern at the distribution boundary

The pattern of plant occurrences at the distribution boundary was studied according to two different scales with respect to two different questions:

- How does population density decrease towards the distribution boundary? - Abruptly or continuously?
- How do individuals occur with respect to each other towards the distribution boundary? - In clumps, randomly or regularly?

The first question was tackled by marking areas of 1 km^2 around plots 1 to 8. The areas were placed in such a way that they contained only forest. At least five subareas of 1 ha each were randomly chosen. The density of *Euphorbia amygdaloides* within each hectare was estimated into six classes.
Since one hectare could be covered by more than 10'000 individuals, much smaller test areas were chosen to answer the second question. In three of the 1 ha subareas with an occurrence of *Euphorbia amygdaloides* a 10 by 10 meter section was randomly chosen. The *Euphorbia amygdaloides* plants within these 100 m^2 were counted and their co-ordinates registered. A test for random, regular or clump distribution was conducted according to Manly (1992) based on the nearest neighbour method.

Measurement of photosynthesis

Plants were again taken from the population near plot 1. Photosynthesis was measured with a compact minicuvette system (CMS 400, Walz) in the laboratory at four different dates.

We studied the response to light and to temperature. Different light intensities (0 to about 550 µmol photons/m^2*s PAR [Photosynthetic active radiance]) were generated using an artificial light source and different grey filters. The temperature of the cuvette was controlled. Due to problems with condensation, however, it could only be varied by 10°C above and below the ambient temperature. The temperature was changed in 4° C steps. For details of measuring photosynthesis see Schulte (1993) or Willert et al. (1995).
Important parameters of photosynthesis were estimated by using a biochemical model of photosynthetic CO_2 assimilation. The extended Arrhenius equation (2) describes the response of A_{max} to temperature (Harley & Tenhunen 1991). The curve shows the temperature optimum for photosynthesis:

$$A_{max} = \frac{e^{c - \frac{dHa}{R*T_{leaf}}}}{1 + e^{\frac{dS*T_{leaf} - dHd}{R*T_{leaf}}}} \qquad [\mu mol\ CO_2/m^2*s] \qquad (2)$$

A_{max}:	Maximum net assimilation rate at irradiation saturation and ambient CO_2-level	[µmol CO_2/m^2*s]
R:	Gas constant R = 0.00831	[KJ/mol*K]
dHa:	Activation energy	[KJ/mol]
dHd:	Energy of deactivation	[KJ/mol]
dS:	Entropy	[KJ/mol]
c:	Constant	
T_{leaf}:	Leaf temperature	[°C]

Results

Temperatures along the transect

The temperatures of all plots during January 1996, which was the coldest month of the year, are shown in Figure 8. Figure 9 shows the temperatures for February 1996, when the lowest temperatures were recorded. It was assumed that the transect would show an oceanic-continental temperature gradient and that the lowest minimum would be observed in the east. This assumption proved to be untrue. In addition to plots 5 and 6 in the eastern part, plot 2 in the west showed the lowest temperature values. The average difference between these three plots and the others was 1.5° C for January 1996 and 1.4°C for February 1996. This same phenomenon was observed throughout the winter (Figure 10).
There was little difference between the plots regarding absolute minimum temperatures. The lowest temperature of all was -18.8°C, recorded on the 7th of February at plot 4. This was followed by -18.7°C at plots 6 and -18.6°C at plot 5. The minimum temperature of the other 4 plots was only 0.9°C to 1.8°C warmer than the temperature of the coldest plot (Table 2).

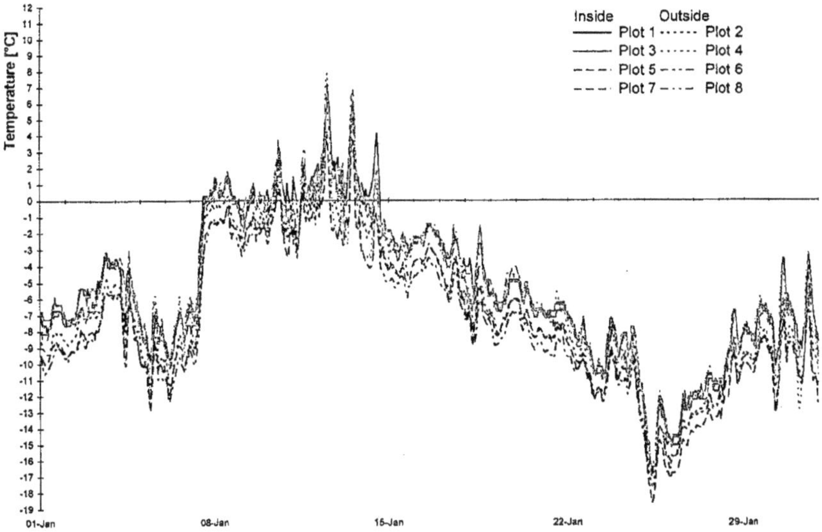

Figure 8: Air temperature along the transect, January 1996, 1.2 h interval, 30cm above ground.

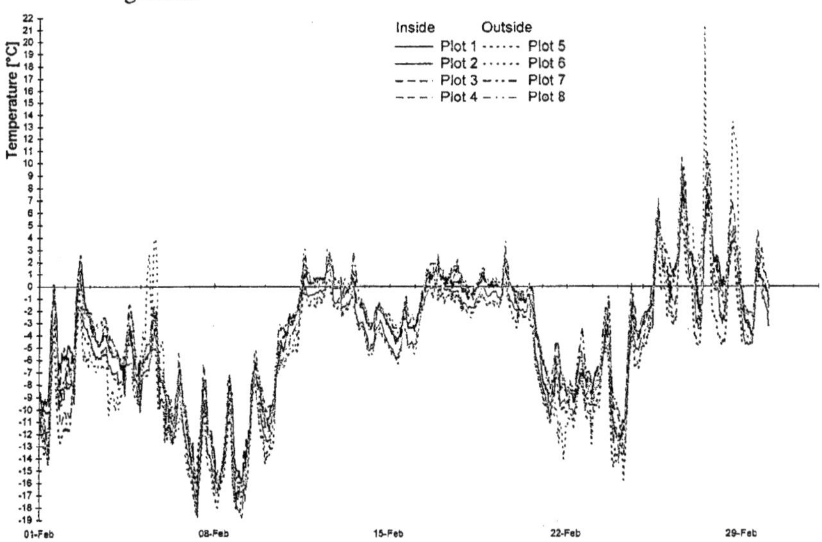

Figure 9: Air temperature along the transect, February 1996, 1.2 h interval, 30cm above ground.
Due to a failure of data loggers at plot 2, 5 and 6 in some periods in January and February and to a replacement by another logger model, the values of these plots may, to a minor degree, vary more than the others.

There were hardly any temperature differences during the summer months (Figure 10). Only plot 4 showed a higher temperature amplitude.

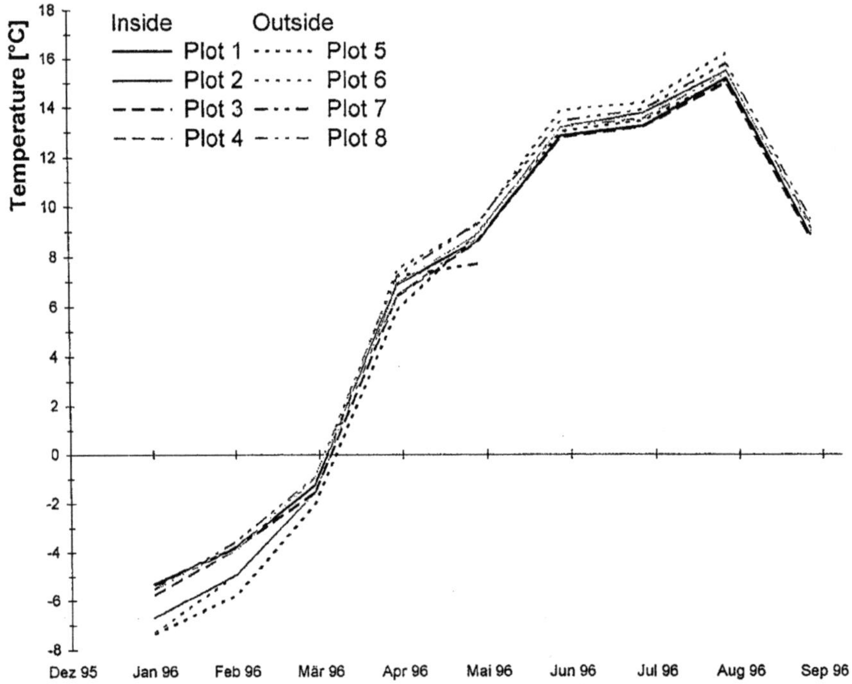

Figure 10: Mean monthly air temperature along the transect, 30cm above ground.

	Inside range of *Euphorbia amygdaloides*				Outside range of *Euphorbia amygdaloides*			
	Plot 1	Plot 2	Plot 3	Plot 4	Plot 5	Plot 6	Plot 7	Plot 8
Absolute minimum temperature [°C]	-17.4	-17.9	-17.0	-18.8	-18.6	-18.7	-17.2	-17.4

Table 2: Absolute minimum temperatures along the transect, 30cm above ground.

Due to a failure of data loggers at plot 2, 5 and 6 in some periods in January and February and to a replacement by another logger model, the values of these plots may, to a minor degree, vary more than the others.

Frost tolerance and frost damage

1. Freeze chamber experiment

Frost damage to leaves after exposure to the six freezing treatments is shown in Figure 11. Only the -24°C and the -30°C treatments could be distinguished from all others within a 5% confidence interval.

Damage to different plant organs is shown in per cent of total damage (Figure 12). Damage among individual plants is highly variable. Therefore, in most cases only two groups can be statistically distinguished. For all organs the -24°C and -30°C treatments differed from all the other treatments. Only the leaves, which are the most delicate plant organs, showed a difference at -18°C as well. This lethal temperature threshold (-18°C) could also be assessed visually. The root colour changed as well, becoming dark when dead.

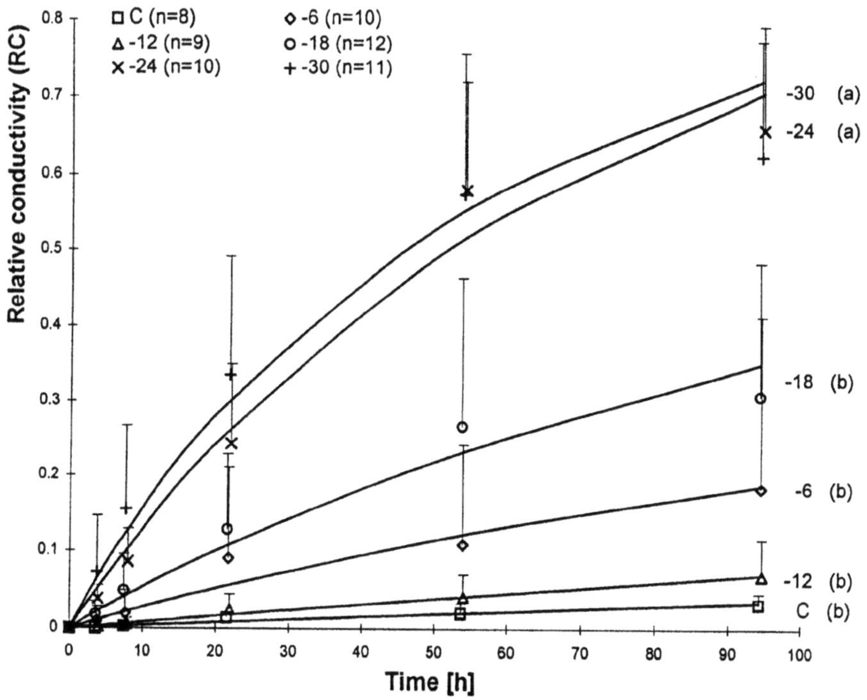

Figure 11: Exponential saturation curve according to equation (1) for the six freezing treatments of the leaves in December 1996. Bars: standard deviation.

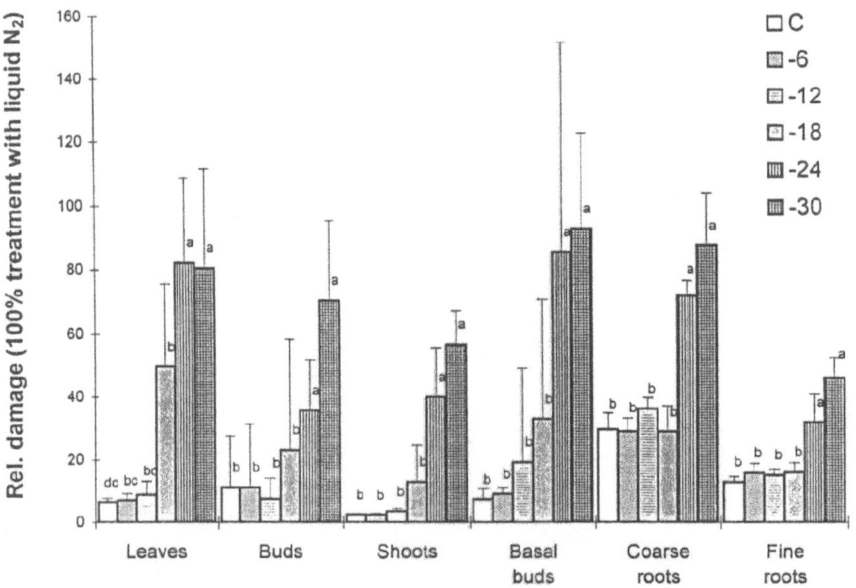

Figure 12: Relative damage to different plant organs in per cent of total damage; 100% damage was caused by treatment with liquid nitrogen, April 1996. Number of replicates (n) see below.

n	Leaves	Buds	Shoots	Basal buds	Coarse roots	Fine roots
°C	10	10	10	8	5	5
-6	10	10	10	9	6	6
-12	9	9	9	8	5	5
-18	10	10	10	8	5	5
-24	10	10	10	10	4	4
-30	10	10	10	6	5	5

2. Field experiment with transplanted individuals along the transect:
   There was no significant difference in electrolyte leakage between the transplanted *Euphorbia amygdaloides* individuals, neither during the winter nor at the end of the winter (Figure 13).

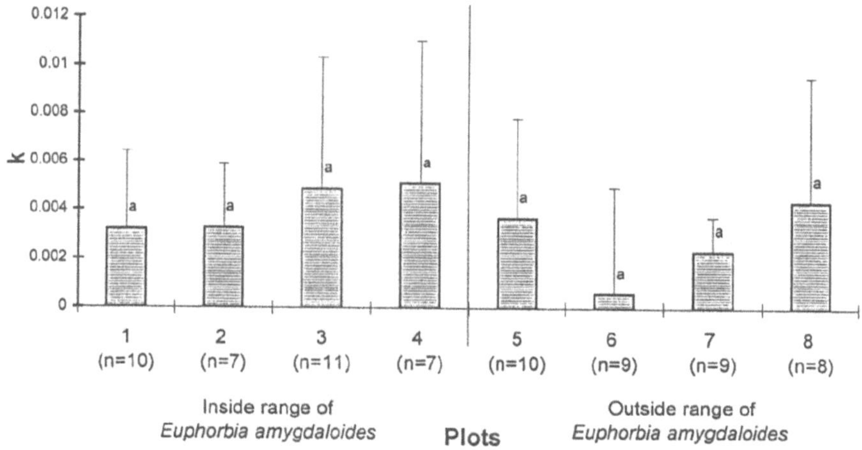

Figure 13: k as a measure of frost damage, leaves from the transplanted *Euphorbia amygdaloides* individuals, April 1996.

3. Snow cover had a significant effect on all investigated organs (Figure 14). k as a measure of frost damage differed in the same order of magnitude as obtained by frost treatments from -12°C to -18°C.

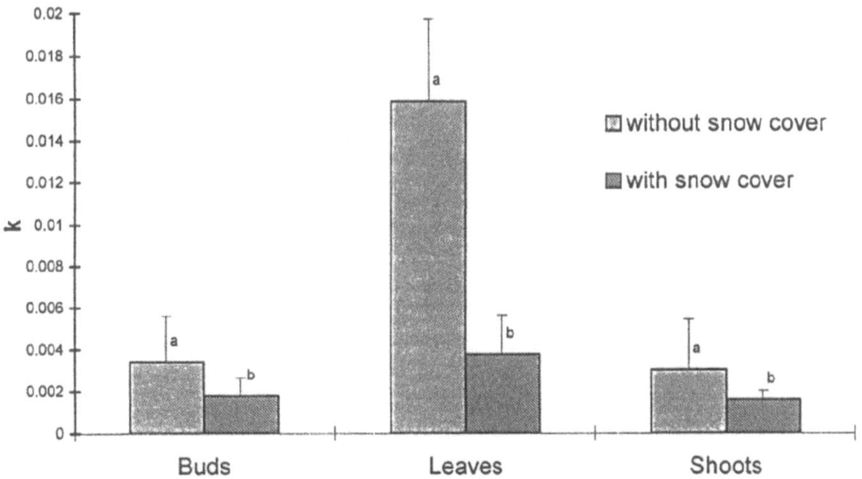

Figure 14: k as a measure of frost damage, *Euphorbia amygdaloides* individuals with and without snow cover, April 1996. (n=20)

Biometrical investigations
Two plants died in each of plots 2, 3 and 7, which was probably not due to frost damage. The plants may have been damaged when put into pots. No significant differences in vegetative growth could be detected during the entire investigation period (Figure 15).

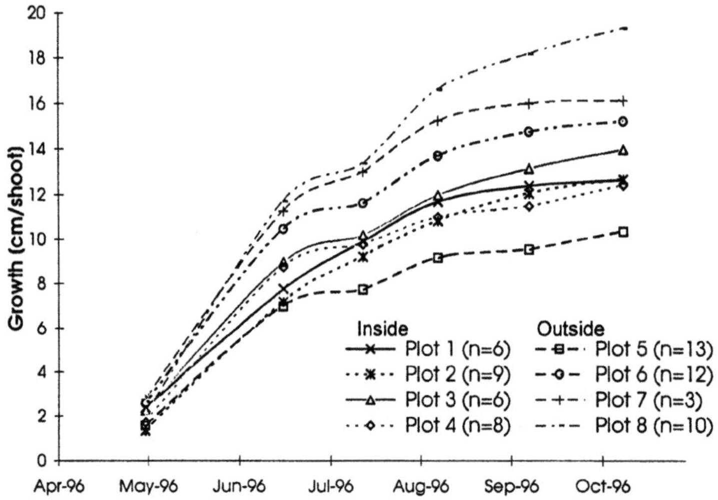

Figure 15: Growth of vegetative shoots. Reference length (0) is December 1995.

Likewise, the germination rate showed no differences (Table 3).

	Inside range of *Euphorbia amygdaloides*				Outside range of *Euphorbia amygdaloides*			
	Plot 1	Plot 2	Plot 3	Plot 4	Plot 5	Plot 6	Plot 7	Plot 8
Germinated seeds (of 50 seeds) May 96	36	26	49	48	34	48	50	43
Survived seedlings (of 50 seeds) Dec. 96	26	1	34	46	16	31	41	34

Table 3: Germination rate and seedling survival rate.

The seedling survival rate was almost equal among the plots except for plot 2, where only one plant survived. No significant differences in fructification were observed (Table 4).

	Inside range of *Euphorbia amygdaloides*				Outside range of *Euphorbia amygdaloides*			
	Plot 1	Plot 2	Plot 3	Plot 4	Plot 5	Plot 6	Plot 7	Plot 8
Average number of fruits per plant	8.0	0.0	13.7	10.0	10.5	7.3	18.0	6.3
Average number of fruits per generative shoot	6.8	0.0	10.2	10.0	7.8	7.3	5.3	5.6
1000-seed weight [g]	4.9		6.3	3.7	5.7	3.9	5.9	5.3

Table 4: Fructification of the transplanted *Euphorbia amygdaloides* individuals, July 1996.

Population pattern

The large-scale population pattern showed a distinct decrease in density towards the distribution boundary (Table 5).

		Inside range of *Euphorbia amygdaloides*				Outside range of *Euphorbia amygdaloides*			
Class	Plants/ha	Plot 1	Plot 2	Plot 3	Plot 4	Plot 5	Plot 6	Plot 7	Plot 8
5	100000	1/5							
4	10000	3/5		1/8					
3	1000	1/5	1/5	2/8	1/5				
2	100		2/5		1/5				
1	10		1/5		1/5				
0	0		1/5	5/8	2/5	5/5	5/5	5/5	5/5

Table 5: Population densities of *Euphorbia amygdaloides* in randomly chosen 1ha squares around the experimental plots.

The numbers indicate the plots. The median density is indicated in black, the range is indicated in grey. The fractions correspond to observations. For example, 100 to 1000 plants per ha were observed two times within $1km^2$ around plot 3. No plants were observed six times.

Black: median, grey: range of density, fractions: fractions of observations.

The highest densities were observed around plot 1. All investigated populations were significantly clumped on the scale of $100\,m^2$ plots. Figure 16 gives an example.

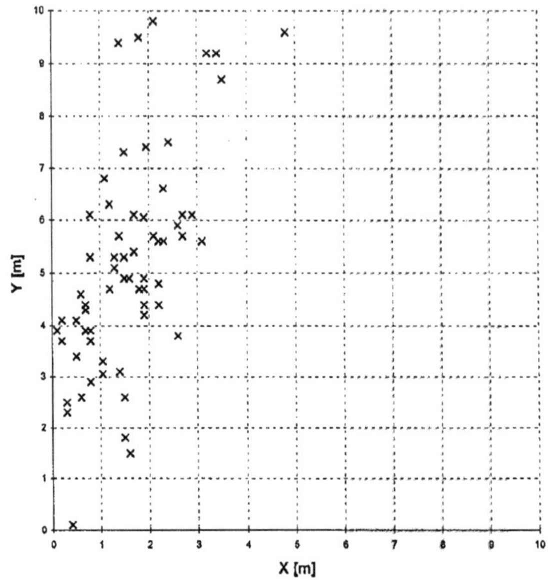

Figure 16: Distribution pattern of all (63) *Euphorbia amygdaloides* plants in a 10 by 10m area, which was randomly located near experimental plot 2.

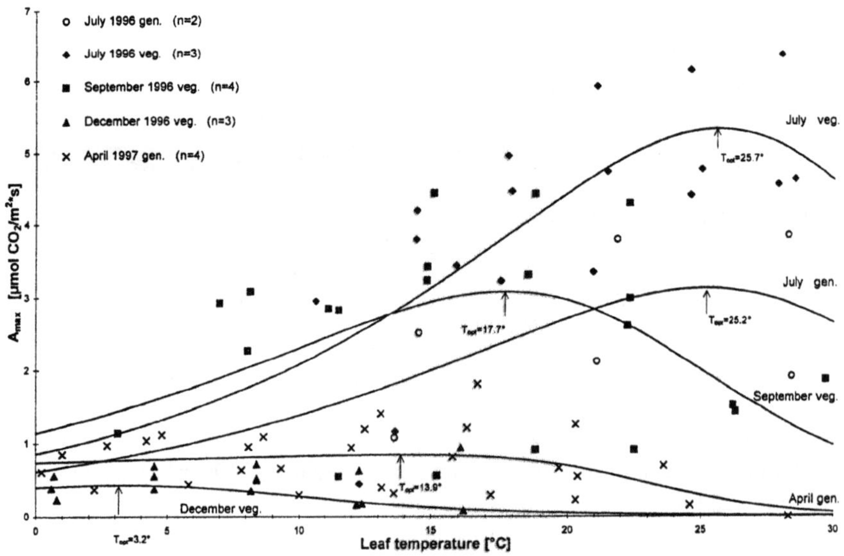

Figure 17: Response of A_{max} (maximum net assimilation rate at irradiation saturation (~550 µmol photons PAR/m^2*s) and ambient CO_2-level) to changes in temperature. Modelled optimum curves according to equation 2.

Measurement of photosynthesis

Figure 17 shows the maximumnet assimilation response (A_{max}) to changes in temperature. Responses of individual plants showed great variability. The acclimation of the plants in colder seasons is obvious, as the temperature optimum of A_{max} drops from 26°C in summer to 18°C in autumn and down to only 3°C in winter. It increases again in April to 14°C.

Discussion

We conclude that winter minimum temperature quite probably determines the north-eastern range limitof *Euphorbia amygdaloides*. Temperatures measured in the field reached values that caused damage to plant tissues in freeze chamber experiments. The threshold temperature was about -18°C for leaves and some degrees lower for buds and shoots (Figure 12). At -24°C and below all plant organs were damaged significantly. These values indicate that the frost tolerance of *Euphorbia amygdaloides* approaches the lower limits for temperate herbs (Till 1956, Bannister 1976, Burke et al. 1976, Larcher & Bauer 1981).

Temperatures below -18°C were also recorded in half of the field plots (Table 2). However, these extremes neither resulted in killing the test plants nor were there significant differences in frost damage among the transplanted individuals (Figure 13). An explanation for the discrepancy between laboratory and field data could be a microclimaticvariability within the plots. The individual plant organs may have experienced higher temperatures than measured in a height of 30 cm. To collect more precise information, sensors would have to be placed in direct contact to the plant's organs. Apart from requiring a much higher expenditure of plot equipment this design would have created much more artificial conditions. For example, fixed sensors could have influenced the shoots' movements and prevented their bending under snow weight.

Assuming that microclimaticvariability is only a minor influence, the question arises as to whether the field conditions met the requirements of a field study. The winter of 1995/1996 might simply not have been severe enough to cause significant damage. The results of our freeze chamber experiments indicate that a winter with extremes of around -24°C would have been necessary to kill the test plants. Nevertheless, the weather conditions met another requirement for frost damage, namely low snow cover. The significant influence of snow cover (Figure 14) indicates that the plants can endure much lower temperatures with sufficient snowfall. A snow cover of more than 20 cm rarely allows the temperature to drop below -5°C (Sakai & Larcher 1987: 16). Since precipitation was exceptionally low in January (Figure 4) the winter can be considered to have offered almost optimum conditions for this field experiment.

The effect of snow on plant distribution has been pointed out by various authors. With respect to the Hainleite area, Köhler (1967) and Meusel et al. (1978: 11) suggested to take snow cover into consideration to understand plant distribution. Uemura (1989) showed that snow exerted a strong influence on the distribution pattern of evergreen plants in Japan.

It should be possible to assess the risk of frost damage at low snow cover by combining both parameters into an index. It is astonishing that such an index has not yet been proposed in plant geography (cf. Tuhkanen 1980). In analogy to the simplest possible humidity index, in which the moisture factor H is defined as a ratio of monthly precipitation in mm to monthly mean temperature in °C, an index for frost risk could be derived with F as a reciprocal of H. F should be formulated for the coldest month and would decrease (negative sign) with increasing frost risk. F should be valid only for mean temperatures below a threshold (e.g. 0°C) and the influence of precipitation should be weakened by adding a constant to monthly precipitation (e.g. +10). It would be worthwhile to test such an index for coincidences with ranges of European plant species.

We would like to include a few comments on using standard climatic data recorded at a height of two meters by weather stations for correlations with plant ranges. In the case of forest plants such as *Euphorbia amygdaloides,* the question arises whether temperatures at the undergrowth level are comparable to temperatures in the open field. It is often stated that the tree canopy of a forest reduces the heat loss of the herbal layer, resulting in more evenly levelled temperatures (Sakai & Larcher 1987: 9). Comparing the minimum temperature in the field of -18.8°C recorded at plot 4 (Table 2) with the minimum temperature of -15.5°C measured by the official weather station of Leinefelde at the standard height of two meters above the ground reveals that the canopy effect does not fully compensate the low temperature closer to the ground. This observation is consistent with values given by Sakai & Larcher (1987: 11) that temperature near to the ground may be 5°C lower than at a height of 2 m and consistent with the measurement in Leinefelde at ground level of -19.1°C. The comparison reveals that forest herbs may experience minimum temperatures of 3.3°C lower than temperatures measured in standard weather stations in the surrounding area. This effect should be taken into consideration when trying to correlate isotherms with plant ranges of forest plants.

Apart from having lethal effects on plants, frosts affect production processes even when they are only moderate. Restriction of photosynthetic production at low temperatures is shown clearly in Figure 17. During seasons of freezing events photosynthesis capacity reaches only 8.0% of the maximum which was found in July. This decrease of photosynthesis can be clearly interpreted as an effect of frost hardening (Larcher & Bauer 1981). In the midst of winter photosynthetic activity is extremely low, indicating that its contribution to the annual dry matter production is negligible. In this respect evergreen leaves do not seem to be advantageous for *Euphorbia amygdaloides*. Of course, this is true for only part of the total range of the species. In regions with mild winters, photosynthetic production in this season will be greater.

The test plants experienced frosts even in April, which might explain their low photosynthetic activity in this month due to an inhibitory effect after freezing, even if the event occurred several nights before (Larcher & Bauer 1981). Nevertheless, a warm spring is likely to support the *Euphorbia amygdaloides* occurrence. Jäger (1970) gives an interpretation of the north-eastern boundary as being coincidental with the 8°C-April isotherm, which indicates a requirement for warm springs. It is quite probable that warm springs are more favourable to respiration processes resulting in a break down of stored carbohydrates for early bud burst and vegetative growth rather than photosynthetic dry matter production.

In summer, the lack of differences in vegetative growth (Figure 15) corresponds to the marginal differences in summer temperature (Figure 10). Other studies showing differences in growth rates along transects have used much steeper climatic gradients (Woodward & Pigott 1975, Friend & Woodward 1992: 89ff.).

Whether a species is able to maintain a viable population at a particular site is not only dependent on its survival and production capacity but also on its ability to propagate and multiply. The observed temperatures had no lethal effects on the plants' buds from the previous year (Figure 12). Freezing events in early spring did not harm the flowers, which can be concluded by the normal seed production of most test plants (Table 4). Germination and the seedling survival were not reduced either (Table 3), with the exception of plot 2. It is likely that the death of all but one seedling in this plot was due to the comparatively low temperatures there, which were probably caused by a more open canopy (Table 1). This observation, together with the direct frost damage to adults, can explain why the species is mostly confined to forests in the study area and only occasionally grows in forest gaps or at forest edges.

From the study's results we conclude that *Euphorbia amygdaloides* could also survive outside of its range at plots 5 to 8, provided that winters are no more severe than during the study period. The question arises whether the actual distribution limit along the transect is identical to the potential limit, which may be located further to the east. Without being able to answer this question we have evidence that it is impossible to answer it on the scale of our study. Variability in microclimate among plots 1 to 4 within the *Euphorbia amygdaloides* range is at least as high or even higher than between plots 5 to 8 outside the range. It can be assumed that at least some sites outside the actual range would meet the requirements for supporting a viable *Euphorbia amygdaloides* population. We suppose that the presence of such sites follows a probability function. For *Euphorbia amygdaloides* the probability of the existence of suitable habitats within the range gradually decreases towards the actual distribution limit. Outside the actual distribution boundary, the probability of encountering suitable habitats greatly decreases. If we assume a lower probability of suitable sites and therefore greater distances between them, it becomes even more improbable that these distances can be spanned by dispersal agents. From this point of view the actual distribution limit is only a sharp bend in the probability function. Consequently, the actual distribution limit would not be a line but a zone with a geographical extension. The potential distribution limit would be the point where probability is zero. This definition differs considerably from the traditional formulation for the potential distribution as the range that could be developed when competition from other plants is excluded (Jäger 1972, Walter & Straka 1970).

Our concept is supported by the results of the distribution pattern along the transect (Table 5). The probability of finding *Euphorbia amygdaloides* within a randomly located square of 1 ha decreases from plot 1 to 4, as do the number of individuals within 1 ha. It is likely that the trend could be demonstrated even more by including further transects in other regions. A consequence of this concept for plant range studies would be to shift the focus from autoecology to population biology.

However, on a scale of 100 m^2 the *Euphorbia amygdaloides* population is unlikely to follow a probability function because all investigated squares revealed a clumped pattern. Most probably, clumping is an effect of the species dispersal biology. *Euphorbia amygdaloides* seeds are spread by ants (van der Pijl 1982). It can be assumed that myrmecochory is responsible for long distance dispersal, ensuring settlement in suitable site conditions. By contrast, autochory may be responsible for a clumped growth pattern at a given site. In order to study the future development of this growth pattern we have marked each of the 100 m^2 squares with magnets.

Although there is evidence that winter temperature is the key factor limiting the *Euphorbia amygdaloides* range in this area there are of course other possible explanations. For example, high summer temperatures may favour an increased floral initiation, resulting in more abundant flowers in the subsequent year (Pigott & Huntley 1981, Owens 1991). Temperature can affect all stages of the life cycle, beginning with the seed bank and ending with seed production. Woodward & Williams (1987) described the transition from one stage to the next as a probability function influenced by temperature. More long-term studies would be useful to assess these effects. Furthermore, temperature can also have an additive quality (Went 1957, Cleary & Waring 1969). The calculation of heat sums could be useful for an interpretation of the north-western boundary of the *Euphorbia amygdaloides* range in Europe.

Finally, the question remains whether this study allows us to make predictions in the larger context of global climatic change. Such predictions are, of course, highly uncertain due to the uncertain trend of all climatic parameters and due to the high acclimation potential of plants (Skre 1993). Further uncertainty factors must be taken into consideration with respect to our probability concept of range limits. However, if absolute temperature minima increase and/or snow in January becomes more abundant, the *Euphorbia amygdaloides* range will probably expand towards the east.

Despite all the difficulties of our approach, we consider further field studies on a local scale in combination with laboratory tests to be useful.

Acknowledgements

We would like to thank Dr. Dirk Gries and Christa Behler for their help with the work on photosynthesis. The field work was supported by the helpful hands of Annette Muhs, Stefan Schrader and Dr. Gustav Schulz. We thank them for their critical discussion as well.
We thank Bernd Raufeisen for the art work and Rainer Kraft for gardening assistance. Many thanks to Judith Weston for the critical reading and revision of the manuscript.
We thank Stephan Pflume for providing the data loggers and their software.
We are grateful to Prof. M. Runge for supporting our ideas in many ways. Prof. E. Jäger and Prof. F. G. Schroeder greatly influenced our ideas on plant distribution and encouraged us to start this work.
We thank Prof. W. Schmidt who put the freeze chamber of the forest department at our disposal.
The importance of plant ranges in understanding the life history of plant species was early recognised by Prof. H. Meusel, who died on the 3rd of January 1997. It is to his memory that we dedicate this paper.

Literature

Bannister P. (1976): Introduction to physiological plant ecology. Blackwell, Oxford.

Begon M., Harper J.L. & Townsend C.R. (1990): Ecology. Individuals, populations and communities. 2nd ed. Blackwell, Boston.

Benkert D., Fukarek F. & Korsch H. (Hrsg.) (1996): Verbreitungsatlas der Farn- und Blütenpflanzen Ostdeutschlands. Fischer, Jena.

Box E.O., Crumpacker D.W. & Hardin D. (1993): A climatic model for location of plant species in Florida, USA. J. Biogeogr. 20: 629-644.

Burke M.J., Gusta L.V., Quamme H.A., Weiser C.J. & Li P.H. (1976): Freezing and injury in plants. Ann. Rev. Plant Physiol. 27: 507-528.

Callauch R. (1986): Die Frosthärte der Stechpalme (*Ilex aquifolium*). Verh. Ges. Ökol. 14: 523-527.

Cleary B.D. & Waring R.H. (1969): Temperature: collection of data and its analysis for the interpretation of plant growth and distribution. Can. J. Bot. 47: 167-173.

Dahl E. (1951): On the relation between summer temperature and the distribution of alpine vascular plants in the lowlands of Fennoskandia. Oikos 3: 22-52.

Deutscher Wetterdienst (1995/1996): Monatlicher Witterungsbericht 43 (12), 44 (1-10). Deutscher Wetterdienst, Offenbach.

Eichler J., Gradmann R. & Meigen W. (1927): Ergebnisse der pflanzengeographischen Durchforschung von Württemberg, Baden und Hohenzollern. Verein für vaterländische Naturkunde in Württemberg, Grüninger, Stuttgart.

Friend A.D. & Woodward F.I. (1992): Evolutionary and ecophysiological responses of mountain plants to the growing season environment. Adv. Ecol. Res. 20: 60-124. London.

Grace J. (1987): Climatic tolerance and the distribution of plants. New Phytol. 106(Suppl.):113-130.

Grime J.P. (1990): Ecological effects of climate change on plant populations and vegetation composition with particular reference to the British flora. In: Jackson M.T., Ford-Lloyd B.V. & Parry M.L. (eds.): Climatic change and plant genetic resources. 40-60. Belhaven Press, London.

Grisebach A. (1880): Gesammelte Abhandlungen und kleine Schriften zur Pflanzengeographie. Engelmann, Leipzig. Reprint of: Vegetationslinien des nordwestlichen Deutschlands, Göttinger Studien 1847: 461-562.

Haeupler H. & Schönfelder P. (1988): Atlas der Farn- und Blütenpflanzen der Bundesrepublik Deutschland. Ulmer, Stuttgart.

Harley P. C. & Tenhunen J. D. (1991): Modeling the photosynthetic response of C_3 leaves to environmental factors. Modeling crop photosynthesis - from biochemistry to canopy. American Society of Agronomy and Crop Science Society, Madison. Special publication no. 19: 17-39.

Hendry G.A.F. (1993): Forecasting the impact of climatic change on natural vegetation with particular reference to the boreal and sub-arctic floras. In: Holten J.I., Paulsen G., Oechel W.C. (eds.): Impacts of climatic change on natural ecosystems, with emphasis on boreal and arctic/alpine areas. 136-150. Trondheim, Norway.

Hintikka V. (1963): Über das Grossklima einiger Pflanzenareale in zwei Klimakoordinatensystemen dargestellt. Ann. Bot. Fenn. 34 (5): 1-64.

Hjelmqvist H. (1940): Studien über die Abhängigkeit der Baumgrenzen von den Temperaturverhältnissen unter besonderer Berücksichtigung der Buche und ihrer Klimarassen. Carl Bloms, Lund.

Holten J.I. (1993): Potential effects of climatic change on distribution of plant species, with emphasis on Norway. In: Holten J.I., Paulsen G., Oechel W.C. (eds.): Impacts of climatic change on natural ecosystems, with emphasis on boreal and arctic/alpine areas. 84-104. Trondheim, Norway.

Huntley B. (1991): How plants respond to climate change: migration rates, individualism and the consequences for plant communities. Ann. of Bot. 67 (Suppl. 1): 15-22. London.

Huntley B., Bartlein P.J. & Prentice I.C. (1989): Climatic control of the distribution and abundance of beech (*Fagus* L.) in Europe and North America. J. Biogeogr. 16: 551-560.

Huntley B., Berry P.M., Cramer W. & McDonald A.P. (1995): Modelling present and potential future ranges of some European higher plants using climate response surfaces. J. Biogeogr. 22: 967-1001.

Iversen J. (1944): *Viscum*, *Hedera* and *Ilex* as climate indicators. Geol. Fören. Förhandl. 66 (3): 463-483.

Jäger E.J. (1968): Die pflanzengeographische Ozeanitätsgliederung der Pflanzenareale. Feddes Repert. 79: 157-335.

Jäger E.J. (1970): Charakteristische Typen mediterran-mitteleuropäischer Pflanzenareale. Feddes Repert. 81 (1-5): 67-92.

Jäger E.J. (1972): Comments on the history and ecology of continental European plants. In: Valentin, D.H. (ed.): Taxonomy, phytogeography and evolution. London. 349-362.

Jäger E.J. (1988): Möglichkeiten der Prognose synanthroper Pflanzenausbreitungen. Flora 180 (1/2): 101-131.

Jäger E.J. (1989): Geobotanik. Areal- und Florenkunde (Floristische Geobotanik). Progr. in Bot. 51: 313-331.

Jäger E.J. (1990): Pflanzenarealgrenzen als Leistungsgrenzen. Biol. Rundschau 28: 295-306.

Jäger E.J. (1992): Die Verbreitung von *Frankenia* in der Mongolei, in Westeurasien und im Weltmaßstab. Flora 186: 177-186.

Jäger E.J. (1995a): Klimabedingte Arealveränderungen von anthropochoren Pflanzen und Elementen der natürlichen Vegetation. Angew. Landschaftsökol. 4: 51-57.

Jäger E.J. (1995b): Geobotany. Plant geography. Progr. in Bot. 56: 396-415.

Kappen L. (1964): Untersuchungen über den Jahreslauf der Frost-, Hitze- und Austrocknungsresistenz von Sporophyten einheimischer Polypodiaceae (Filicinae). Flora 155: 123-166.

Klötzli F., Walther G.-R., Carraro G. & Grundmann A. (1996): Anlaufender Biomwandel in Insubrien. Verh. Ges. Ökol. 26: 537-550.

Köhler H. (1967): Vegetationskundliche Untersuchungen der natürlichen Waldgesellschaften des oberen und mittleren Eichsfeldes und der Randgebiete des Thüringer Beckens. Diss. Univ. Halle.

Larcher W. & Baue, H. (1981): Ecological significance of resistance to low temperature. In: Lange O.L., Nobel P.S., Osmond C.B., Ziegler H. (eds.): Physiological plant ecology. I. Encyclopedia of plant physiology 12 A: 403-437. Berlin, Heidelberg.

Manly B. (1992): The design and analysis of research studies. Cambridge University Press.

Meusel H. & Jäger E.J. (1989): Ecogeographical differentiation of the submediterranean deciduous forest flora. Pl. Syst. Evol. 162: 315-329.

Meusel H., Jäger E., Rauschert S. & Weinert E. (1978): Vergleichende Chorologie der zentraleuropäischen Flora. Band 2. VEB Gustav Fischer Verlag, Jena.

Murray M. B., Cape J. N. & Fowler D.(1989):Quantification of frost damage in plant tissues by rates of electrolyte leakage. New Phytol. 113: 307-311.

Owens J.N. (1991): Measuring growth and development of reproductive structures. In: Lassoie J.P., Hinckley T.M.: Techniques and approaches in forest tree ecophysiology 423-452. Boca Raton, Ann Arbor, Boston.

Pigott C.D. & Huntley J.P. (1981): Factors controlling the distribution of *Tilia cordata* at the northern limits of its geographical range. III. Nature and causes of seed sterility. New Phytol. 87: 817-839.

Sakai A. & Larcher W. (1987): Frost survival of plants. Responses and adaptation to freezing stress. Springer, Berlin.

Schulte M. (1993): Saisonale und interannuelle Variabilität des CO_2-Gaswechsels von Buchen (*Fagus sylvatica* L.). Bestimmung von C-Bilanzen mit Hilfe eines empirischen Modells. Diss. Universität Göttingen. Shaker, Aachen.

Skre O. (1993): Consequences of possible climatic temperature changes for plant production and growth in alpine and sub-alpine areas in Fennoscandia. In: Holten J.I., Paulsen G., Oechel W.C. (eds.): Impacts of climatic change on natural ecosystems, with emphasis on boreal and arctic/alpine areas 125-135. Trondheim, Norway.

Till O. (1956): Über die Frosthärte von Pflanzen sommergrüner Laubwälder. Flora 143: 499-542.

Tukahnen S. (1980): Climatic parameters and indices in plant geography. Acta Phytogeogr. Suecica 67: 1-110.

Uemura S. (1989): Snowcover as a factor controlling the distribution and speciation of forest plants. Vegetatio 82: 127-137.

Van der Pijl L. (1982): Principles of dispersal in higher plants. 3rd ed. Springer, Berlin.

Walter H. & Straka H. (1970): Arealkunde (Floristisch-historische Geobotanik). 2. Aufl. Ulmer, Stuttgart.

Weinitschke H. (1959): Die Waldgesellschaften der Hainleite. Diss. Univ. Halle.

Went F.W. (1957): The experimental control of plant growth. Waltham, Massachusetts (USA).

Willert D.J. von, Matyssek R. & Herppich W. (1995): Experimentelle Pflanzenökologie. Grundlagen und Anwendungen. Thieme, Stuttgart.

Woodward F.I. & Williams B.G. (1987): Climate and plant distribution at global and local scales. Vegetatio 69: 189-197.

Woodward F.I. & Pigott C.D. (1975): The climatic control of the altitudinal distribution of *Sedum rosea* (L.) Scop. and *S. telephium* L. I. Field observations. New Phytol. 74: 323-334.

Woodward F.I. (1987): Climate and plant distribution. Cambridge University Press, Cambridge.

Plenary discussion

Statement: N. KRÄUCHI

I have a minor remark with respect to the setup of your meteorological instruments. I strongly suggest to measure temperature according to standardized manuals. Temperature is commonly measured at 2 meters above ground. If you measure at different heights within the forest stand and in the open field you need to transform the data for comparison. I would further suggest to add meteorological stations in the open field as close as possible to the sites in the forest.

Statement: H. BRUELHEIDE

Exactly this has been tried by forest-ecologists in Göttingen. They had to establish a net of weather stations inside the forests in the Harz Mountains, because they could not compare any station outside the forest with the climatic conditions inside. Additionally, they established stations nearby in the open field.

Statement: N. KRÄUCHI

These microclimatic differences of the forested and non-forested nearby site was the main reason to use two meteorological stations for each plot of the Long-term Forest Ecosystem Research (LTFER) Programme in Switzerland.

Statement: F. KLÖTZLI

I also wanted to point out the fact. I'm not against the fact that you measure it near the plant or on the plant but you should have a reference point on the international level of measurements. What I want to say is also endorsing your conclusion: one year is not enough, that's clear, we don't have to discuss that. But you will see *Euphorbia amygdaloides* is behaving in a rather crazy way. It is very resistant, it is not browsed by roe-deer or so, it's not attacked by insects, it's influenced by pathogens e.g. from fungi such as ascomycetes. So you get jumps in its appearance, that's our experience. You have chosen a "funny plant". It's not because you have done a funny work, no, you have just chosen a plant which in an absolutely unpredictible way is very resistant, very stable and all of a sudden it is gone.

Statement: H. BRUELHEIDE

We have not known before our studies that *Euphorbia amygdaloides* tolerates such low temperatures, so this is one of the main results of our work.

Statement: F. KLÖTZLI

That's what I meant to say, it's a very stable plant but all of a sudden it's not here anymore. And then it jumps and that's why you get such "funny" boundaries. The plant is well chosen in a way as you have said, the criteria are clear, but if you continue you will have a lot of surprises with it.

Statement: A. BORHIDI

I would like to suggest, that perhaps it would be useful to make additional temperature measuring points on the surface of the leaves.

Answer: H. BRUELHEIDE

Yes, of course, but you have to remember that we already used 32 data loggers for this experiment (8 for air temperature, 24 for soil temperature). Since there is the risk of theft if you just put them in the field, increasing the number of plots could be rather expensive.

Question: N. KRÄUCHI

Have you tried to figure out for what reason so many plants disappeared in some plots? Were there fungi?

Answer: C. SCHULZ

Unfortunately, we don´t have any evidence for parasites influencing the survival of our plants. It is also possible that plants at plot 2 suffered low temperatures.

Answer: H. BRUELHEIDE

This was probably due to the open canopy. The plot with the largest damage to seedlings was plot 2, where we recorded just 8 % relative light intensity. Maybe the seedlings experienced frost due to heat emission in the night (radiation cooling) which we were unable to detect by measuring air temperature. We did not recognise this inappropriate feature of plot 2 because we chose all the plots during the winter.

Recent shifts in vegetation boundaries of deciduous forests,
especially due to general global warming, pp. 151–166
edited by F. KLÖTZLI and G.-R. WALTHER

Lianisation and Therophytation as complementary processes of Laurophyllisation

A. BORHIDI,[1,2] T. MORSCHHAUSER[2] and E. SALAMON-ALBERT[2]
[1] Institute of Ecology and Botany, Hungarian Academy of Sciences, H-2163 Vácrátót, Hungary
[2] Botanical Department of the Janus Pannonius University, H-7624 Pécs, Hungary

Introduction

Climatic changes and the shifting of vegetation are permanent and continuous processes. The reason why we have to study these processes is their elevated speed, the acceleration of the movement. The special characteristics of the actual vegetation shift are its multivectorial nature and the multifactorial control. Multivectorial nature means, that several important changes are going on simultaneously, like climate change, acidification, eutrophication, desertification. Multifactorial control means that all these processes are controlled by a large set of environmental factors acting in different ways in time and space.
A relevant aspect of the climatic change and the vegetation shift has been demonstrated by Klötzli and collaborators, in the area of the Insubrian Alps, canton Tessin (Ticino), southern Switzerland. An invasion of far eastern laurophyllous evergreen elements into the indigenous winterdeciduous beechwoods started several years ago. This process was named by Klötzli laurophyllisation of the temperate deciduous forests and interpreted as an increasing warm-temperate–subtropical influence enhanced by the global warming of the climate (Klötzli et al. 1996, Gianoni et al. 1988, Klötzli 1988).
In Hungary, a country of SE Middle-Europe, with a slightly cooler but also notably drier climate with more unbalanced temperatures, the global warming of the climate provoked other types of vegetation responses. These responses may be interpreted as lianisation and therophytation which are complementary phenomena to the laurophyllisation of the more humid and atlantic Insubrian Alps.
During the studies of vegetational changes occurring in the southernmost mountainous areas of Hungary, the Mecsek and Villányi Mts. (S of the city Pécs) the investigations have been concentrated on these two kinds of structural and compositional vegetation changes.

Lianisation

Lianisation has been observed and studied on the southern slopes of the Mecsek Mts. between 400 and 600m a.s.l. in the hairy-oak woods. Ten relevés taken by A.O. Horvát between 1949 and 1960 (Horvát 1972) were used as comparative material, published under the name Orno-Quercetum pubescentis mecsekense (A.O. Horvát 1946, Soó 1961) (see Table 1). Subsequently B. Kevey and T. Morschhauser studied the same area and ten relevés which were taken at almost the same sites as before were selected. These relevés are listed in Table 2 as Tamo-Quercetum virgilianae (Borhidi and Morschhauser 1996). The two different association names may suggest, that two different communities are compared, but that is not the case. The reasons for the different names are partly taxonomic, forty years ago *Quercus pubescens* and *Q. virgiliana* were not distinguished by the Hungarian phytosociologists and even the microspecies of the *Quercus petraea* group (*Q. dalechampii, Q. polycarpa*) has been differentiated according to the article of Borhidi (1969). Another reason for different names is nomenclatural'; it is explained in detail by Borhidi & Kevey (1996).

The most appealing phenomenon one may observe, is the increased density of the shrub layer interwined by several climber species. These lianas are *Lonicera caprifolium* and *Tamus communis* climbing up to the higher shrub layer up to a height of 3-4m and *Hedera helix* which has not been detected earlier in this community, but acts as a widely distributed liane in the meso- and hygrophilous forest communities in the area. Comparing the cover values of these 3 species we find that the average cover of the climbers was 2.1 % and their maximum height 2m in the relevés of 1949-1960(Table 1), while in the years 1991-1994 their average cover surpassed 37% and their maximum height reached 4.5m. *Tamus communis* did not only climb up to the branches of shrubs but also climbed the trunks of the trees (Table 2).

Simultaneously to this process another can be detected by comparing the two tables, which show the increasing number of the geophytes in the relevés of the nineties. Both the number of the geophyte species (12 new geophyte species that had not been detected in the earlier relevés) and their cover are considerably higher than in the relevés of the forties and fifties.

Both structural changes are indicating a warming up of the climate, since both wines and geophytes are common structural elements of the mediterranean vegetation and less characteristic to the submediterranean thermo-xerophilous winterdeciduous forests.

Table 1: Orno-Quercetum pubescentis mecsekense (A.O.Horvát 1946, Soó 1961)

	Relevés		1	2	3	4	5	6	7	8	9	0
Acer	*campestre*	A	-	-	+	-	-	+	-	1	1	-
		B	-	-	+	+	-	-	-	-	-	-
		C	+	1	-	+	-	-	-	-	-	-
Fraxinus	*ornus*	A	-	-	+	1	1	1	3	1	1	1
		B	2	2	-	2	1	1	1	-	-	-
		C	-	1	-	-	-	-	-	-	-	-
Quercus	*cerris*	A	-	-	+	+	1	-	-	-	-	-
Quercus	*petraea & dalechampii*	A	-	-	3	-	-	1	-	-	-	-
		B	-	-	+	-	-	-	-	-	-	-
		C	-	-	-	-	-	-	-	-	+	-
Quercus	*pubescens & virgiliana*	A	4	4	3	3	4	1	5	3	4	4
		B	1	-	+	+	-	-	+	-	-	-
		C	+	+	-	1	-	-	-	-	-	-
Sorbus	*torminalis*	A	-	-	-	-	-	-	+	-	-	-
		B	-	-	+	-	-	-	+	-	-	-
		C	-	+	+	-	-	-	+	-	-	-
Tilia	*tomentosa*	A	-	-	+	-	1	+	+	1	1	+
Ulmus	*minor*	A	-	-	-	-	+	-	-	-	-	-
Acer	*tataricum*	B	-	-	-	-	-	-	-	1	1	-
Cornus	*mas*	B	3	2	1	2	2	1	1	1	1	1
Crataegus	*monogyna*	B	1	1	-	1	1	1	1	1	1	1
Euonymus	*verrucosa*	B	-	-	+	+	-	-	-	1	-	1
Ligustrum	*vulgare*	B	-	-	+	+	-	-	1	1	1	1
Prunus	*spinosa*	B	-	+	-	+	-	+	+	-	-	-
Rosa	*arvensis*	B	-	-	-	-	-	-	-	1	-	1
Rosa	*canina*	B	+	1	+	+	-	-	1	-	-	1
Rosa	*gallica*	B	-	-	+	+	+	-	-	1	1	-
Lonicera	***caprifolium***	**B**	**+**	**+**	**+**	**+**	**+**	**1**	**-**	**1**	**1**	**1**
Tamus	***communis***	**C**	**1**	**+**	**+**	**+**	**+**	**+**	**+**	**-**	**+**	**1**
Aconitum	*anthora*	C	-	-	-	+	-	-	-	-	-	-
Arabis	*turrita*	C	-	+	-	-	-	-	+	+	-	-
Brachypodium	*pinnatum*	C	-	-	-	-	-	1	-	-	-	-
Brachypodium	*sylvaticum*	C	3	3	1	1	1	2	-	3	1	1
Bromus	***pannonicus***	**C**	**-**	**2**	**-**	**-**	**-**	**-**	**-**	**-**	**-**	**-**
Calamintha	*clinopodium*	C	-	-	+	-	-	+	+	-	-	-
Calamintha	*officinalis*	C	+	-	-	+	+	-	-	-	-	-
Campanula	*bononiensis*	C	-	-	+	-	-	-	-	-	-	-
Campanula	*persicifolia*	C	-	-	-	-	+	-	-	-	-	-
Campanula	*rapunculoides*	C	-	-	-	-	-	-	+	-	-	-
Carex	***flacca***	**C**	**2**	**3**	**3**	**-**	**3**	**3**	**-**	**-**	**-**	**-**
Carex	***humilis***	**C**	**-**	**+**	**-**	**-**	**-**	**3**	**1**	**-**	**4**	**-**
Chrysanthemum	*corymbosum*	C	-	-	-	-	-	-	+	-	-	-
Coronilla	*varia*	C	+	1	+	-	+	+	+	-	+	-
Cytisus	*supinus*	C	-	-	+	-	+	-	+	-	-	-
Dactylis	*aschersoniana*	C	1	1	1	+	+	+	+	1	1	-
Dictamnus	*albus*	C	-	+	-	+	+	-	+	-	1	-

Dorycnium	*herbaceum*	C	+	+	+	-	+	+	-	-	-	-
Epipactis	***microphylla***	**C**	-	+	+	-	-	-	-	-	-	-
Erysimum	*odoratum*	C	+	+	+	+	+	+	-	-	-	-
Euphorbia	*cyparissias*	C	-	+	+	+	+	+	-	1	1	1
Euphorbia	*polychroma*	C	-	-	-	-	-	-	-	-	1	-
Festuca	*heterophylla*	C	-	+	+	+	+	+	-	-	1	1
Festuca	*valesiaca*	C	-	-	-	-	-	-	-	3	5	-
Filipendula	*vulgaris*	C	+	-	+	+	+	-	-	1	-	1
Fragaria	*moschata*	C	-	-	-	-	+	-	-	-	-	-
Galium	*lucidum*	C	+	1	+	+	1	+	+	1	-	1
Genista	*nervata*	C	-	-	-	-	+	-	-	-	-	-
Geum	*urbanum*	C	-	-	+	+	-	-	-	-	-	-
Glechoma	*hirsuta*	C	-	+	+	1	-	-	1	1	-	1
Helleborus	*odorus*	C	+	+	+	1	+	+	+	1	1	1
Helianthemum	*ovatum*	C	+	+	-	-	-	+	-	-	-	-
Hieracium	*bauhinii*	C	-	-	+	-	+	-	-	-	-	-
Iris	***graminea***	**C**	-	-	-	-	-	-	+	-	+	-
Lactuca	*quercina*	C	-	-	-	-	+	-	-	-	-	-
Lathyrus	*niger*	C	-	-	-	+	+	-	2	-	+	-
Lithospermum	*purp.-coerul.*	C	-	-	-	-	+	+	-	1	1	+
Melampyrum	*nemorosum*	C	+	-	-	-	-	+	-	-	-	-
Melica	*uniflora*	C	-	-	-	+	-	-	+	-	3	-
Melittis	*grandiflora*	C	-	-	-	-	+	+	+	-	-	-
Platanthera	***bifolia***	**C**	+	+	+	+	-	+	-	-	-	-
Polygonum	*dumetorum*	C	-	-	+	+	+	+	-	-	+	-
Sedum	*maximum*	C	-	-	-	+	+	+	-	-	-	-
Silene	*viridiflora*	C	-	-	-	-	-	+	+	-	-	-
Stellaria	*holostea*	C	-	-	-	1	-	-	-	-	-	-
Symphytum	*tuberosum*	C	+	-	+	+	+	-	+	-	-	-
Teucrium	*chmaedrys*	C	+	-	+	+	+	-	-	-	1	-
Trifolium	*alpestre*	C	-	-	-	-	+	+	-	-	-	-
Verbascum	*austriacum*	C	-	-	-	-	-	+	+	-	-	-
Veronica	*chamaedrys*	C	+	-	+	+	+	-	-	-	1	-
Vincetoxicum	*hirundinaria*	C	+	-	-	-	-	-	+	-	-	-
Viola	*alba*	C	+	+	+	+	+	+	1	-	-	-

Table 2: Tamo-Quercetum virgilianae Borhidi & Morschhauser (1996)

		1	2	3	4	5	6	7	8	9	0
Fraxinus ornus	A1	+	+	1	1	2	2	3	2	2	1
	A2	2	2	1	1	+	+	+	+	+	2
	B1	2	2	+	2	1	1	1	+	+	+
	B2	+	+	2	+	+	2	2	2	1	2
	C	+	+	+	+	1	+	+	+	1	+
Quercus cerris	A1	-	+	+	+	1	1	+	+	-	+
	A2	+	-	-	-	-	-	-	-	+	-
	B1	-	-	+	-	-	+	-	-	-	-
	B2	+	+	-	-	+	+	-	-	-	+
	C	+	+	+	+	1	1	+	+	+	3
Quercus dalechampii	A1	1	1	2	2	+	3	-	1	+	-
	A2	+	-	-	-	1	-	-	+	-	-
	B1	1	+	1	-	-	+	-	-	+	-
	B2	-	1	+	1	-	1	-	2	-	-
	C	+	+	+	+	-	-	-	+	-	-
Quercus virgiliana	A1	4	4	3	3	4	1	5	3	4	4
	A2	+	-	-	1	-	1	-	-	-	-
	B1	-	+	+	-	-	-	-	-	-	-
	B2	+	-	-	1	1	2	1	+	+	+
	C	+	+	+	+	+	+	+	1	1	+
Quercus pubescens	A1	1	-	+	+	-	-	+	-	1	-
	B2	+	1	1	1	-	-	+	-	+	2
	C	+	+	-	1	+	-	-	-	-	-
Sorbus domestica	A2	+	-	+	-	-	-	-	-	-	-
	B1	-	+	-	+	-	-	-	-	-	-
	B2		-	+	-	+	+	-	-	-	-
	C	+	+	+	+	-	-	-	-	-	-
Sorbus torminalis	A1	-	-	-	-	-	-	+	-	-	-
	A2	-	+	-	+	-	-	-	-	-	-
	B1	+	-	+	-	+	+	+	-	-	-
	B2	-	+	+	+	+	+	-	+	+	-
	S	+	+	+	+	+	+	+	+	+	-
Tilia tomentosa	A1	-	-	-	-	1	-	-	1	1	-
	A2	-	-	+	-	+	1	1	-	-	-
	B1	-	-	1	-	+	+	+	+	-	1
	B2	-	-	+	-	2	1	1	1	2	+
	C	+	+	-	1	+	-	-	+	+	1
Acer campestre	B1	-	+	+	-	-	+	-	1	1	-
	B2	-	+	+	+	-	+	-	1	1	1
Cornus mas	B1	3	2	1	2	2	2	2	1	1	2
	B2	+	+	2	+	1	+	+	2	2	+
	C	+	+	+	+	+	+	1	1	2	+
Crataegus monogyna	B1	1	1	1	1	1	1	1	1	-	1
	B2	2	1	2	2	2	2	1	1	1	2
Euonymus verrucosa	B2	+	+	1	1	1	1	+	+	-	1
	C	+	+	+	+	+	+	+	+	-	+
Ligustrum vulgare	B1	+	+	1	1	1	+	+	+	+	1
	B2	1	1	2	+	2	2	+	+	+	+
	C	+	+	+	+	+	+	+	+	-	+
Prunus spinosa	B1	-	+	-	+	-	+	+	-	-	-
	B2	+	+	+	+	-	-	+	+	+	-
	C	1	1	+	+	-	-	1	+	1	-
Pyrus pyraster	B1	+	+	-	-	+	-	-	+	-	-
	B2	+	+	+	-	+	-	-	-	+	+
	C	-	-	+	-	+	-	-	+	+	+
Rosa arvensis	B1	+	+	+	+	-	-	-	1	-	1

	B2	+	+	+	-	-	+	-	+	-	-
	C	+	+	+	-	-	1	+	+	-	+
Rosa canina	B1	+	1	+	+	-	-	1	-	-	1
	B2	-	+	+	1	+	-	-	-	1	+
	C	1	1	+	+	+	1	+	-	+	+
Rosa gallica	B2	-	-	-	+	+	-	-	1	1	-
Viburnum lantana	B1	1	1	2	2	-	-	-	1	2	2
	B2	2	+	+	+	-	-	-	+	+	+
	C	+	+	+	+	-	-	-	+	+	+
Hedera helix	**B2**	**+**	**1**	**1**	**-**	**-**	**+**	**-**	**-**	**-**	**1**
	C	**+**	**+**	**1**	**+**	**-**	**+**	**-**	**-**	**-**	**+**
Lonicera caprifolium	**B1**	**1**	**1**	**+**	**+**	**1**	**-**	**-**	**-**	**-**	**1**
	B2	**2**	**2**	**1**	**3**	**2**	**1**	**1**	**2**	**2**	**1**
	C	**+**	**1**	**2**	**+**	**-**	**2**	**2**	**-**	**-**	**-**
Tamus communis	**B1**	**2**	**1**	**1**	**2**	**1**	**1**	**2**	**2**	**1**	**1**
	C	**1**	**2**	**2**	**2**	**3**	**3**	**1**	**1**	**+**	**2**
Aconitum anthora	C	-	-	+	1	-	-	-	+	-	+
Adonis vernalis	C	-	-	-	-	-	-	-	+	1	1
Ajuga genevensis	C	+	-	+	+	-	-	-	+	+	-
Alliaria petiolata	C	+	+	1	1	+	+	+	+	-	+
Allium oleraceum	**C**	**+**	**+**	**-**	**-**	**-**	**+**	**-**	**-**	**-**	**-**
Anacamptis pyramidalis	**C**	**-**	**-**	**-**	**-**	**+**	**+**	**-**	**-**	**-**	**-**
Anthericum ramosum	**C**	**-**	**+**	**+**	**-**	**-**	**-**	**-**	**1**	**+**	**+**
Arabis turrita	C	+	+	+	+	+	+	+	-	-	-
Arum maculatum s.str.	**C**	**+**	**+**	**-**	**-**	**1**	**1**	**+**	**-**	**-**	**-**
Brachypodium pinnatum	C	2	2	3	1	-	-	1	+	-	1
Brachypodium sylvaticum	C	1	1	+	+	1	+	+	-	-	+
Bromus pannonicus	**C**	**-**	**1**	**-**	**2**	**2**	**1**	**1**	**-**	**-**	**1**
Campanula bononiensis	C	+	+	+	+	+	+	-	-	-	+
Campanula persicifolia	C	+	+	+	+	1	1	+	-	-	+
Campanula rapunculoides	C	+	+	+	-	+	+	-	-	-	-
Carex humilis	**C**	**-**	**+**	**-**	**-**	**-**	**3**	**1**	**-**	**4**	**-**
Carex flacca	**C**	**2**	**2**	**3**	**+**	**3**	**3**	**+**	**+**	**-**	**+**
Chamaecytisus supinus	C	-	+	-	+	-	+	-	-	-	-
Chrysanthemum corymbosum	C	+	+	+	-	+	+	1	-	-	1
Clinopodium vulgare	C	+	+	+	+	+	+	-	-	-	+
Coronilla varia	C	+	1	+	-	-	-	+	-	-	-
Corydalis solida	**C**	**+**	**+**	**1**	**1**	**-**	**+**	**+**	**-**	**-**	**-**
Dactylis polygama	C	1	+	+	1	1	+	+	1	1	-
Dictamnus albus	C	+	+	+	1	1	+	1	1	2	1
Erysimum odoratum	C	+	+	+	1	1	1	+	+	-	+
Euphorbia cyparissias	C	-	+	+	+	+	+	-	1	1	1
Euphorbia polychroma	C	+	+	+	+	-	-	-	+	1	1
Fallopia dumetorum	C	+	+	+	+	+	+	-	-	-	+
Festuca heterophylla	C	-	-	-	-	+	+	-	-	-	-
Festuca rupicola	C	-	-	-	-	-	-	-	3	5	-
Festuca valesiaca	C	-	-	-	1	1	-	-	-	-	-
Filipendula vulgaris	C	+	-	+	+	1	1	-	1	-	1
Fragaria vesca	C	-	+	+	+	+	1	+	-	-	-
Fragaria viridis	C	+	-	-	-	+	1	-	-	-	-
Galium lucidum	C	+	1	+	+	1	1	+	1	1	1
Genista ovata ssp. nervata	C	-	-	-	+	+	+	-	-	-	-
Geranium sanguineum	C	+	1	1	+	-	-	-	-	-	-
Geum urbanum	C	+	+	+	+	+	+	-	-	-	+
Glechoma hirsuta	C	+	+	+	1	+	+	1	1	-	1
Helleborus odorus	C	+	+	+	1	+	+	+	1	1	1
Hieracium bauhini	C	+	+	+	+	+	-	-	-	-	-
Inula spiraeifolia	C	1	1	-	-	+	-	-	-	-	1
Iris graminea	**C**	**-**	**+**	**-**	**+**	**-**	**-**	**+**	**-**	**+**	**-**

Iris variegata	**C**	-	-	-	-	-	-	-	**1**	**+**	**+**
Lactuca quercina ssp. sagittata	C	+	+	+	+	1	+	-	-	-	1
Laser trilobum	C	+	+	+	+	-	-	-	-	-	-
Laserpitium latifolium	C	-	-	-	-	+	+	+	-	-	-
Lathyrus niger	C	-	-	+	+	+	-	2	-	-	+
Lathyrus pannonicus	C	+	+	-	-	-	-	-	-	1	-
Lathyrus venetus	C	+	-	+	-	-	-	-	-	-	-
Lithospermum purpureo-coeruleum	C	+	+	1	1	1	1	+	2	2	1
Melampyrum nemorosum	C	+	+	-	-	+	+	+	-	-	-
Melica uniflora	C	+	+	+	+	2	2	+	-	-	3
Melittis carpatica	C	-	-	-	+	+	+	+	-	-	+
Muscari tenuiflorum	**C**	**+**	**+**	-	**+**	-	-	**+**	**+**	**+**	**+**
Orchis simia	**C**	**+**	**+**	-	-	**+**	-	-	-	-	**+**
Ornithogalum sphaerocarpum	**C**	-	-	-	-	**+**	**+**	-	**+**	-	**+**
Peucedanum cervaria	C	+	-	-	-	-	-	-	1	1	+
Platanthera bifolia	**C**	**+**	**+**	**+**	-	-	**+**	-	-	-	-
Poa pratensis	C	+	-	+	+	1	1	-	-	-	-
Polygonatum odoratum	**C**	**1**	**1**	-	-	-	-	-	**1**	**1**	-
Ranunculus illyricus	**C**	-	-	-	-	-	-	-	**+**	**+**	-
Ruscus aculeatus	C	-	+	1	-	-	-	-	2	-	1
Sedum maximum	C	+	+	+	+	1	+	1	+	+	+
Silene viridiflora	C	-	-	-	-	+	+	-	-	+	-
Stellaria holostea	C	+	+	+	+	-	+	-	-	-	-
Symphytum tuberosum	C	1	1	+	1	+	+	+	-	-	+
Teucrium chamaedrys	C	1	1	+	+	+	1	1	2	2	+
Thalictrum aquilegiifolium	C	-	-	+	-	+	+	-	-	-	-
Trifolium alpestre	C	-	-	-	-	+	+	+	-	-	-
Verbascum austriacum	C	-	-	-	-	-	+	+	-	-	-
Veronica chamaedrys	C	+	-	+	+	+	-	-	-	1	-
Vincetoxicum hirundinaria	C	1	1	+	+	1	1	+	+	-	+
Viola alba	C	+	1	+	+	1	1	1	+	-	+
Viola odorata	C	1	+	+	1	+	-	-	-	-	1
Waldsteinia geoides	C	+	-	-	-	-	-	-	-	-	2

Therophytation

For testing the changes in the composition of the vegetation caused or stimulated by the global warming several standard plots have been selected in the extremely exposed southern slopes of the southernmost mountain and hilly areas of Hungary six years ago. Rocksward-rockbush-woodland ecotones were sampled at least twice a year, in early springtime and midsummer. Standard plots have been situated on the following places:

1. Jurassic hard limestone in the Villány hills at an altitude of 350 m a.s.l., exposition S 35°. Ecotone formed by *Sedo sopianae-Festucetum dalmaticae* rocksward and *Inulo spiraeifoliae-Quercetum pubescentis* rockbush-woodland. Important species: *Agropyron intermedium, Artemisia alba ssp. saxatilis, Carex liparicarpos, Ceterach javorkaeanum, Chrysopogon gryllus, Convolvulus cantabrica, Crupina vulgaris, Dianthus giganteiformis, Dictamnus albus, Euphorbia cyparissias, Festuca dalmatica, Geranium molle, Melica ciliata, Orlaya grandiflora, Phleum phleoides, Potentilla arenaria, Sedum neglectum ssp. sopianae, Stipa capillata, Teucrium chamaedrys,*

Thymus praecox, Trigonella gladiata in the rocksward and *Quercus pubescens, Q. virgiliana, Fraxinus ornus, Cornus mas, Euonymus verrucosa, Alliaria petiolata, Arum maculatum, Dictamnus albus, Fallopia convolvulus, Geranium molle, Inula spiraeifolia, Lithospermum purpureo-coeruleum, Polygonatum odoratum, Ruscus aculeatus, Tamus communis, Veronica hederifolia* in the bush-woodland.
Transects were made of 6 quadrats measuring 4m x 4m each, and of 28 quadrats measuring 1m x 1m each.

2. Steep loess slope on the Misina in Mecsek Mts. Nature Reserve Area of Pintér Kert, 300m a.s.l., exposition SW 40°. Ecotone formed by *Serratulo radiatae-Brometum pannonici* closed grassland, *Inulo spiraeifoliae-Quercetum pubescentis* bush-woodland, a transitional forestedge vegetation and by the subclimax hairy oak forest association, *Tamo-Quercetum*. Important species in the grassland are: *Allium flavum, Artemisia alba* ssp. *saxatilis, Bromus pannonicus, Cleistogenes serotina, Convolvulus cantabrica, Erysimum odoratum, Galium lucidum, Inula spiraeifolia, Melica ciliata, Serratula radiata, Teucrium chamaedrys, Thymus degenianus*. In the bush-woodland and hairy oak forest *Quercus pubescens, Fraxinus ornus, Cornus mas, Cerasus mahaleb, Prunus spinosa, Staphylea pinnata, Alliaria petiolata, Brachypodium pinnatum, Dictamnus albus, Euphorbia cyparissias, Glechoma hirsuta, Laser trilobum, Lithospermum purpureo-coeruleum, Melampyrum nemosorum, Melittis carpatica, Tamus communis, Teucrium chamaedrys, Viburnum lantana* and *Viola alba* are of major importance.
A transect was made with 4 plots of 4m x 4m each and with 16 plots of 1 square meter each.

3. Moderately steep limestone slope on the Tubes Mt. about 450m a.s.l., exposition SW 15°. Ecotone is formed by *Serratulo radiatae-Brometum pannonici* closed rocksward, a forest edge community and the *Tamo-Quercetum virgilianae* shrubforest. Important species in the grassland community are: *Artemisia alba* ssp. *saxatilis, Bromus pannonicus, Carex humilis, Centaurea triumfettii, Convolvulus cantabrica, Dictamnus albus, Eryngium campestre, Euphorbia cyparissias, Festuca rupicola, Galium lucidum, Helianthemum ovatum, Koeleria cristata, Melica ciliata, Potentilla arenaria, Salvia pratensis, Sedum neglectum, Serratula radiata, Teucrium chamaedrys, Thymus degenianus, Verbascum phoeniceum, Veronica austriaca, Vinca herbacea.* In the forest association *Quercus virgiliana, Fraxinus ornus, Cornus mas, Euonymus verrucosa, Ligustrum vulgare, Aconitum anthora, Anthericum ramosum, Arabis turrita, Carex humilis, Dictamnus albus, Helleborus odorus, Inula spiraeifolia, Mercurialis ovata, Polygonatum odoratum, Sedum maximum, Stellaria holostea, Tamus communis, Trifolium alpestre, Vincetoxicum hirundinaria, Viola alba* are the most important species.

A transect was made with 4 plots of 4m x 4m metres each and with 16 $1m^2$ plots.

4. Acidophilous grassland-shrubwoodland ecotone on red Permian sandstone on the southern slopes of the Jakabhegy, West of the city Pécs, exposition SSW 25°. In the acidophilous grassland the most important species are: *Agrostis tenuis, Aira caryophyllea, Genista pilosa, Hieracium bauhiniformis, Jasione montana, Polygonatum odoratum, Rumex acetosella, Viscaria vulgaris, Veronica dillenii, Vulpia myuros, Cladonia pyxidata, C. rangiferina, Ceratodon purpureus, Polytrichum juniperinum*. In the shrubwood association (*Genisto pilosae-Quercetum polycarpae*) the *Quercus dalechampii, Q. polycarpa, Q. pubescens, Fraxinus ornus, Juniperus communis, Anthericum ramosum, Genista pilosa, Hieracium bauhiniformis, H. sabaudum, H. umbellatum, Luzula campestris, Polygonatum odoratum, Rumex acetosella, Sedum maximum, Solidago virga-aurea, Veronica dillenii, Viola tricolor, Viscaria. vulgaris, Hypnum cupressiforme,* and *Polytrichum attenuatum* are the most importants species.
A transect is composed of 4 plots, 4m x 4m each, and of 16 $1m^2$ plots.

In each square plot structure, composition and cover of the plant communities were investigated twice a year, in spring and summer, starting in 1991. The changes in the composition were ecologically interpreted by using the ecological indicator values of Ellenberg (1974, 1992) revised and completed for the Pannonian flora by Borhidi (1993, 1995). From the phytocenological cover data mean values were calculated for each ecological factor in each plot for every year.
As for the ecological values the extremely exposed grassland-shrubwood ecotone of the Szársomlyó Hill showed the most explicit changes. Figure 1. shows, that – including some fluctuations – the mean temperature (mT) values have increased from 6 to 6.5, the humidity values (mW) have decreased from 4 to 3.5, the soil reaction values (mR) have increased from 7.2 to 7.6, nutrient values (mN) have fluctuated within a rather wide range, continentality values (mK) have increased half a degree from 3.9 to 4.6 and light intensity values (mL) have increased from 6.3 to 6.9. These changes may be related to the general decrease of the plant cover in the grassland communities.

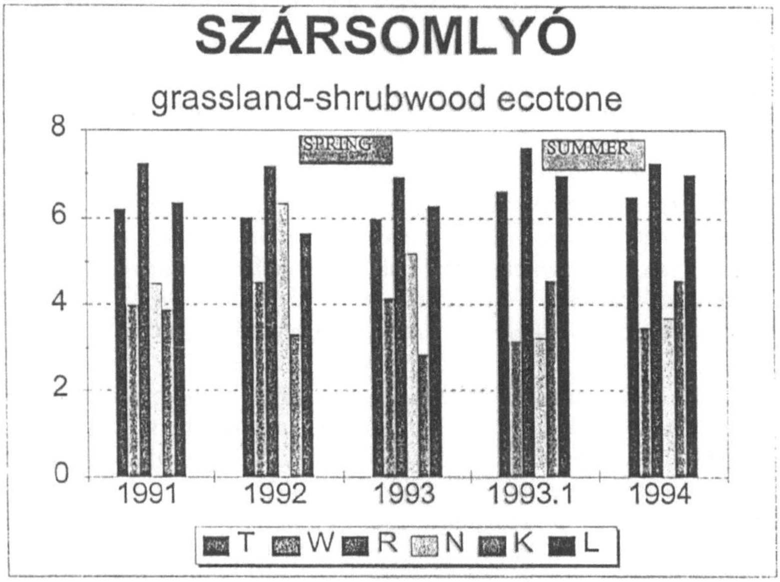

Figure 1: Changes in a grassland-shrubwood ecotone of the Szársomlyó Hill

The decrease of the plant cover has been combined with a structural change of the grassland vegetation, namely with a change of the amount of the annual and perennial forbs and grasses. Figure 2 presents the changes in the relative cover of the annual vs perennial plants in four different habitats. Including the fluctuations of different years, the general trend is obvious. The cover of perennial plants is decreasing while the free surfaces are more and more covered with annual plants. These annuals are the natural pioneer elements of the open grassland. That is the reason why we call this process therophytation, which is not a degradative process but a repeated new start of the succession from a bottom level with indigenous flora elements.

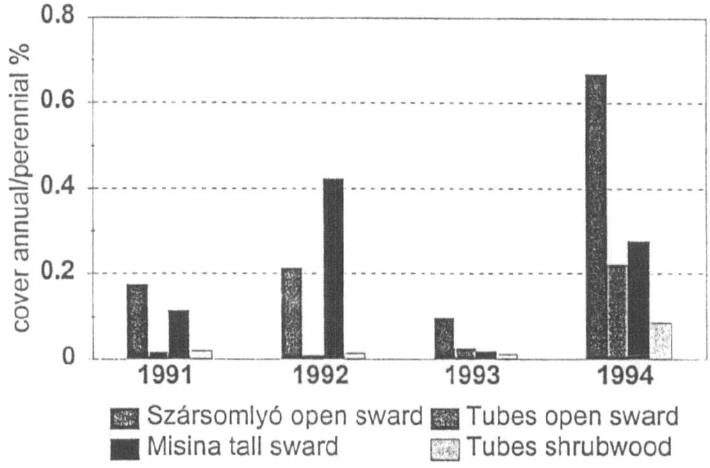

Figure 2: Changes in cover of annual vs. perennial plants

In order to get an idea of the specific process on population level, Figure 3 shows the decreasing trend of the perennial dominants, like *Festuca dalmatica, Melica ciliata* and the generalist *Helianthemum ovatum*, simultaneously with the massive appearence of the annual *Medicago minima* in the open sward plots of the Szársomlyó Hill. More concrete changes may be seen on Figure 4. which shows population dynamics of some species from the plot no. c67 of the Szársomlyó Hill. The slow increase of the geophyte *Iris variegata*, the rapsodic changes of the small-shaped spring therophyte *Lithospermum arvense*, and the consistant trends of increasing populations of the annuals *Medicago minima* and *Orlaya grandiflora* are in balance with the decrease of the perennial dominant *Melica ciliata*. At last the descending dynamism of the different perennial life forms – the chamaephyte dominant *Potentilla arenaria* and *Thymus praecox*, the perennial *Sanguisorba minor* and *Stipa capillata* and the perennial succulent *Sedum neglectum* ssp. *sopianae* – can be observed in contrast with the strong increase of the annual populations of *Orlaya grandiflora* in an average of ten plots of the open sward community in the Szársomlyó Hill (Figure 5).

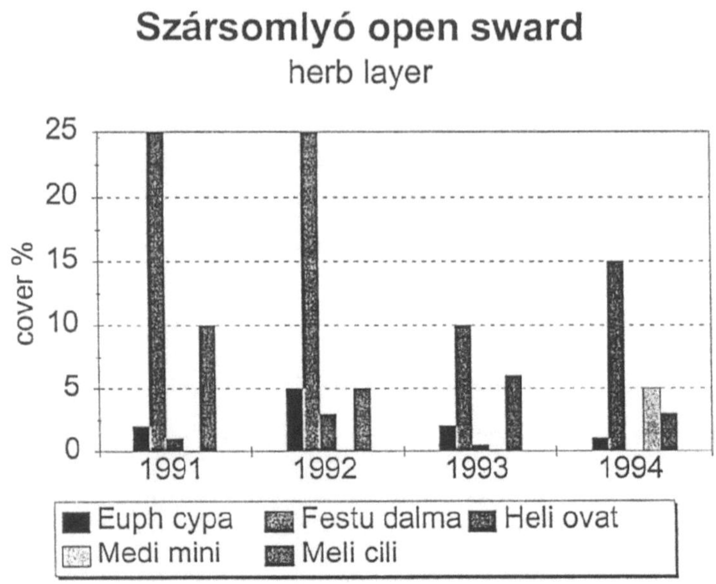

Figure 3: Decreasing trend of the perennial dominant species

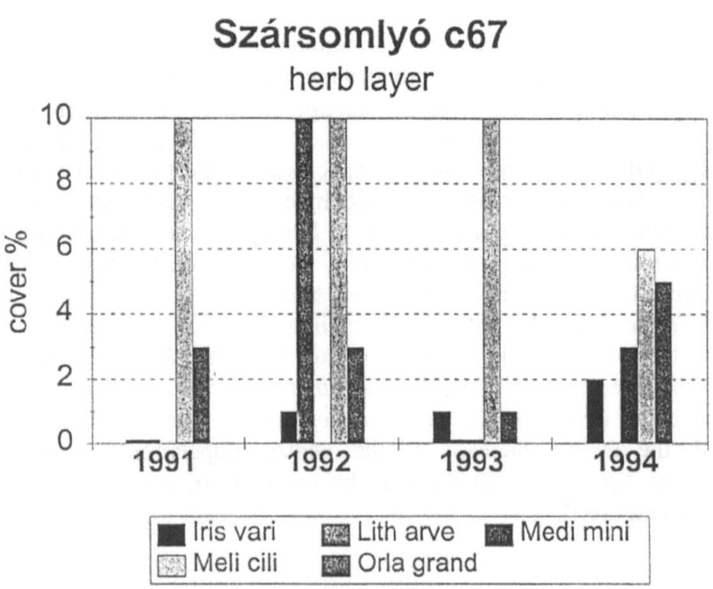

Figure 4: Population dynamics of some species from the plot no. c67

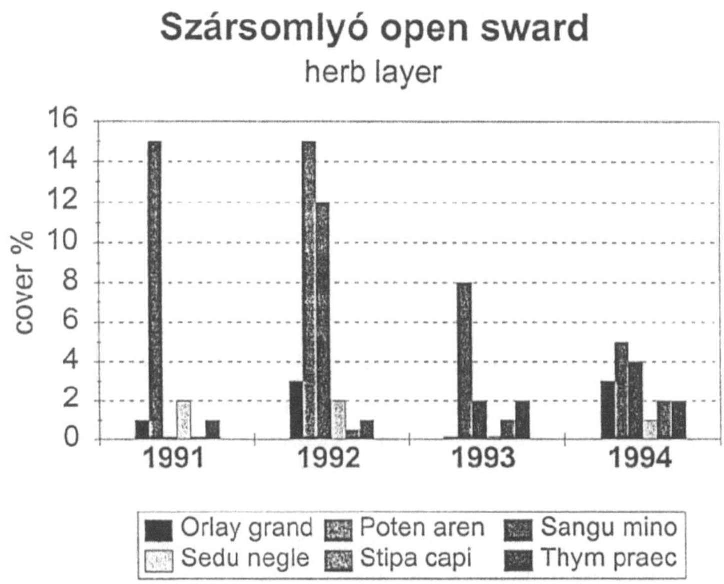

Figure 5: Descending dynamism of different perennial life forms in contrast with the strong increase of annual populations in the open sward community in the Szársomlyó Hill.

Conclusions

It is obvious, that 4 years of observation is a short period for drawing important conclusions about the trend of changes. However, the increase of therophytes and geophytes is clearly manifested and this change is in concordance with the general characteristics of the east-submediterranean-mediterranean vegetation. The climate in the Pannonian Basin does not only tend to be warmer but also drier. During the eighties the annual average precipitation of one year has been lost within a ten years period. At the same time, no alien species have appeared, neither new nor old ones. The only exception – for the time being – is the escape of *Lunaria annua,* an ornamental garden-therophyte, to the herb layer of the *Tamo-Quercetum virgilianae* in the Mecsek Mts and to one *Asperulo taurinae-Carpinetum* stand in the Villányi Mts, both rather local. Consequently, the native communities are still not very vulnerable to the opening of gaps for alien elements.

From 1997 onwards the field observations will be continued, so further hypotheses may be formed.

Acknowledgements

The authors are very thankful to the organizers of the "Laurophyllisation" symposium, especially to Prof. F. Klötzli, G.-R. Walther and G. Carraro and to the ETH Zürich for the ideal conditions of the symposium. The field studies presented at the symposium and published here, were supported by the Hungarian National Research Fund OTKA 1298.

References

Bartha D., Kevey B., Morschhauser T. & Pócs T. (1995):Hazai erdõtársulásaink (Hungarian forest associations, in Hungarian). TILIA 1: 8-86.

Borhidi A. (1993): A Magyar Flóra Szociális Magatartás Típusai, Természetességi és Relativ Ökológiai Értékszámai. Social Behaviour Types of the Hungarian Flora, its naturalness and relativ ecological indicator values. Janus Pannonius Tud. Egy Kiadványai, Pécs.

Borhidi A. (1995): Social behaviour types, their naturalness and relative ecological indicator values of the higher plants of the Hungarian Flora. Acta Bot. Hung. 39: 97-182

Borhidi A. (1996):An annotated checklist of the Hungarian plant communities, I. The non-forest vegetation. In: Borhidi,A. (ed.): Critical Revision of the Hungarian Plant Communities. Janus Pannonius Univ. Pécs, 43-94.

Borhidi A. & Dénes A. (1997): A Mecsek és a Villányi hegység sziklagyep társulásai (The rock sward communities of the Mecsek and Villány Mts.). Studia Phytologica Jubilaria, Pécs, 45-66.

Borhidi A. & Kevey B. (1996):An annotated checklist of the Hungarian plant communities, II. The forest vegetation. In: Borhidi A. (ed.): Critical Revision of the Hungarian Plant Communities. Janus Pannonius Univ. Pécs,95-138.

Borhidi A. & Morschhauser T. (1995):Social Behaviour Types (SBTs) a new tool for assessing naturalness and ecology of plant communities. Eureco 95 Conf. Proceedings.

Borhidi A., Morschhauser T. & Albert É. (1995): Global change induced vegetation dynamics in extreme exposed grassland-bush-woodland ecotones. Eureco 95 Conf. Proceedings.

Ellenberg H. (1974): Zeigerwerte der Gefäßpflanzen Mitteleuropas. Scripta Geobotanica 9: 1-97.

Ellenberg H., Weber H.E., Düll R., Wirth W., Werner W. & Paulissen D. (1991): Zeigerwerte der Gefäßpflanzen in Mitteleuropa. Scripta Geobotanica 18: 1-248.

Gianoni G., Carraro G. & Klötzli F. (1988): Thermophile, an laurophyllen Pflanzenarten reiche Gesellschaften im hyperinsubrischen Seenbereich des Tessins. Ber. Geobot. Inst. ETH, Stiftung Rübel, Zürich, 54: 164-180.

Horvát A.O. (1946): A pécsi Mecsek természetes növényszövetkezetei. (Phytoassociationes naturales montis Mecsek ad civitatem Pécs in Hungaria) Pécs.

Horvát A.O. (1972): Die Vegetation des Mecsekgebirges und seiner Umgebung. Akadémiai Kiadó, Budapest.

Klötzli F. (1988): On the global position of the evergreen broad-leaved (non-ombrophilous) forest in the subtropical and temperate zones. Veröff. Geobot. Inst. ETH, Stiftung Rübel, Zürich, 98: 169-196.

Klötzli F., Walther G.-R., Carraro G. & Grundmann A. (1996): Anlaufender Biomwandel in Insubrien. Verhandl. Ges. f. Ökologie 26: 537-550.

Landolt E. (1977): Ökologische Zeigerwerte zur Schweizer Flora. Veröff. Geobot. Inst. ETH, Stiftung Rübel, Zürich, 64: 1-208.

Morschhauser T. (1995): A mecseki Tubes növényzete (Vegetation of the Mt. Tubes in the Mecsek Mts) In Hungarian. TILIA 1: 199-210.

Plenary discussion

Question: F. KLÖTZLI

I've got one question, a more organisational one: Now that you have shown that there are certain shifts, directed shifts, you have an other political position. Unfortunately, after four years only you can't really be sure of your results, because you need more years. Do you get the necessary organisation and financing for the next years? Because you should continue.

Answer: A. BORHIDI

Yes, I have again. That's a good question because two years ago I haven't, but the problem has been solved and from 1996 it starts again.

Question: F. KLÖTZLI

How about the quantity of certain species that might have changed considerably as well? It is not only just a matter that species appeared but the quantitative parts are essential too.

Answer: A. BORHIDI

Yes, it is different because we studied it in four different ecotones which are going from the open rocks sward to the closed grassland and to the forest edge and to the interior of the forest. So practically the bigger change is in the open rocks sward. It has the least stability. Practically the perennial plants and the chamaephytes are restricted more and they are replaced by therophytes but not weedy therophytes. I think it is something which is similar to the east mediterranean xerotherm places where practically the geophytes and therophytes play a greater role, because there are practically two periods which are untolerable and to be survived and therefore geophytes and therophytes are most able to avoid both winter frost and summer drought.

Statement: F. KLÖTZLI

I think the most considerable change or, let's say, if you look at the weight of the change, is this climbing up of the lianas into the tree layer.

Answer: A. BORHIDI

Yes, it is very important because it is a structural change.

Statement: G. DOYLE

I should like to thank Professor Borhidi for presenting us with this multi-faceted approach to the problem of laurophyllisation. It is abundantly clear that there are many reasons for change. The variety of examples that you have identified demonstrate the complexity of the problems associated with an understanding of laurophyllisation.

Recent shifts in vegetation boundaries of deciduous forests,
especially due to general global warming, pp. 167–179
edited by F. KLÖTZLI and G.-R. WALTHER

Evergreen broad-leaved vegetation in the Colchis

Georgi NAKHUTSRISHVILI
Institute of Botany, Georgian Academy of Sciences
Tbilissi 380007, Georgia

Abstract

In Colkheti relict (tertiary) evergreen broad-leaved vegetation exists, which used to be widespread all throughout western Eurasia in the past and can nowadays be found as undergrowth in forests. Its representatives seek shelter in the relatively warm and humid climate of the Colkheti. The snow cover plays the crucial role in those species' survival. These plants have mainly semi-prostrate growth form, although single individuals that reach 10-12m do exist. A number of syntaxons of broad-leaved and coniferous forests, including the Colkheti evergreen undergrowth, have mainly been marked out by Dolukhanov, and partially by Kolakovski.

If temperature rises and the precipitation declines due to the expected global change of climate, we assume that Mediterranean hard-leaved forests and shrubs will prevail on Colkheti foothills.

Einleitung

"Kolchis" ist die uralte (XII - XIII Jh. v. u. Z.) Bezeichnung des georgischen Staates im süd-östlichen und östlichen Schwarzmeergebiet. Die Kolchis gehört zum Schwarzmeerbecken. Ihre Grenzen nehmen den Anfang im Norden des Großkaukasus und verlaufen an der Wasserscheide dieses Gebirgsrückens entlang über den Meßcheti-Paß zur Adshara-Imereti- und Schawscheti Bergkette, weiter längs des Pontusgebirges ziehend - reichen sie ungefähr bis Trapezunt (zur Zeit: Trabzon) (Figur 1).

Nicht selten wird die Kolchis viel enger gefaßt und sogar auf den Küstenlandstrich von Batumi beschränkt.

Die Kolchis stellt eine einheitliche botanisch-geographische Naturprovinz dar. Trotzdem sind ihre physikalisch-geographischen Bedingungen ziemlich mannigfaltig.

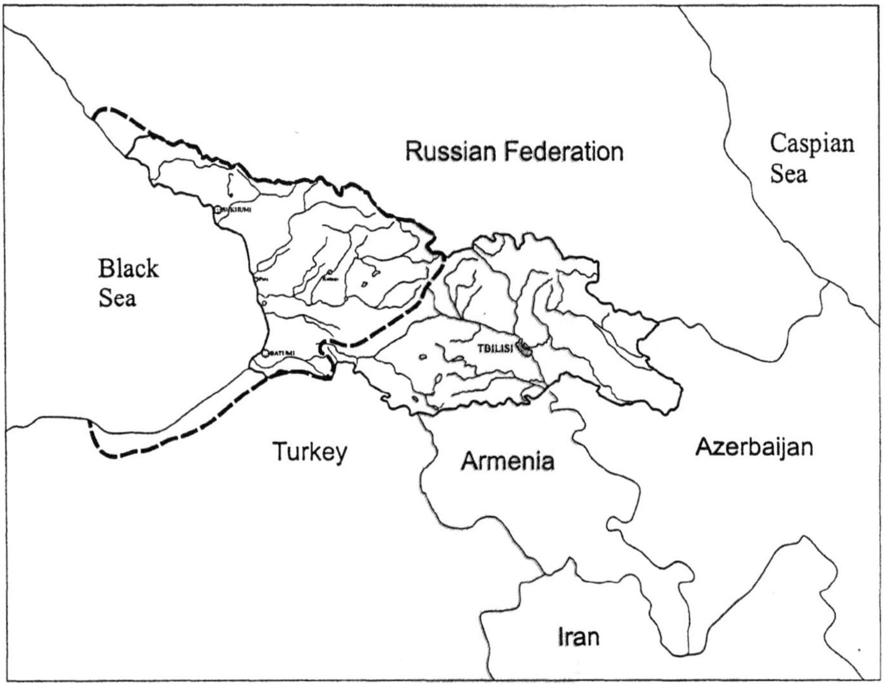

Figur 1: Kolchische botanisch-geographische Provinz (unterbrochene Linie)

Geologie, Geomorphologie

Während des Miozäns gab es anstelle der gegenwärtigen Kolchis zwei Höhenabschnitte des Festlandes: einen nördlichen und einen südlichen, die durch die Meeresstraße voneinander getrennt waren. Aber bereits schon gegen Ende des Miozäns bestand jene allgemeine geographische Konfiguration der Kolchis, die bis zur Gegenwart in etwa erhalten geblieben ist.

Die ältesten (vorkimmerischen) kristallischen Gesteinsschichten des Großkaukasus (Gneisse, Granite) stiegen hoch hinauf. In den Bergen der Kolchis sind Granite mit paläozoischen Schiefern bedeckt. Ihnen folgen jurassische Kalk-und Dolomitsedimente usw.

Für die Zone tertiärer Vorgebirge der Kolchis sind die Sedimente in Gestalt von Schieferton, Mergel u.a. aus dem Paleozän und übrigen Jungtertiär charakteristisch.
Der jüngste Abschnitt (aus der Quartärzeit) des Festlandes auf dem Territorium der Kolchis ist die Küsten- oder die Kolchisniederung.
Das Relief der Kolchis ist außerordentlich zergliedert und kompliziert. Eine Ausnahme bildet nur die Südkolchis, die sich durch die Weichheit der Umrisse und mindere Ausprägung der glazialen Formen des Reliefs auszeichnet.
Die Kolchis ist vom Norden, Osten und Süden von dem Dreieck der Gebirgsketten geschützt. Seewärts ist sie geöffnet, was ihre günstigen Naturbedingungen bestimmt (Kolakowsky 1961).

Klima

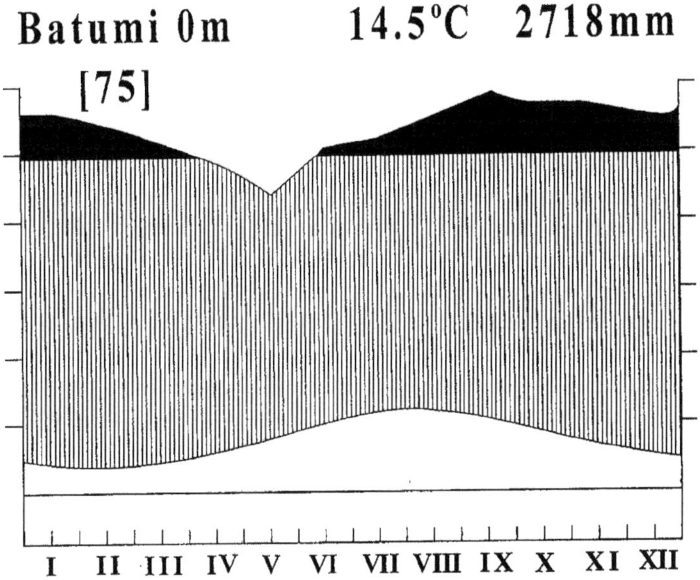

Figur 2: Klimadiagramm der Südkolchis

Das Klima der entferntesten geologischen Vergangenheit der Kolchis hat sich niemals durch Rauhheit, insbesondere durch kontinentalen Charakter ausgezeichnet. Das heutige Klima der Kolchis ist von subtropischem Typus, vor allem zeichnet es sich durch reiche Niederschläge (> 4500mm im Jahr) aus, welche die Westwinde bringen. Die orographische Abgeschlossenheit des Territoriums begünstigt die Kondensation der Niederschläge in der Niederung und auf den angrenzenden Berghängen.

Hauptzüge dieses "feucht-subtropischen" Klimas sind ein hohes Jahrestemperaturmittel mit unbedeutenden Schwankungen, reiche Niederschläge (Figur 2), mäßig warme Sommer mit einer mittleren Julitemperatur von 23-25°C und milde Winter mit einer mittleren Januartemperatur von 3-6°C.
Wie Kolakowski (1961) bemerkt, erinnert das Klima der Kolchis in seinen allgemeinen Zügen an das von Süd-Japan und der Berggebiete Süd-Chinas, allerdings treten im Kolchischen Klima auch Merkmale des Mittelmeerklimas auf (z.B. Niederschlagsabnahme in der Sommerzeit).

Refugialgebiet der Kolchis

Die Kolchis ist ein Refugialgebiet der subtropischen Flora des Neogens, also jener Pflanzenwelt, die in der Vergangenheit in West-Eurasien sehr verbreitet war (Dolukhanov, 1980).
In der Gegenwart hielten sich die Verwandten dieser Pflanzen in Ostasien und an einigen Stellen Nord-Amerikas.
In West-Eurasien befinden sich nur zwei größere und reichere Refugialgebiete, in denen kompakte Überbleibsel der uralten mesophilen Pflanzenwelt dieses Erdteils erhalten geblieben sind. Das eine befindet sich in der Kolchis, das andere - am nördlichen Abhang der Gebirgskette des Elburs, dessen kleinerer Teil nach Aserbeidschan eindringt (nach Talisch), bekannt als Hyrkanisches Refugialgebiet. Der größere Teil des Kolchischen Refugialgebiets befindet sich in Georgien, ein kleinerer Teil in der Türkei (Schiechtl u.a. 1965).
Beide Refugialgebiete haben eine Reihe von allgemeinen Zügen, unterscheiden sich aber auch durch stark ausgeprägte Eigenarten.
Im Hyrkanischen Refugialgebiet ist das Klima im Unterscheid zur Kolchis etwas wärmer und trockener. Starke Niederschläge fallen nur auf verhälthismäßig niederen gipsometrischen Anhöhen. In der Kolchis sind alle hohen Berggürtel mit reichen Niederschlägen versorgt, die sich im Verlaufe des Jahres gleichmäßig verteilen. Für die Hyrkanik ist das Vorhandensein von engendemischen Arten wie *Quercus castaneifolia, Parrotia persica, Alnus subcordata, Acer velutinum, Albizzia julibrissin, Gleditschia caspia, Hedera pastuchovii, Danaë racemosa* besonders charakteristisch. Interessant ist, daß keine einzige dieser Arten in den Kolchischen Refugialgebieten wachsen und ihrer Natur nach näher zu den Hirkanischen stehen. *Pterocarya pterocarpa*, sehr nahe stehende vikariante Arten *Zelcova hircanica* und Z. *carpinifolia.*
Zu den Kolchischen Refugialgebieten stehen folgende Arten näher als zum hirkanischen Gebiet: *Vaccinium arctostaphylos, Laurocerasus officinalis, Daphne pontica, Hypericum androsaemum* und *Arachne colchica.*
Im Verlaufe fast des ganzen Miozäns, waren auf dem Gebiet der gegenwärtigen Kolchis zwei voneinander durch eine Meerenge getrennte Berglandflächen.
In wärmeren Perioden des niederen und mittleren Miozäns herrschte an den Kolchischen Küstenstreifen ein feucht-subtropisches Klima, das zur Existenz

von immergrünen Wäldern mit Lauraceae, immergrünen Eichen und anderen typisch subtropischen Pflanzen verholfen hat. Am Ende der Tertiärzeit trat Kälte ein, die den Untergang vieler wärmeliebender immergrüner Formen hervorrief. Wahrscheinlich waren an den Küstenstreifen zeitweilig Perioden mit weniger Niederschlag im Sommer, das Klima veränderte sich in dieser Zeit allmählich und näherte sich dem Klima der Mittelmeerländer.

In der Periode der Quartär-Vereisung waren die Temperaturverhältnisse in der Kolchis nicht so katastrophal niedrig, aber aufgrund der sehr starken Niederschläge war die Vergletscherung an manchen Stellen sehr stark.

Die Kolchis blieb schon ab dem frühen Miozän ein gutes Refugialgebiet für die relikten Arten der wärmeliebenden mesophilen Pflanzenwelt miozänischer und sogar oligozänischer Wälder, die ausserhalb der Kolchis, bereits Millionen Jahre früher ausgestorben sind.

Die Hauptrolle im Prozeß der Erhaltung des subtropischen Klimas in der Kolchis spielte zweifellos die mächtige Bergschranke des Großen Kaukasus, die Transkaukasien vor nördlichen, kalten Luftmassen schützte. Außer der Gebirgskette des Kaukasus bestimmen vor allem die vielen Niederschläge das Klima der Kolchis, welche westlichen Winde in das Gebiet eintragen. Dank der orographischen Geschlossenheit des Territoriums konzentrieren sich die Niederschläge auf die Ebene und die anliegenden Abhänge der Berge. Die meisten Niederschläge fallen in der oberen Bergzone. Die größte Feuchtigkeit tritt im Süden der Kolchis (Adscharien) auf.

Vegetation

In einem verhältnismässig kleinen Gebiet Georgiens entwickelten sich die verschiedenartigsten Pflanzenformationen, angefangen von Wüsten in den Vorgebirgen im Osten, über üppige Wälder im Gebiet mit feuchtem, fast subtropischem Klima bis zu der eigenartigen kryophilen Vegetation der Hochgebirge der Kolchis.

Dieser wesentliche Unterschied zwischen dem Klima Ost- und Westgeorgiens (Kolchis) bestimmte die Mannigfaltigkeit der Pflanzendecke, was sich auch auf die Struktur der Höhenstufen auswirkte. In Westgeorgien fehlt der trockene, waldlose Gürtel der Vegetation vollständig; Täler und Berghänge sind bereits ab Meereshöhe mit Wäldern bedeckt. Deshalb ist das Profil der vertikalen Zonierung einfacher und durch drei Vegetationsgürtel dargestellt; Waldgürtel (von Ufer bis zu 1900m), subalpine (1900-2500m),alpine Zone (2500-2900m) (Figur 3).

In der Kolchis beginnnen stellenweise bereits vom Meeresstarnd an Wälder mit relativ hoher Produktivität - das Alnetum (*Alnus barbata*) und Pterocaryetum (*Pterocarya pterocarpa*) auf der versumpften kolchischen Niederung, Quercetum (*Quercus pedunculiflora, Q. dschorochensis*), Carpinetum (*Carpinus caucasica*), Castanetum (*Castanea sativa*) mit einer Beimischung

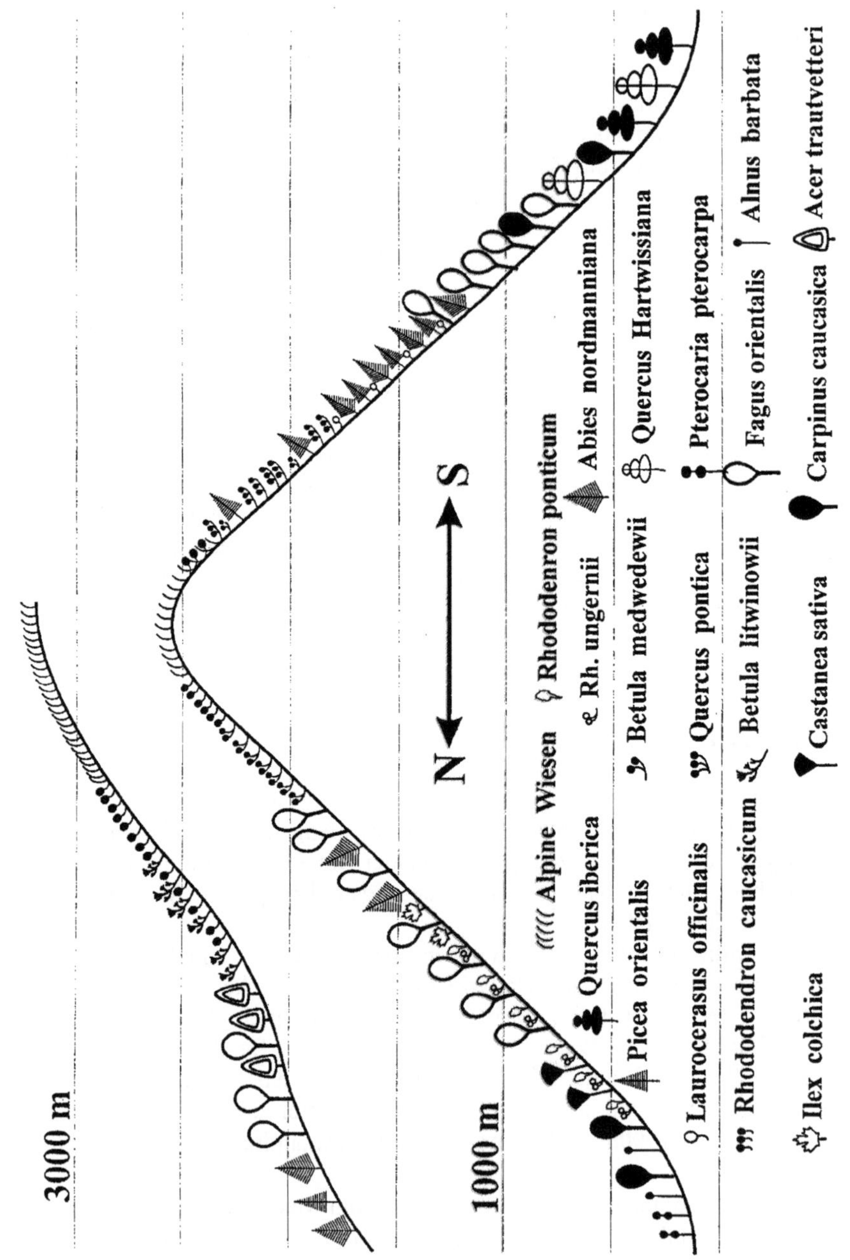

Figur 3: Vegetationsprofil der Südkolchis

anderer Arten sind an wenigen feuchten Orten verbreitet. Hier entwickeln sich üppig Lianen (*Smilax excelsa, Hedera colchica* u. a.).
Die Botaniker des Kaukasus (Dolukhanow 1980) sind mit Riklis Meinung nicht einverstanden, der die Kolchischen Wälder für subtropisch hielt.
Die Kolchis unterscheidet sich durch andersartige, spezifische Eigenschaften und hebt sich auch durch das Relief von allen anderen Waldlandschaften West-Eurasiens hervor.
Am Schwarzmeerstrand Abchasiens wächst *Pinus pithyusa.* Der unikale Hain auf der Pizunda-Landzunge wird als wertvolles Denkmal der Natur geschützt. Dort trifft man auch andere Vertreter der Mittelmeerflora, darunter die für Georgien seltenen Arten (*Arbutus andrachne, Cistus, Erica arborea, Pancratium maritimum* u.a.).
Die Vorgebirgslandschaft der Kolchis ist durch die Tätigkeit des Menschen stark verändert; von ihrer natürlichen Vegetation ist fast nichts erhalten geblieben. In den letzten Jahrzehnten wurden dort auf den Ebenen (mit Ausnahme einiger Bereiche) und in den Vorgebirgen subtropische Früchte eingeführt und Kulturen angelegt (Tee, Zitrusgewächse, Tung, Lorbeer, Pflanzen zur Gewinnung ätherischer Öle u.a.). Zu dekorativen Zwecken werden Eukalyptus, Palmen, Oleander und viele andere Zierpflanzen kultiviert.
Die Bergwälder an den Hängen im unteren Teil des Berggürtels (bis 500-600m), welche nicht durch Kulturen verdrängt wurden, bestehen meist aus Eichen- (*Quercus iberica, Q. hartvissiana*), Kastanien- und Buchenformationen; in höher gelegenen Gebieten hauptsächlich aus Buchen, ab 1000m mit *Picea orientalis* und *Abies nordmanniana* (Figur 3). Der grösste Teil dieser Wälder wurde intensiv abgeholzt, sodass gegenwärtig natürliche Waldflächen nur an äusserst beschränkten Stellen in Schutzgebieten und an schwer zugänglichen Bergen erhalten geblieben sind. Für die Waldvegetation der Kolchis ist das an manchen Stellen entwickelte dichte Unterholz sehr charakteristisch. Ein Dickicht halbkriechender Sträucher, unter denen relikte Arten dominieren, wie beispielsweise der immergrüne, pontische u.a. *Rhododendron, Laurocerasus, Ilex.* In einigen Gebieten (besonders in den Kalksteinbergen Abchasiens und Megreliens) gibt es viel *Buxus*, welcher grösstenteils in der oberen Schicht des Baumbestandes wächst. In den Bergwäldern Georgiens (wie im westlichen so auch im östlichen) findet man des öfteren solch wertvolle und heute seltene Holzgewächse wie *Taxus baccata.* Der Taxus-Hain in der Bazara-Schlucht (Oberlauf der Alasani) ist einer der am besten erhaltenen im ganzen westlichen Eurasien. Einzelne Taxusbäume erreichen hier die Höhe von 30 m mit einem Stammdurchmesser bis zu 1,5m.
Die Pflanzendecke der subalpinen Stufe ist am artenreichsten. Ihre untere Grenze liegt in einer Höhe von 1800-1900m.Von diesem Niveau an lichtet sich der Wald allmählich, und die Vegetation anderer Gesellschaften erscheint: Lichte Wälder, Krummholz, Hochstaudenfluren und Wiesen. Eigenartig und sehr charakteristisch für diese Stufe der Kolchis ist die Krummholzvegetation mit lokalendemischen Arten: *Betula medwedewii,* B. *megrelika, Quercus*

pontica und Kniehholz-Arten: *Rhamnus imeretina, Sorbus subfusca, Corylus colchica, Daphne alboviana* und dem in den Bergen des Kaukasus weit verbreiteten *Rhododendron caucasicum.*
Der größte Teil der Unterholzvegetation des Kolchischen Waldes besteht aus immergrünen Sträuchern und Kniehholzarten: *Rhododendron ponticum. Rh. ungernii, Rh. smirnowii, Epigaea gaultherioides, Laurocerasus officinalis, Ilex colchica, Ruscus colchicus, Rhododendron caucasicum.* Das dichte Gestrüpp der immergrünen Sträucher und Knieholzbäume bekam den Namen "Schkeriani" - "Schkeri" - georgisch - *Rhododendron ponticum.* Der berühmte Botaniker Golizin versteht unter dem Namen "Schkeriani" den ganzen Komplex der immergrünen Sträucher.
Alle diese Pflanzen waren in West-Eurasien im Neogen stark verbreitet. Sie haben eigenartige biomorphologische Eigenschaften: halbes Knieholz und säbelhaft gekrümmte Schößlinge. Ein solcher Wuchs hilft ihnen am Leben zu bleiben und begünstigt die Einwurzelung der Schößlinge. Im feuchten Klima der Kolchis haben die Mehrheit dieser Pflanzen eine sehr breite vertikale Verbreitung - über mehr als 2000m. Das Element der Pflanzenwelt, das gerade der Waldpflanzenwelt der Kolchis ein subtropisches Äußeres verleiht, erreicht die beste Entwicklung nicht nur unter den feucht-tropischen Bedingungen des Kolchischen Gebirgsvorlandes, sondern auch in den verhältnismäßig hohen gipsometrischen Berghöhen - in den Wäldern des kalten Klimas mit mächtiger Schneedecke.
Außer den obenerwähnten Arten wächst in der Kolchis eine Reihe anderer Pflanzen, die zu den Relikten gehören: *Buxus colchica, Ruscus ponticus. Arachne colchica, Arbutus andrachne, Hypericum androsaemum, Philadelphus caucasicus, Staphylea colchica, St. pinnata* (Dolukhanow 1980). Ausführlicher behandeln wir Wälder mit immergrünem Kolchischen Unterholz.

Buchenwälder (*Fagus orientalis*) mit immergrünem Kolchischen Unterholz

Im Unterschied zu den Buchenwäldern Westeuropas ist im Fagetum der Kolchis oft dichtes Unterholz vorhanden, das von halbprostratem immergrünen Buschwerk (Bestand) gebildet wird.
Im Fagetum kommen alle Hauptvertreter des immergrünen Kolchischen Unterholzes vor, für welches ein außerordentliches Bedürfnis nach Feuchte und gleichmäßiger Versorgung mit Niederschlägen während des ganzen Jahres kennzeichnend ist.

Fageta Rhododendrosa (*Rhododendron ponticum*) A. Dolukhanow (1989)

Diese Gesellschaften sind in Westgeorgien sehr verbreitet. Nur in Kalkgebirgen sind sie selten anzutreffen und in den verhältnismäßig trockenen Schluchten sind sie überhaupt nicht vorhanden. Dort, wo die Jahresniederschläge 1400mm übersteigen, sind sie bis auf 1250m.ü.M. verbreitet. Es ist von Interesse, daß in

sehr feuchten Schluchten (> 2000mm) *Rhododendron ponticum* sogar die obere Waldgenze (2100 - 2200m.ü.M.) erreicht.
Fageta rhododendrosa entwickeln sich sowohl auf fast ebenen Stellen, als auch auf steilen Berghängen.

Fageta rhododendrosa (*Rhododendron ungernii*) **A. Dolukhanow (1989)**
Die Gesellschaften der höchstinteressanten Kombination der Buche und der uralten lokalendemischen Reliktpflanze *Rhododendron ungernii* sind in der Kolchis in den an Niederschlägen besonders reichen Orten verbreitet (> 3000mmim Jahr). Diese Gesellschaften kommen hauptsächlich auf der Höhe von 1200 bis 2000m.ü.M. vor. Aber in den Schluchten, wo die Niederschläge mehr als 3200 - 3500mmerreichen und beim Vorhandensein häufiger Nebel im Sommer sinken sie stellenweise bis 800 oder sogar bis 600m.ü.M. herab. *Rhododendron ungernii* gedeiht in den verhältnismäßig lichten Wäldern und erreicht eine Höhe von 3 - 4m. Im Unterschied zu *Rhododendron ponticum* ist *R. ungernii* weniger schattentolerant und deshalb kommt *Rhododendron ungernii* in dichten Wäldern oft nicht vor. Jedoch verträgt *Rhododendron ungernii* die volle Auslichtung besser als *Rhododendron ponticum.*

Fageta lauricerasosa A. Dolukhanow (1989)
Die Verbreitung der schon erwähnten sowie der vorher angegebenen Gesellschaften ist mit der hohen Feuchtigkeit des Klimas eng verbunden. Bei einem Jahresmittel von über 2000mm Niederschlag entwickeln sich die Kirschlorbeergesellschaften bis zu 2200 - 2250m.ü.M. In Höhen von 700 bis 2000m sind sie typisch.
In großen Höhen werden *Rhododendron ponticum* und *Laurocerasus officinalis* besonders anspruchsvoll bezüglich der Menge an Winterniederschlägen für das Überleben in harten Winterbedingungen.
Fageta lauricerasosa gedeihen im Unterschied zu *Rhododendron* Arten in Kalkgebirgen und unter feuchten Klimaverhältnissen gut. Sie wachsen besser auf stark ausgelichteten, südexponierten Hängen. Von den *Rhododendron*-Waldtypen unterscheiden sie sich dadurch, daß Fageta lauricerasosa bedeutend weiter verbreitet sind und auch außerhalb der Grenzen der Kolchis vorkommen.
In den feuchtesten und wärmsten Gebieten Westgeorgiens wächst *Laurocerasus officinalis* als gerader Hochstamm bis in eine Höhe von 14 - 18m, mit einem Stammdurchmesser von 60 - 70cm, mitunter sogar von 1m (Golizin 1935, datiert nach Dolukhanow 1989). Solche Bäume wachsen nur auf den niedrigsten Bergstufen.

Fageta ilicitosa (*Ilex colchica*) **A. Dolukhanow (1989)**
Die Gesellschaften von *Fagus orientalis* und *Ilex colchica* sind etwas stärker verbreitet als Fageta rhododendrosa, ihr Areal fällt beinahe mit dem Areal von Fageta lauricerasosa zusammen. Stellenweise dringen sie bis Ostgeorgien vor.

Ilex colchica gedeiht im feuchtesten Klima ca bis 2150 - 2180m.ü.M., allerdings sind sie selten unter 500m.ü.M. oder über 2000m.ü.M. zu finden. Meistens sind sie auf Hängen nördlicher Exposition verbreitet und für tiefe, schattige Schluchten besonders charakteristisch. *Ilex colchica* ist eine typische Schattenpflanze und braucht mehr Feuchte als der ihm nah verwandte *Ilex aquifolium*. Hülsenfrüchtler bilden keine kompakten Bestände auf großen Flächen. Am häufigsten liegen Dickichte fleckenweise vor. Der Wuchs von *Ilex colchica* variiert stark je nach den gegebenen Klimaverhältnissen (Abb. 1 & 2). Im untersten Teil des Waldgürtels richtet sich diese Pflanze oft gerade auf und wächst wie ein kleiner Baum bis zu einer Höhe von 6 - 7m – selten 10 - 12m – und einem Stammdurchmesser von 20 cm. In Höhen von 1000 - 1200m.ü.M. bilden sie halbkriechende Bestände von 1 - 1,5m Höhe. Im oberen Teil des Waldgürtels (1800 - 2000m) kriechen sie noch stärker.

Fageta ruscosa (*Ruscus colchicus*) **A. Dolukhanow (1989)**

Die Gesellschaften von Fageta ruscosa kommen im Vergleich mit anderen Fagetum mit kolchischem Unterholz bedeutend seltener vor. Man findet sie als kleine Abschnitte in verschiedenen Wäldern der Kolchis.

Der bekannte Forscher der kolchischen Pflanzenwelt Kolakowsky (1961) führt Bezeichnungen für eine Reihe von Waldgesellschaften mit immergrünem kolchischen Unterholz an. Aber die Angaben über ihre Verbreitung, Pflanzensoziologie und Ökologie sind überaus spärlich. Deswegen führen wir nur ihr Verzeichnis mit Angabe der Höhen (in m.ü.M.) ihrer Vorkommen an:

Querceta (*Quercus iberica*) **rhododendrosa** (*Rhododendron ponticum*) 20-700m,

Querceta ruscosa (*Ruscus ponticus*),

Carpineta (*Carpinus caucasica*) **rhododendrosa** (*Rhododendron ponticum*),

Carpineta lauricerasosa (*Lanrocerasus officinalis*) 0 - 1000m,

Carpineta ruscosa (*Ruscus ponticus*),

Carpineta ilicitosa (*Ilex colchica*),

Carpineta buxosa (*Buxus colchica*),

Castaneta (*Castanea sativa*) **rhododendrosa**(*Rhododendron ponticum*) 200 - 1400m.

Dunkle Nadelwälder mit immergrünem kolchischen Unterholz

Im Unterschied zu den Buchenwäldern mit immergrünem kolchischen Unterholz sind die dunklen Nadelwälder mit gleichem Unterholz nicht so sehr verbreitet. Abgesehen davon bilden dunkle Nadelwälder günstige Vorkommen für die Vegetation von fast allen immergrünen Pflanzen der Region.

A. Dolukhanow (1989) sondert Assoziationsserien der Nadelwälder aus, wie z.B.

Piceeta rhododendrosa (*Picea orientalis*),

Picceto-Abieta rhododendrosa (*Abies nordmanniana*),

Abb. 1
Ilex colchica in den
Carpineta caucasici
(Südkolchis, 800 m)

Abb. 2
Ilex colchica
mit Früchten

Fageto-Abietarhododendrosa,
Piceeta lauricerasosa,
Fageto-Abietalauricerasosa,
Piceeta ilicitosa,
Piceeto-Abieta ilicitosa,
Fageto-Abietailicitosa.
Das Unterholz mit *Rhododendron ponticum* entwickelt sich vorwiegend in Buchen- und Fichtenwäldern und kommt auf allen Stufen der dunklen Nadelwälder vor. Höher als 1900m.ü.M. steigt es nur in niederschlagsreichen Bergregionen. Für die Vegetation dieser Wälder sind generell hohe Feuchtigkeit während des ganzen Jahres und mäßig feuchte Böden notwendig. Die natürliche Regeneration in diesen Wäldern ist sehr schwach.
Die Assoziationen der Dunkelnadelwälder mit *Laurocerasus officinalis* sind auf allen Höhenstufen verbreitet, die durch Niederschläge intensiv befeuchtet sind. Im Unterschied zu den Gesellschaften mit *Rhododendron ponticum* bevorzugen die Nadelwälder mit Lorbeerkirsche die Südhänge und meiden die Kalkstandorte nicht. *Laurocerasus officinalis* ist weniger schattentolerant als *Rhododendron ponticum*, deshalb kommt er verhältnismäßig selten in den gedrängten Abieta vor. Piceeta laurocerasosa sind an steinigen Standorten anzutreffen. Das Unterholz aus Lorbeerkirsche wird oft von *Ilex colchica* begleitet.
Unter den dunklen Nadelwäldern mit dem Unterholz *Ilex colchica* sind diejenigen Fageto-Abieta-ilicitosa am besten entwickelt, welche den mittleren Teil des Waldgürtels einnehmen (1200 - 1700m.ü.M.).
In den dichten Beständen von *Ilex colchica* keimen keine Holzarten.

Prognose

Im Zusammenhang mit den globalen klimatischen Veränderungen ist nur bekannt, daß in der Kolchis in der wärmsten Periode die Temperaturen um maximal 0,5°C ansteigen und in den kältesten Perioden um maximal 0.5°C abnehmen. Vorläufig ist es nicht bekannt, wie sich Niederschlagsmengen verändern. Wenn die Niederschläge mit der Temperatur ansteigen, was anzunehmen ist, wird sich in der Kolchis üppige subtropische Vegetation (besonders immergrünes Unterholz) entwickeln und intensive Laurophyllierung (Klötzli et al. 1996) einsetzen.

Literatur

Dolukhanow A. (1980): Kolchische Unterholz, Metsniereba, Tbilissi, (Russisch).

Dolukhanow A. (1989): Waldvegetation Georgiens, Metsniereba, Tbilissi (Russisch).

Klötzli F., Walther G.-R., Carraro G. & Grundmann A. (1996): Anlaufender Biomwandel in Insubrien. Verh. Ges. Ökol. 26: 537-550

Kolakowsky A. (1961): The Flora and vegetation of Colchida. The publisch. House of the Moscow university, Moscow, (Russisch).

Schiechtel, H., Stern, R. & Weiss E. (1965): In Anatolischen. Gebirgen, Verlag des Geschichtvereines für Kärnten, Klagenfurt.

Recent shifts in vegetation boundaries of deciduous forests,
especially due to general global warming, pp. 181–197
edited by F. KLÖTZLI and G.-R. WALTHER

On the Colkhic vegetation

Nugzar ZAZANASHVILI
Institute of Botany, Georgian Academy of Sciences
Tbilissi 380007, Georgia

Abstract

This paper is a short review of Colhkic vegetation wich is very rich in relics and endemics. The general distribution and contemporary ecological conditions of the vegetation are presented, with emphasis on evergreen Colkhic relicts. Apart from this summarised data for natural and introduced vegetation the relation to the global climatic change is given. The present conservation activities as well as perspectives to preserve this unique vegetation are discussed in the light of establishing a new system of protected areas in Georgia.

Introduction

Colkheti (Colkhida) is one of the historical-geographical provinces of the Caucasus, surrounded by the Great Caucasus and Small Caucasus range systems and belonging to the Black Sea catchment basin (western Transcaucasus). This region is one of the shelters for tertiary flora relics in western Eurasia. Here a lot of representatives of evergreen and deciduous relics are spread, which often determine phythosociological and spatial structures of contemporary plant communities. Prostrate and half-prostrate species define the uniqueness of this region. Since they have active vegetative reproduction ability, in most cases they also define dynamics of plant communities. The following species are typical examples: *Quercus pontica, Betula medwedewii, B. megrelica, Rhamnus imeretina, Corylus colchica, Sorbus subfusca, Vaccinium arctostaphylos, Viburnum orientale, Rhododendron luteum;* evergreens - *Rh. ponticum, Rh. ungernii, Rh. smirnowii, Rh. caucasicum, Laurocerasus officinalis, Ilex colchica, Ruscus colchicus*. Besides those many other evergreen and deciduous relic trees, shrubs, grasses and lianas occur (*Quercus hartwissiana, Q. imeretina, Buxus colchica, Staphylea colchica, Philadelphus caucasicus, Hypericum inodorum, Daphne pontica,*

D.alboviana, Hedera colchica, Primula magaseifolia, etc.). We have to point out five half-prostrate evergreen widespread species.

Physical-geographical factors of the Colkheti region

Relief

Colkheti borders are given by the Black Sea in the west and the Great Caucasus ranges system (in this part the highest peak is Shkhara - 5068m) in the North. The Small Caucasus ranges (Mepistskaro - 2850m) in the South and the Likhi low range, located between Great and Small Caucasus, building the eastern boundary. In the central part of the region Colketi lowland is situated. Here, due to the high soil moisture, Colkhic relic species are rare. In the relief of the Great Caucasus predominate tectonic-erosions, karst and glacial forms, on the Small Caucasus tectonic-erosions and volcanic forms.
Representatives of the Colkhic evergreen flora are distributed from sea level to the timberline - 2200-2300m a.s.l. (see Table 1), with exception of *Rhododendron caucasicum*, which is spread from 2000 to 2800-2900 (3000)m a.s.l. In most cases these species spread in valleys open for humid air masses, mainly on the shadowy slopes. *Rhododendron ponticum* and *Rh. caucasicum* grow mainly on the northern slopes, *Rh. ungernii* prefers less shadowy slopes, *Laurocerasus officinalis* southern slopes and *Ilex colchica* the prominent elements of the relief.

Geology

The greatest part of the Colkhetian section of the Great Caucasus is structured from Jurassic slates and sandstone, Cretaceous limestone, marls, sandstone and Paleoezoic granites. On the Small Caucasus Eocene volcanic bed-rocks (mainly andesites), Paleocene-Low Eocene clays, marls and sandstone (flysch) are most frequent. *Rhododendron ponticum* (and *Quercus pontica*) avoid calcareous biotops; at the same time, under relevant climate conditions, on limestone its competitor *Laurocerasus officinalis* is promoted (from Colkhic relics clearly distinguished calciphile plant is *Corylus colchica*).

Climate

Colkheti is the most humid region in the Caucasus. The annual precipitation falls rarely short of amount to 1000mm, on the greatest part of the Colkheti area it amounts to 1500-2500mm with a maximum of over 4000mm, which is characteristic for the middle mountains of the southern Colkheti seaside. Besides, the distribution of precipitation throughout the year is mostly equal. In some areas on the plains autumn maxima and summer minima occur. In the mountains an intense precipitation is a necessary condition for evergreen Colkhic relics to survive the winter successfully. On an altitude of 1000-2200m

a.s.l. the snow cover lasts 90-196 days (Kordzakhia 1961). Among evergreen Colkhic species the most in need for humidity is *Rhododendron urgernii* (see Table 1). In places were the annual precipitation drops below 1200mm Colkhic evergreen relics rarely develop.
The annual air temperature in the plains varies between 12 and 14°C. The altitudinal thermal gradient is in the limits of 0,5-0,6°C/100m.

Soils

The greatest part of the soils is represented by alluvial soils, yellow, yellow-podzolic, red earth (up to 500-600m a.s.l.), rendzina, forest brown, mountain meadow soils and andosols (the availability of trailing-shrubs *Rhododendron caucasicum* is responsible for developing mountain peat soils). Colkhic evergreen species are spread on all these soil types. The soil pH and moisture are of greater importance for plants, which avoid wet as well as dry soils.

Natural vegetation and its contemporary ecological condition

Figure 1 shows the distribution of natural vegetation in Colkheti. *Alnus barbata* forests are spread in the coastal part of the Colkheti lowlands. Nowadays only their separate groves are still remaining in the extreme western part of the Colkheti lowlands in wet and damp biotops, wetlands, etc. The riverside forests with *Pterocarya pterocarpa* and some other types of alder forests have disappeared. Evergreen underwood is not characteristic for lowlands alder forests, its representatives rarely occur here.
Mixed broad-leaved forests are characteristic for foothills (up to 500-600m a.s.l.) with dominant endemic relics like *Quercus imeretina*, *Zelkova carpinifolia* (northern Colkheti), *Quercus hartwissiana*, as well as *Carpinus caucasica, Castanea sativa, Fagus orientalis*, *Tilia caucasica, Fraxinus excelsior,* in moist biotopes - *Alnus barbata*, etc., a second layer is represented by *Staphylea colchica, S. pinnata, Buxus colchicus,* etc. For such forests the existence of Colkhic underwood (*Rhododendron ponticum, Rh. luteum, Laurocerasus officinalis, Ruscus colchicus, Ilex colchica, Vaccinium arctostaphylos*) and the diversity of lianas is typical. These wonderful and unique forests were damaged by human impact. In the lowlands they are totally destroyed (except *Quercus imeretina* grove in the Ajameti reserve). They are also damaged on foothills and protected only in the Bichvinta-Mjusera reserve in north-western Colkheti (Abkhazia).

Figure 1: Map of natural vegetation of Colkheti, sc.1:2 500 000

1. Subnival open vegetation (*Primula bayernii, Cerastium undilatifolium,* etc.).
2. North Colkhetian alpine grasslands (*Geranium gymnocaulon, Festuca djimilensis, Woronowia speciosa, Carex pontica,* etc.), mats (*Sibbaldia semiglabra, Ranunculus brachylobus,* etc.) and *Rhododendron caucasicum* communities.

2a. South Colkhetian alpine grasslands (*Carex tristis, Agrostis lazica, Festuca ovina,* etc.), mats (*Campanula tridentata, Gentiana pontica,* etc.) and *Rhododendron caucasicum* communities.

3. North Colkhetian subalpine crooked-stem forests and woodlands (*Fagus orientalis, Acer trautvetteri, Betula litwinowii, partly B. megrelica, Quercus pontica*), shrub communities (*Rhododendron caucasicum, Rh. ponticum, Laurocarasus officinalis, Rhamnus imeretina, Corylus colchica,* etc.), tall herbaceous vegetation (*Heracleum ponticum, H. aconitifolium,* etc.) and grasslands (*Poa iberica, Calamagrostis arundinacea,* etc.). *planifolia, Geranium platypetalum,* etc.).

3a. South Colkhetian subalpine crooked-stem forests and woodlands (*Fagus orientalis, Acer trautvetteri, Betula litwinowii, B. medwedewii, Quercus pontica*), shrub communities (*Rhododendron caucasicum, Rh. ponticum, Rh. ungernii, Laurocerasus officinalis*, etc.), tall herbaceous vegetation (*Ligusticum alatum, Milium schmidtianum*, etc.) and grasslands (*Agrostis planifolia, Geranium platypetalum*, etc.).

4. Fir, spruce-fir and beech-fir forests (*Abies nordmanniana, Picea orientalis, Fagus orientalis*).
5. Fir, spruce-fir and beech-fir forests (*Abies nordmanniana, Picea orientalis, Fagus orientalis*) often with evergreen underwood (*Rhododendron ponticum, Laurocerasus officinalis, Ilex colchica*).
6. Pine forests (*Pinus kochiana*) alternating with crooked-stemed birch (*Betula litwinowii*) and spruce-fir forests.
7. Beech forests (*Fagus orientalis*) mainly with Colkhic underwood (*Rhododendron ponticum, Rh. luteum, Laurocerasus officinalis, Ilex colchica, Vaccinium arctostaphylos*, etc.).
8. Beech-chestnut forests (*Castanea sativa, Fagus orientalis*) alternating with oak forests (*Quercus iberica*) with Colkhic underwood (*Rhododendron ponticum, Rh. luteum, Laurocerasus officinalis, Ilex colchica, Vaccinium arctostaphylos*, etc.).
9. Lowland and submontane mixed broad-leaved forests (*Quercus imeretina, Q. hartwissiana, Zelkova carpinifolia, Carpinus caucasica, Castanea sativa, Fagus orientalis*) with Colkhic underwood (*Rhododendron ponticum, Rh. luteum, Laurocerasus officinalis, Vaccinium arctostaphylos*, etc.).
10. Swamp alder forests (*Alnus barbata*).
11. Alder forests alternating with *Sphagnum spp*. bogs.
12. Pitsundian pine forests (*Pinus pithyusa*) with *Carpinus orientalis, Cistus ponticus*.
13. Ajarian oak-pine forests (*Pinus kochiana, Quercus dshorochensis*).

The lower mountain belt (500-1000m a.s.l.) is characterised by chestnut, beech-chestnut and beech-hornbeam-chestnut forests (Figure 1, #8). In the "chestnut zone" of north Colkheti, in relatively dry biotops (southern slopes, etc.), oak and hornbeam-oak forests (*Quercus iberica, Carpinus caucasica, C. orientalis*) are spread. Evergreen species are not specific for such oak forests. In this zone forests are also destroyed. Best preserved are the forests with evergreen understorey, since they are unapproachable for the forest industry. They are protected in Bitchvinta-Mjusera and Kintrishi reserves.
Beech forests dominate between 1000 and 1500m a.s.l. (Figure 1, #7). In southern Colkheti (Ajara) beech forests with evergreen underwood occupy approximately 50% of the beech zone (it is supposed that in the northern Colkheti this indicator should not be significantly different). In the first instance beech forests set up monodominant communities. In the Caucasus beech forests are main objects for forest industry, but at the same time are beech forests with primary structure preserved in some inaccessible gorges. They are

protected in the Kintrishi reserve and in the reserves located in the north-western Caucasus.

Dark-coniferous forests (fir, spruce-fir, beech-fir; Figure 1, #4-5) are mainly spread within boundaries of 1400-2000m a.s.l. In some places they form the timberline and climb up to 2250m a.s.l. These forests are also under anthrophogenic influence. At the same time they are well preserved in some inaccessible gorges (especially in northern Colkheti) and protected in reserves situated in the north-western Great Caucasus (also in Borgomi reserve, in the northern part of the Small Caucasus).

Subalpine forests are coenotically most diverse (Figure 1, #3,3a). Lower subalpine elfin woods formed with endemic relics are also unique (mainly within 1700-2000ma.s.l.). *Betula medwedewii* forests are spread only in southern Colkheti (including Turkish Lazistan), *B. megrelica* only in northern Colkheti and *Quercus pontica* in both areas. An evergreen understorey is typical for such forests. They differ in physiognomy: in most cases dominants grow nearly vertically. In Colkheti crooked-stem beech forests are widespread (mainly within 1900-2200m a.s.l.). Evergreen underwood is also typical for these forests. For the upper subalpine belt (2200-2500m a.s.l.) crooked-stem birch forests with *Rhododendron caucasicum* layers are frequent. Unfortunately, subalpine forests like broad-leaved forests are extremely damaged and spread only on 15 % of their natural distribution area. Subalpine forests are protected in Kintrishi, Pskhu-Gumista, Ritsa reserves and others located in the north-western Great Caucasus (*Betula medwedewiii* and *Quercus pontica* are only protected in the Kintrishi reserve). Subalpine forests are replaced by various meadows, which are changed by alpine vegetation - from 2500ma.s.l. onwards. In the alpine belt *Rhododendron caucasicum* communities are spread up to 2800-2900m a.s.l. - this is considered a post-wood process in relation to historical changes of climate (Dolukhanov 1980, Kolakovsky 1961).

Evergreen Colkhic relics: geographical distribution, ecology and coenotical position

As we stated above, we should pay attention to 5 species, which have a common Colkhic origin and mostly half-prostrate growth form.

Laurocerasus officinalis

This species is widespread in Western Georgia (Colkheti) and partly, in the north-western and south-eastern Great Caucasus, the Small Caucasus near the Colkheti region (Borgomi velley), the Talish mountains (southern Azerbaijan), Pontic mountains, northern Anatolia (north-eastern Turkey) and in the Gilian mountains (northern part of Iran, see Figure 2); it occurs also, separated from the main area, in the north-western part of the Balkan peninsula.

L. officinalis, during periods of high precipitation, is distributed from sea level to 2200-2300ma.s.l. In regions where the annual precipitation is less than 1200mm it does not occur (Table 1). In humid and warm areas of the foothills (up to 500-600m a.s.l.) where snow cover is limited and does not press young branches of trees to the ground, *L. officinalis* may grow as a 14-16m high straight tree, with 50-100cm diameter (Golitsin 1935; Mandjavidze 1982).

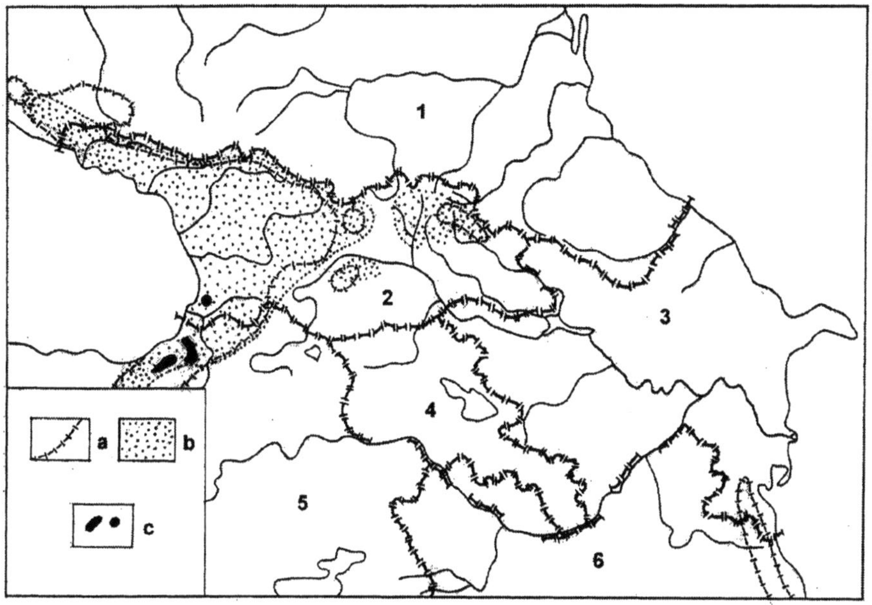

Figure 2: Distribution of *Laurocerasus officinalis* (a), *Ilex colchica* (b), *Epigaea gaultherioides* (c) in the Caucasus isthmus.
1. Russian federation, 2. Georgia, 3. Azerbaijan, 4. Armenia, 5. Turkey, 6. Iran.

However, for Colkheti the growth form of half-prostrate bush (thickets) is typical. In the middle part of the forest belt (1000-1400m a.s.l.) their height reaches 3-4,5m. At the same time does the length of branches exceed the brush height 2-4 times. The lower part of the branches are prostrate and rooted and other parts have a sickle-like form with a straight top. The diameter of the branches does not exceed 18-20cm. With higher altitudes the strength of the bushes is reduced: on 2000m a.s.l. the height is not more then 2,5m, on 2200m a.s.l. it stays below 1,5m.

L. officinalis mainly forms layers in broad-leaved as well as in dark-coniferous forests. It is especially characteristic for forests where *Tilia caucasica* is dominant or co-dominant. Authors of the Russian school of forest typology are considering all forests with such underwood as one type with 2-3 subtypes differentiated altitudinally. Dolukhanov (1989) disagrees with this formulation. He described the following associations: Castanetum laurocerasosum (1953),

Fageta laurocerasosa (Fagetum laurocerasosum typicum, F. ilicitoso-laurocerasosum), Piceeta laurocerasosa, Piceeta-Abieta laurocerasosa, Fageto-Abieta laurocerasosa.)

Species	Laurocerasus officinalis	Rhododendron ponticum	Rh. ungernii	Rh. caucasicum	Ilex colchica
Altitude					
0-500	+	+	-	-	+
500-1000	+	+	-	-	+
1000-1500	+	+	-+	-	+
1500-2000	+	+	+	-+	+
2000-2200	+	+	-+	+	+
2200-2350	+-	-+	-	+	-
Precipitation(mm)					
<1100	-	-	-	+	-
1100-1400	+-	-+	-	+	+-
1400-1800	+	+	-	+	+
1800-2200	+	+	-	+	+
>2200	+	+	+	+	+
Soils					
Calcareous	+	-+	+?	+	+
Non-calcareous	+	+	+	+	+
Forest types					
Mixed broad- leaved	+	+	-	-	+-
Beech	+	+	+	-	+
Dark coniferous	+-	+-	+	-+	+
Elfin woods	+	+	+	+	+
Crooked - steam	+-	+-	-+	-	+-
Density of canopy					
0,3	+	+-	+	+	-+
0,3-0,4	+	+	+	+	+
0,5-0,6	+-	+	+-	-+	+
0,7-0,8	-+	-+	-	-	+-
0,8-0,9	-	-	-	-	-+

Table 1: Some biotopological and coenotical dates
(+ frequently, well development; + - moderate frequency, partly insufficient development; - + fragmentary distribution, insufficient development).

Under a canopy of a density of 0,6 in beech forests *L. officinalis* stops fructification. In beech forests, on the northern slopes, where the value of the density is over 0,7 it does not occur. On the southern slopes with the extreme density of 0,85 it occurs as solitary weak individuals. In fir and spruce forests it cannot live under canopies with such a density. *Betula medwedewii* and *Quercus pontica* elfin woods with evergreen underwood are structurally peculiar: here in most cases a layer of bush is higher then elfin wood. *L. officinalis,* like *Rhododendron ungernii,* often occurs outside the forest canopy, where it forms strong, up to 4-4,5 m high bushes.

Rhododendron ponticum

It spreads mainly in neighbourhood regions of the catchment basin of the Black Sea (Figure 3). Beyond its main area, it occurs in the eastern part of the Balkan peninsula, separately - in the southern part of Iberia and Lebanon mountains.

With its growth form and ecological peculiarities it resembles *Laurocerasus officinalis*. Normally it develops in the mountains where the annual rainfall is over 1300mm and does not grow in areas with precipitation lower than 1100mm. Altitudinally it spreads from sea level to the timberline (2200-2300m a.s.l.). In the foothills, where winter is quite warm, it seldom grows as a low, straight tree. In the understorey *Rh. ponticum* is strictly linked with fresh soils and often occurs in combination with *Rubus hirtus, Trachystemon orientale, Hedera colchica or Viburnum orientale* (Dolukhonov 1980).

Rh. ponticum is especially typical for beech, chestnut and hornbeam-beech-chestnut forests, although it occurs in other types as well. They are described as: Quercetum pontico-rhododendrosum (Maleev 1936), Fagetum rhododendrosum (Povarnitsin 1936), Castanetum rhododendrosum (Dolukhanov 1953), Fageta rhododendrosa, Fagetum rhododendrosum typicum, Piceeta rhododendrosa, Abieta rhododendrosa (Dolukhanov 1980), etc. Moreover, on the timberline it occurs together with *Laurocerasus officinalis, Ilex colchica* and *Ruscus colchicus* (partly with *Rhododendron ungernii*) and is typical for trailing elfin woods (*Quercus pontica, Betula medwedewii*).

It develops best on the northern slopes of humid mountains under the forest canopy with a density from 0,3 to 0,6, in less moist conditions - with densities of 0,5-0,65.

In forests with *Rh. ponticum* underwood the natural forest regeneration is suppressed, this particularly influences the dynamics of many forests types and in some cases forests are replaced by *Rhododendron* bushes.

Rhododendron ungernii

It is a local endemic spread in the south Colkheti (including Lazistan, the north-west Turkey, Figure 3).

Rh. ungernii develops in the upper belt of the forest zone (mainly 1400-2000m a.s.l., 2200m a.s.l.), where the annual precipitation is over 2500mm. Physiognomically the bush resembles *Rh. ponticum*. In undergrowth the height of the bushes on 1700ma.s.l. reaches 2,5m and in glades 4-4,5m.The length of the stem is 2-3 times more then bush height. This species is more light-requiring and often occurs beyond the forest canopy.

Rh. ungernii forms strong bush (underwood) in beech, beech-fir and spruce-fir forests. Seldom it occurs under a canopy with a density of about 0,5 and does not develop when this value is over 0,6. Underwood formed by *Rh. ungernii* prevents natural forest from regeneration and often replaces them for a long time.

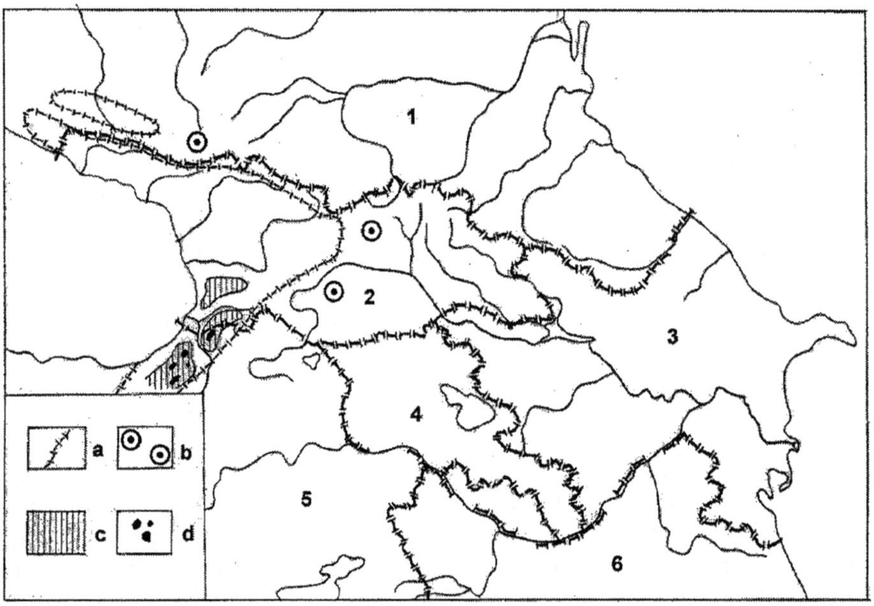

Figure 3. Distribution of *Rhododendron ponticum* (a,b), *Rh. ungernii* (c) and *Rh. smirnowii* (d) in the Caucasus isthmus.
1. Russian federation, 2. Georgia, 3. Azerbaijan, 4. Armenia, 5. Turkey, 6. Iran

Rhododendron caucasicum

This species is more widespread in the Caucasus than other Colkhic relics, though it's whole area is narrower than that of *Rh. ponticum* and *Laurocerasus officinalis*, which are also distributed beyond the Caucasus isthmus (everywhere except driest part).

Rh. caucasicum communities are mainly situated between 2000 and 2900ma.s.l. They are drawn to the northern slopes. The stem of this species is developed from the base forming quite far-stretching branches. They either lie on the peat soil or are covered with it and are rooted by a great number of surface-lying roots. In the underwood the height of this plant is 60-100cm; in the open areas - two or three times less. The distribution of this species beyond forests is generally considered as secondary process due to historical changes of climate and anthropogenic impact. Remains of crooked-stem forests located within *Rh. caucasicum* communities spread on 2600-2650m a.s.l.

Rh. caucasicum mainly developes in crooked-stem forests and elfin woods (*Betula litwonowii, B. raddeana, B. medwedewii, Fagus orientalis, Quercus pontica*). Some communities have been described: Rhododendretum caucasici, Fagetum caucasico-rhododendrosum, Betuletum rhododendrosum, B. herboso-rhododendrosum (Bush & Bush 1936). It is more light-requiring than *Rh. ponticum;* e. g. in the beech forests it grows only under a canopy with a density of less than 0,5, often accompanied by various grasses, mosses and subshrubs. It permanently occurs together with *Gymnocarpium dryopteris, Huperzia selago, Vaccinium myrtillus, Empetrum hermaphroditum,* etc.

Ilex colchica

Ilex colchica is quite widespread in the Caucasus areas (Figure 2), but in compact thickets it is only distributed in Colkheti.

In Colkheti it spreads from the sea level to the subalpine belt, but it especially is typical for altitudes of 1000-2000m a.s.l. (up to 2150m a.s.l.). It grows mostly under fir and beech-fir, partly also beech and spruce forests. Some communities have been described: Piceeta ilicitosa, Piceeto-Abieta ilicitosa, Abieta ilicitosa, Fageto-Abieta ilicitosa, Fagetum ilicitosum typicum, F. ilicitosum superior, F. festucoso-ilicitosum (Dolukhanov 1989). A monodominant underwood of *I. colchica* can be seen in small areas only (about 0,2ha). It often occurs in combination with *Laurocerasus officinalis* and *Ruscus colchicus*. *I. colchica* develops better in well drained, fresh and moderately dry soils. In forests it may develop on cobbly biotops, limestone screes, etc. *I. colchica* is a highly shadow resistant species: in fir forests with a 0,6 density of canopy it begins fructification and occurs even under a density of 0,9. The height of *I. colchica* underwood is seldom over 1,5 m.

Dynamic tendencies in relation with global change

Natural vegetation

In the Caucasus investigations of vegetation dynamics in relation to global climatic changes is a new enterprise. In Georgia an appropriate governmental programme, supported by GEF in 1997-98, has been launched. The team of colleagues from the Institute of Botany of the Georgian Academy of Sciences is also participating in this project. In 1996 the high-mountain vegetation was emphasized, but in 1997 the same activities are planned in the Colkheti region (Ajara). According to some preliminary data, collected during the compilation of vegetation maps of Ajara and through decoding of space photo-information as well as analysis of other available information, some tendencies have been shaped out. However, further confirmation is necessary.

This area is characterised by an increase of the upper distribution border of some Colkhic evergreen relics. According to some references and our present investigations they are in some areas distributed up to 2400 m a.s.l., and *Rhododendron caucasicum* is quite frequent on heights of 3000 (3100) m a.s.l., occasionally surrounded by subnival vegetation, which is evidently unusual and therefore at least needs special investigation. It is difficult to estimate this occurrence as a clearly evident tendency. In any case we found out (for the end of the 80ies) that some individuals and groups of *Quercus macranthera*, which is quite frequent in the eastern and southern Caucasus but unusual for the western Caucasus (Colkheti), were distributed between 2000-2200m a.s.l. This was estimated earlier (e.g. Gagnidze *et al.* 1985) but not on similar elevations; at the same time the positive (developing) tendency for this species could be presumed.

Adventitious/introduced species

The introduction of species in the Colkheti region should have been started in ancient times - considering the fact that the Greek already have established town-colonies in this area - but it is difficult to estimate clearly these first steps of introduction (it is still a matter of discussion if species like *Ficus carica, Laurus nobilis* and others are native or introduced and later distributed in Colkheti). The second wave of introductions occurred during the colonising of Georgia by Russia in the beginning of the XIX century.

Presently, e.g. in the southern Colkheti (Ajara), on an area of 3000km^2, about 1650 species are found, among them 7% are adventitious, in the coastal area about 14% (Manjavidze 1982, etc.). The adventitious flora of Colkheti is well studied (Memiadze 1971, Iabrova-Kolakovsky 1977, etc.), but geobotanically introduced or adventitious vegetation is investigated in much less detail. Amongst others the contemporary geobotanical position of herbaceous adventitious vegetation is relatively well studied, like *Paspalum dilatatum* of Brazilian origin which was introduced in about 1910 as a forage grass and presently is one of the main dominants of secondary grasslands in the Colkheti lowland. The same could be applied to the east Asian *P. digitaria*, the north American *Andropogon virginicus* and others.

Considering trees and shrubs, it should be mentioned that in spite of the diversity of introduced species a trace of "Laurophyllisation" is not yet found in Colkheti and the prime reason for this is probabely the fact that native Colkhic relics (among them evergreen ones) are still keeping powerful coenotic positions in the natural vegetation. However, it must be mentioned also that sufficient attention to the process of Laurophyllisation has so far not been paid. One of the most aggressive introduced species appears to be the North American *Robinia pseudoacacia*, which is widely distributed throughout the whole Colkheti region. It builds monodominant forests up to the lower mountain belt (1000-1200m a.s.l.) on deforested areas and frequently participates in the structure of primary forests. Especially some species should be mentioned

which are mixing with aboriginal species and in particular cases even replace them. Such species are: the Japanese *Cryptomeria japonica*, the North American *Pinus taeda*, the Australian *Acacia dealbata*, the Chinese *Paulownia tomentosa*, as well as *Ailanthus altissima, Asimina triloba, Catalpa speciosa, Acer palmatum, Rhus chinensis, Deutzia scabra, Hydrangea macrophylla, Pueraria hirsuta*, etc. often species of laurineous forests.
The gradual expansion of some species (particularly *Robinia pseudoacacia*) could be linked with a general tendency of global climate changes, but this subject needs special investigation.

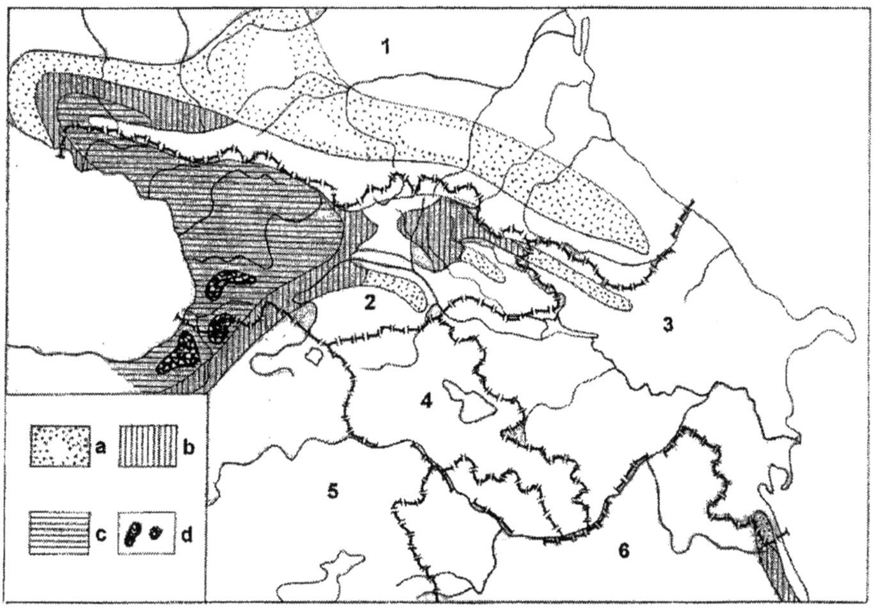

Figure 4: Distribution of half-prostrate Colkhic relics in the Caucasus isthmus (*Laurocerasus officinalis, Rhododendron ponticum, Rh. luteum, Rh. ungernii, Rh. smirnowii, Vaccinium arctostaphylos, Epigaea gaultherioides, Viburnum orientale, Ruscus colchicus, Ilex colchica, I. stenocarpa, I. hyrcana,* including *Rhododendron X sochadze,* except *Rhododendron caucasicum*):
a - areas where spread 1-2 species; b - 3-4; c - 5-8; d - 9-11
(by Dolukhanov 1980)
1. Russian Federation, 2. Georgia, 3. Azerbaijan, 4. Armenia, 5. Turkey, 6. Iran

Diversity and conservation

As it is shown on Figure 4 the maximalquantity of Colkhetian half-prostrate relics occurs in the south Colkheti (Ajara and adjacent Turkish territory, in the Small Caucasus and Pontic mountains). The same region is characterised by the highest value of diversity of trees and shrub species in the Caucasus. Particularly in the lower mountain belt (200-1200m a.s.l.) of south Colkheti on an area of 100-200km^2 over 100 species of trees and shrubs are present - among them many relics of Colkhic and other origin: widespread are: *Abies nordmanniana, Picea orientalis, Alnus barbata, A. incana, Fagus orientalis, Fraxinus excelsior*, as well as *Taxus baccata, Buxus colchica, Quercus pontica, Q. hartwissiana, Pterocarya pterocarpa, Periploca graeca, Evonymus latifolia, E. leioplea, Daphne pontica, Diospyros lotus, Hedera colchica, Hypericum androsaemum, H. inodorum, Ilex colchica, Laurocerasus officinalis, Mespilus germanica, Ostrya carpinifolia, Rhododendron ponticum, Rh. luteum, Ruscus colchicus, Smilax excelsa, Staphylea colchica, S. pinnata, Tilia caucasica, Ulmus elliptica, Vaccinium arctostaphylos, Viburnum orientale, V. opulus*, local endemics - *Betula medwedewii, Epigaea gaultherioides, Rhododendron smirnowii, Rh. ungernii, Phillyrea medwedewii*, and some species occurring rarely in the Caucasus as e.g. *Arbutus andrachne*, etc. Some of the aforementioned species are not characteristic for the northern Colkheti, but some of the species are still relatively widespread such as *Zelkova caprinifolia, Quercus imeretina*, in limestone biotops *Corylus colchica* (it should be mentioned that the grade of endemism in the Colkhetian limestone rock-scree vegetation reaches up to 80% - Sokhadze 1982). The coastal area is characterised by *Pinus pithyusa*, etc.

The total quantity of tress and bushes gradually decreases with altitude. Between 1100-1500m a.s.l. there are 75 species, in the range of 1500-2100m a.s.l. 55 species, over 2000m a.s.l. 43 species (Manjavidze 1982). However, this general rule does not apply to the Colkhetian relics. The majority of which is concentrated in the lower subalpine sub-belt (1700-2200m a.s.l.).

The lower forest and subalpine belts are distinguished by a high phytocoenotic diversity; in the northern Colkheti, e.g. in the Bichvinta-Mjusera reserve, in the range of 0-100 m a.s.l. and on an area of 0,5 km^2 there is a hornbeam-oak forest (*Quercus iberica, Carpinus orientalis*) on the top of ridges. The southern and eastern slopes are occupied by another type of oak forest (with *Carpinus caucasica*), the northern and western slopes by a beech-chestnut forest (*Castanea sativa, Fagus orientalis*) with an evergreen understorey. The coastal sand plain is characterised by a *Pinus pithyusa* forest with *Cistus tauricus*, the coastal rocky abrasive biotops by *Pinus pithyusa* in a complex with rock-scree vegetation. The same area with relatively developed soils includes communities with *Arbutus andrachne*, apart from this these small areas are characterised by *Alnus barbata* forests in the bottoms of gorges, *Erica arborea* communities, etc.

The diversity of Colkhetian vegetation is best evident in the above-mentioned reserve, as well as in Kintrishi and Pskhu-Gumistra reserves (Figure 5).

Through the initiative of the World Wide Fund for Nature (WWF) a new system of protected areas is presently on the way to be established. This project proposes the creation of areas with several protection categories (national park, managed nature reserve, protected landscape, etc.); the establishment of Colkheti National Park is also planned. Both areas generally include Colkhetian evergreen vegetation. In 1997 a WWF Project Office in Georgia (Director - Dr. G. Sanadiradze) launched the planning activities for a National Park in south Colkheti (Ajara) using GIS methods. The overall goal of the establishment of this national park is the conservation of diversity of relic Colkhic vegetation.

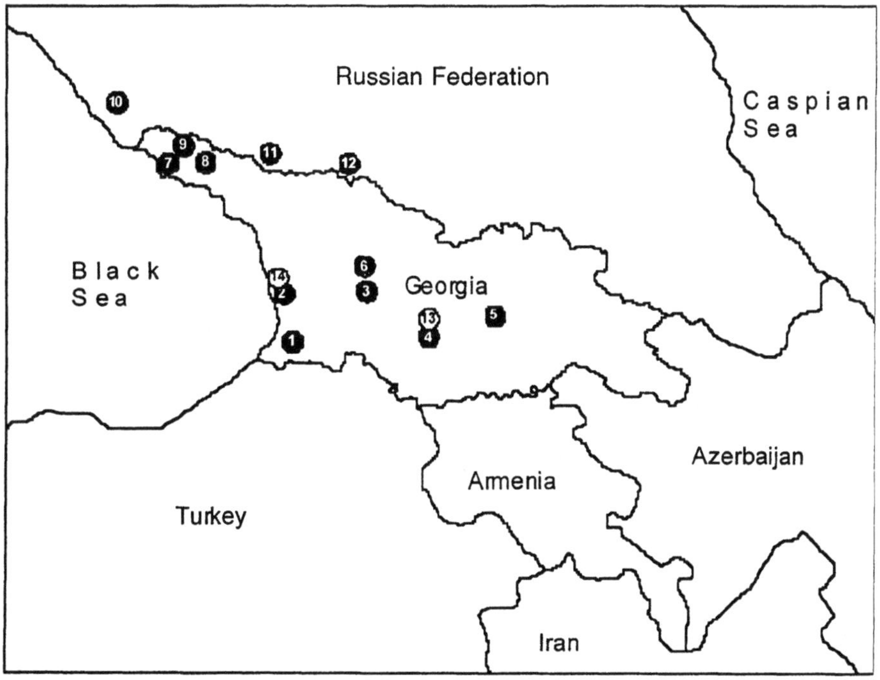

Figure 5: Location of nature reserves where Colchic evergreen vegetation is protected in the Caucasus.

EXISTING STRICT NATURE RESERVES	ALTITUDINAL LOCATION (m)	AREA (ha)	MAIN OBJECTS FOR PROTECTION
1. Kintrishi	450-2600	13893	Colkhic mountain forests, elfin woods
2,6. Sataplia-Colkheti	0-500	854	Colkheti lowland wetlands
3. Ajameti	100-200	4845	*Quercus imeretina* forests
4. Borjomi	600-2200	17948	Fir-spruce, pine, beech forests
5. Saguramo	500-1400	5359	Beech forests with some Colkhic relics
7. Pitsunda-Mjusera	0-100	3645	*Pinus pithyusa* , Colkhic foothill forests
8. Pskhu-Gumista	300-2850	40819	Beech, spruce-fir, subalpine forests
9. Ritsa	2000-2200	16289	Subalpine dark-coniferous forests
10. Caucasian Biosphere	500-3360	267277	North-west Caucasian mountain forests
11. Teberda	1500-4040	84996	North-west Caucasian mountain forests
12. Kabardino-Balkerian	1000-5000	74081	North-west Caucasian mountain forests
PLANNED NATIONAL PARKS			
13. Borjomi-Kharagauli	1000-2600	50400	Fir-spruce, pine, beech forests
14. Colkheti wetlands	0-20	15000	*Alnus barbata* forests, *Sphagnum* bogs

References

Bush E.A. & Bush N.A. (1936): *Vegetation of the eastern South Oseti.* Publishing House of Academy of Sciences of USSR, Moscov-Leningrad. (In Russian).

Dolukhanov A.G. (1953): Chestnut forests of Georgia. *Transactions of Tbilisi Botanical Institute* 15: 339-364. (In Russian).

Dolukhanov A.G. (1980): *Colkhidian underwood.* "Metsniereba", Tbilisi. (In Russian).

Dolukhanov A.G. (1989): *Forests of Georgia, Part 1.* "Metsniereba", Tbilisi. (In Russian).

Gagnidze R., Mtskhvetadze D., Mukbaniani M. & Chelidze D. (1985): Botanical geography and flora of Svaneti. *Transsactions of Institute of Botany of the Georgian Academy of Sciences* 30: 3-115 (In Georgian).

Golitsin S. (1939): Shceriani communuties in humid woody mountains of Ajara. *Bulletin of Vorohej University* 11: 13-31. (In Russian).

Iabrova-Kolakovsky V.S. (1977): *Adventitious flora of Abkhazia.* "Metsniereba", Tbilisi. (In Russian).

Kolakovsky A.A. (1961): *Plant world of Kolkhida.* Publishing House of Moscow University. (In Russian).

Kordzakhia M. (1961): *Climate of Georgia.* Publishing House of Academy of Sciences of Georgian SSR, Tbilisi. (In Georgian).

Maleev V.P. (1936): Oak forests of Abkhazia. *Abkhasia: geobotanical and forestry sceth.* Publishing House of Academy of Sciences of USSR, Moscov-Leningrad. (In Russian).

Manjavidze D.V. (1982): *Relic forests of Ajara.* "Metsniereba", Tbilisi. (In Russian).

Memiadze V.M. (1971): Adventitious flora of Kintrishi valley. *Bulletin of Batumi Botanical Garden* 16: 20-29. (In Georgian).

Povarnitsin V.A. (1936): Forest types of Abkhazia. *Abkhasia: geobotanical and forestry sceth.* Publishing House of Academy of Sciences of USSR, Moscov-Leningrad. (In Russian).

Sokhadze E.V. (1982): *Limestone and vegetation.* "Metsniereba", Tbilisi. (In Russian).

Recent shifts in vegetation boundaries of deciduous forests,
especially due to general global warming, pp. 199–216
edited by F. KLÖTZLI and G.-R. WALTHER

Arealogical review of Colchic evergreen broad-leaved mesophyllous dendroflora species

Revaz GAGNIDZE
Institute of Botany, Georgian Academy of Sciences
Tbilissi 380007, Georgia

Abstract

The botanico-geographical character of the Caucasian and especially specific evergreen broad-leaved woody species of the Colchic mesophyllous floristic complexes and their florogenetic connections are considered. The complexes are referred to as "Colchian mesophyllous dendroflora" in the Caucasian botanical literature. Phytocoenologically they are called "Colchian subprostrate evergreen underwood".

Introduction

The article deals with the character of the Caucasian and especially with areas of mesophyllous evergreen broad-leaved woody species of the Colchican floristic complexes and with their florogenetic connections. In the Caucasian botanical literature the above mentioned complex is known as Colchican mesophyllous dendroflora. Phytocoenologically it is called a Colchian subprostrate evergreen underwood.

From the botanico-geographical point of view the Colchis belongs to the Eastern Euxinian or Colchican Phytogeographic Province and expands in the territory from the north-western part of the Caucasus, especially from Tuapse - Novorosiysk to the north-eastern part of Anatolia (Maleev 1940, Kolakovsky 1960, Kharadze & Gagnidze 1970, Gagnidze 1974, 1996) the so-called "Floristic Diagonal of Anatolia", which crosses the Meret - Ordu Watershed Range (Davis 1971) (Figure 1).

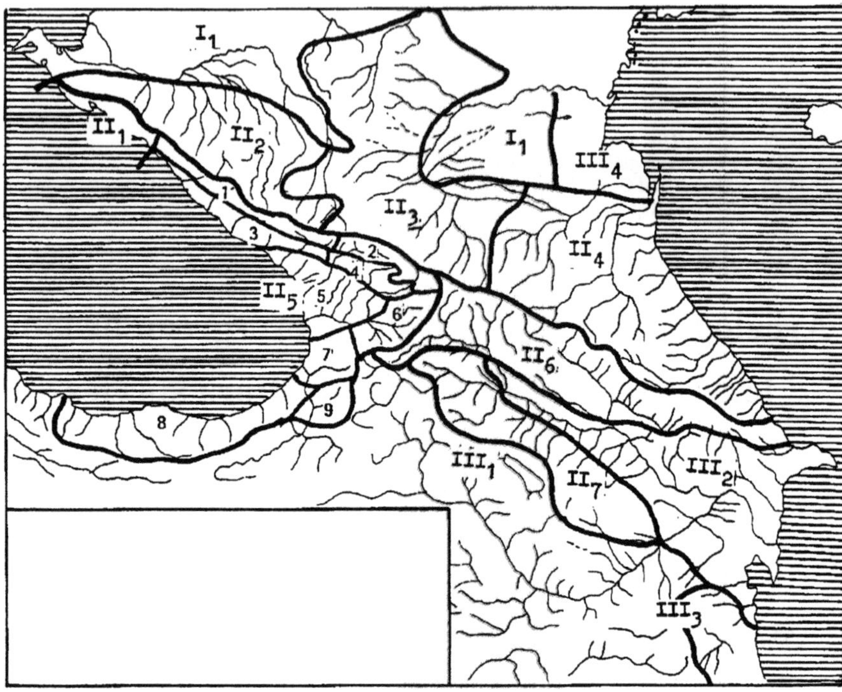

Figure 1: Phytogeographical division of the Caucasus (Gagnidze 1974, 1996). I. Eurasian (Euro-Siberian) Steppen Region: I_1 - Pontic Province; II. Mediterranean Region: Eumediterranean Subregion - II_1. Krym-Novorossiysk or Northern Euxinian Province. Submediterranean or Southeuropemountain-Caucasian Subregion: II_2 - Kubanian, II_3 - Elbrus Kazbekian, II_4 - Dagestanian, II_5 - Colchian or Eastern Euxinian [Districts: 1 - Abkhazian part of Greater Caucasus (Kavkasioni), 2 - Svaneti-Racha, part of Greater Caucasus (Kavkasioni), 3 - Cherkesia - Abkhazeti (Abkahazia) - Samegrelo (part of limestone's Western Transcaucasia), 4 - Racha-Lechkhumi (part of limestone's Western Transcaucasia), 5 - Kadori-Rioni, 6 - Guria-Imereti, 7 - Adjara-Shavsheti, 8 - Lazistan, 9 - Artwin], II_6 - East-Transcaucasian, II_7 - Minor Caucasian Provinces. III. Irano-Turanian Region: III_1 - Iran-South-Transcaucasian, III_2 - Kura-Araxian, III_3 - Hyrcanian, III_4 - Turanian or Aralo-Caspian Provinces

The positive balance of atmospheric moisture throughout the year is one of the main factors that enhance the evergreen mesophyllous dendroflora's species growing, development and spreading. The altitudinal and latitudinal distribution of the Colchian mesophyllous evergreen broad-leaved underwood depends on this factor as well. These conditions cause its relict character (Dolukhanov 1980).

In the western part of Eurasia (western Palaearctic) the largest and the richest refuge of the relicts appears in the Transcaucasus - Colchis, in the eastern and north-eastern part of the Black Sea Basin, especially in districts of Adjara-Guria and the Lazistan, and in the north-western part of the Great Caucasus, which is phytogeographically influenced by the Colchis (Maleev 1940), and situated on mountain ridges of the Minor Caucasus bordering on the mountains of Anterior Asia (Western Asia) (the plateau of Iran) from the north, as are the subcaspian slopes of Mt. Elburs in the northern part of the Iranian Phytogeographical Province.

In comparison with the Colchis, the Hyrcanian Refuges' climate is warm and dry. The surplus precipitation mostly appears on low hypsometric heights (1300-1500ma.s.l.). Because of that the evergreen dendroflora is concentrated on the low and middle part of the broad-leaved forests' belt, contrary to the Colchian mountains where the surplus precipitations occur in every belt (Figure 2) and are equally distributed throughout the year. So the two centres - Colchian and Hyrcanian - do differ from each other floristically.

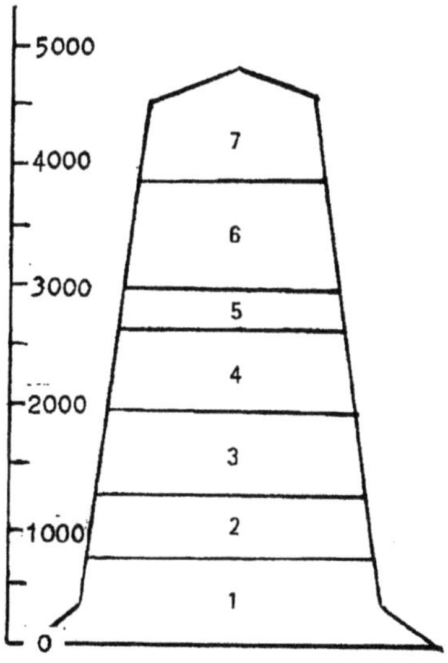

Figure 2: Colchian type of altitudinal zonality
Altitudinal belts of:
1 - mixed-leaved forest,
2 - deciduous broad-leaved forest,
3 - dark coniferous forest,
4 - subalpine,
5 - alpine,
6 - subnival,
7 - nival.

Although the floristic composition of the Colchian evergreen underwood is not rich, the complex is highly specific.

Certain species of relict genera such as *Rhododendron*, *Epigaea*, *Ruscus*, *Ilex*, *Daphne*, *Hedera*, *Laurocerasus* are phytogeographically and arealogically most interesting in the complexes of the Colchian broad-leaved evergreen mesophyllous dendroflora.

It should be mentioned that some species have only a Colchian area and others are characterised by a disjunctive spreading in the Colchis, in Mediterranean and Western Europe (Maleev 1941). However, outside of the Colchis the significance of these plants in the formations of broad-leaved and dark coniferous forests considerable decreases. Here they appear in a state of "regressive relicts" and have quite a different position in the scheme of general areas and in biocoenosis.

As a floristic peculiarity of the Colchian mesophyllous dendroflora; this complex is a characteristic phenomenon of the Colchis.
Another botanico-geographical peculiarity of the Colchian mesophyllous dendroflora is the wide amplitude of its latitudinal distribution although species may have stenochoric areas. Most of the species grow on the forest profile, at 2300-2400m a.s.l. In cold climate conditions, the lower limit of latitudinal distribution shifts to 2000-2200m a.s.l. At these elevations evergreen woody species have to endure temperatures below zero under a deep snow cover.
It should be noted that a certain complex of evergreen plants which consists of *Buxus colchicus*, *B. hyrcanus*, *Ruscus ponticus*, *R. hyrcanus*, *Danaë racemosa*, *Arbutus andrachne* etc. and species of other genera, which belong to the number of relicts, but ecologically don't come from the same florogenetic complex, is met in the Caucasus. These findings are confirmed by the following observations: the peculiarity of their latitudinal and geographical distribution; their growing and development; the effect of total temperature on mesophyllous species; the appearance of these species mainly in the lower mountain belt.
In winter their development does not depend on snow cover. Some species are less shade enduring than others and because of that they belong to habitats of other ecosystems. The ecology of these species corresponds with the Mediterranean conditions. This is the difference between the complexes of these two evergreen woody plants.
Botanico-geographical relations of the Colchic mesophyllous dendroflora with other centers have been detected. Using the arealogical analysis the concentration centres of species closely related to the Colchian ones have been established (Figure 3).

Species connected with centre I

One of the main genera of the Colchis complex is *Rhododendron*. In the Colchis it is represented with 5 species: *R. ponticum*, *R. ungernii*, *R. smirnowii*, *R. caucasicum*, *R. sochadze*.
R. ponticum is characterized by disjunctial spreading (Figure 4, 5). The main areas are the Colchis, the north-western part of the Caucasus and the seashore slopes of Anatolia. It spreads in the Balkanus as well, the second fragment - in the Lebanon and Antilebanon Mounts, where it runs the south-eastern boundary of the Mediterranean Phytogeographical Region (Gagnidze & Margalitadze 1981). The third fragment lies in the Iberian Peninsula. The race of the SW-Iberia, *R. baeticum* is regarded as an independent species or subspecies.

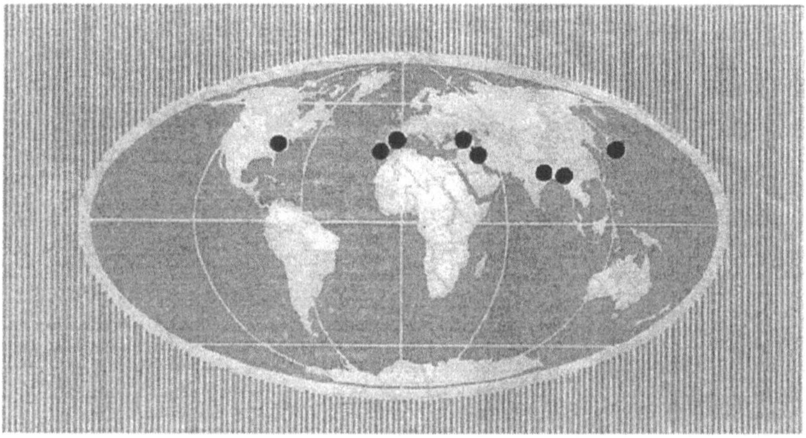

Figure 3: Concentration centres of Colchic close-related species

I The centre of West-Europe-Talysh
Nowadays these closely related species occur on disjunctal habitats in Mediterranean countries. Atlantic Europe, the Macaronesian Region on one hand and in the Talysh-subcaspian slopes of the Elburs in North Iran on the other hand.

II The centre of East and partly South Eastern Asia, including Himalaya.

III The centre of the Appalachian Mountains - North Atlantic America.

These closely related species are concentrated either in the Appalachians or in South-Eastern Asia including Himalaya

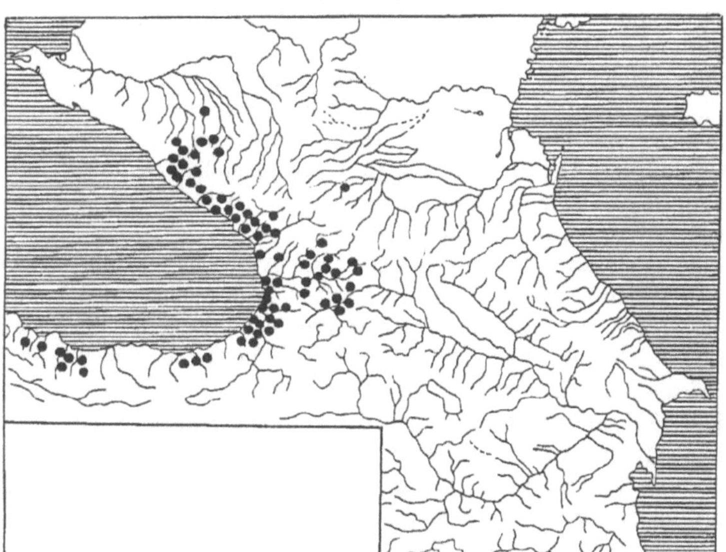

Figure 4: *Rhododendron ponticum* in Caucasus and Anatolia

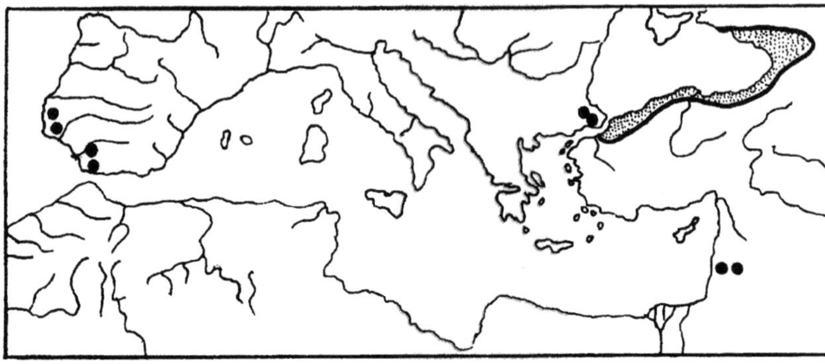

Figure 5: Total area of *Rhododendron ponticum*

In the tertiary period *R. ponticum* was widely spread in mountainous southern Europe. In the glacial period *R. ponticum* disappeared from the flora of southern Europe. The palaeobotanical material of the species was discovered in the Mendel-Riss interglacial deposits of Tyrol and the Alpes at the lakes Como, Iseo and Lugano (Shatalina 1964, Dolukhanov 1980).
The reason for the elimination of *R. ponticum* from the flora of southern Europe might be the increase of aridity in the Mediterranean area during summer.
The Caucasian-Mediterranean relationship group *R. ponticum* - *R. baeticum* (or subsp. *baeticum*, Stevenson 1947, 1972) - *R. algabriense* (or var. *algabriense*) (*R. baeticum* and *R. algabriense* are not accepted as independent species in the "Flora Europaea", 1972) is related with the species of North America - the Appalachian Mountains Phytogeographical Province and the Pacific Ocean coast: *R. maximum*, *R. catawbiense* (the Appalachian Mountains); *R. macrophyllum* (*R. californicum*) (Sect. *Ponticum* subsect. *Pontica*) (Chamberlain 1982). The allied species to *R. ponticum* live on rocky slopes, 50-1000m a.s.l. (*R. catawbiense*); Upland woods, 300-1700m a.s.l. (*R. maximum*) and forest margins, 150m a.s.l. (*R. macrophyllum*).
R. ponticum is closely related to the Caucasus-Anatolia subalpine species *R. caucasicum* which has been formed in habitats with mountainous conditions. On the calcareous ridges of the Colchian Province (Western Georgia and NE Anatolia, 1700-2400m a.s.l.) above the timberline, rarely found in *Fagus* forests *R. caucasicum* and *R. ponticum* create hybrids (*R. sochadze*) (Kharadze & Davlianidze 1967, Stevens 1978) (Figure 6). On the other hand *R. caucasicum* is morphologically and coenologically related to the East-Asian *R. aureum* (*R. chrysanthum*). (Central and Eastern Siberia, Kamchatka, Sakhalin, Kuriles, Korea, Sri Lanka, Japan, alpine slopes, 1500-2700ma.s.l.) (Chamberlain 1982; Sect. *Ponticum*, subsect. *Pontica*). The coenosis of both species are rich in arcto-alpine species (Dolukhanov 1980). However, *R. caucasicum* is sensitive to winter frosts. That is the reason why it spreads in places where the snow cover

is deep in winter and spring. *R. aureum* is more cryopilic (Aleksandrova 1975). It is difficult to tell in which stage of the evolution the divergation of their common ancestor took place.

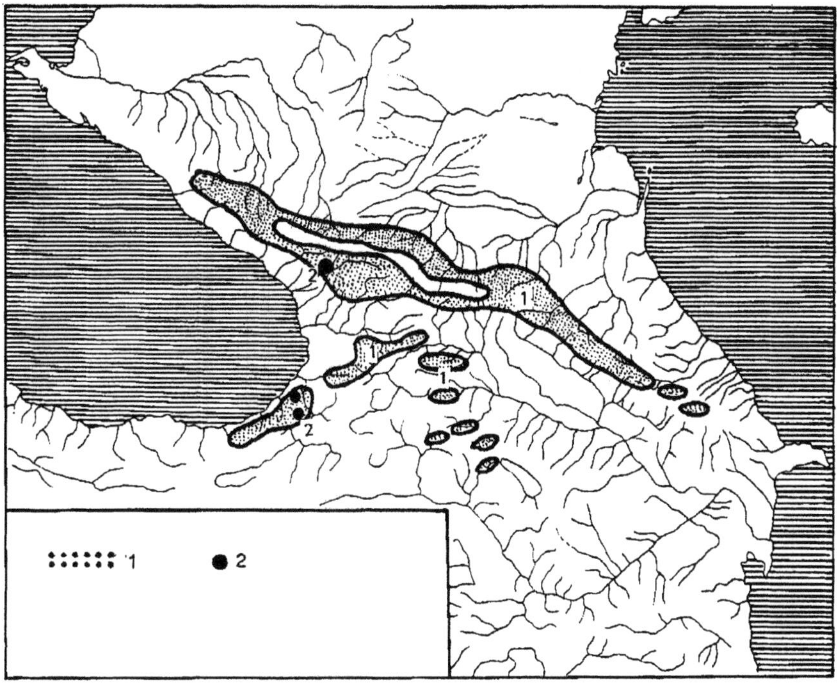

Figure 6: 1 - *Rhododendron caucasicum*, 2 - *R.* x *sochadze*

R. ungernii (Figure 7) and *R. smirnowii* (Figure 8) are considered to be rather peculiar, both, morphologically and florogenetically. These local Colchic-Lazistanian species (Grossheim 1967) are mutually related and are isolated in the system of genera. They are likely to be related with *R. arboreum* of the Himalaya and have remote ancestors in East-Asia (Sect. *Ponticum*, subsect. *Arborea*; Chamberlain 1982). They occur in the following habitats: in open or mixed forests (Klötzli 1988), on rocky slopes and in upland regions 1850-3650m a.s.l.; (a subsection of uncertain affinities though probably allied to subsection *Argyrophylla* (particularly *R. ririei*, open rocky slopes and endemic in China) (Chamberlain 1982:328) (Figure 12; scheme 1).
Laurocerasus officinalis establishing the Caucasian southern European group spreads not only in the Caucasus and in southern Europe but also in northern Anatolia, subcaspian Iran and the Balkans. In the territory of Europe it creates different populations. The closely related species *L. luzitanica* is distributed in the north-western part of the Pyrenees and Macaronesia. Close-related species are spread in south-eastern Asia as well as in equatorial and South Africa.

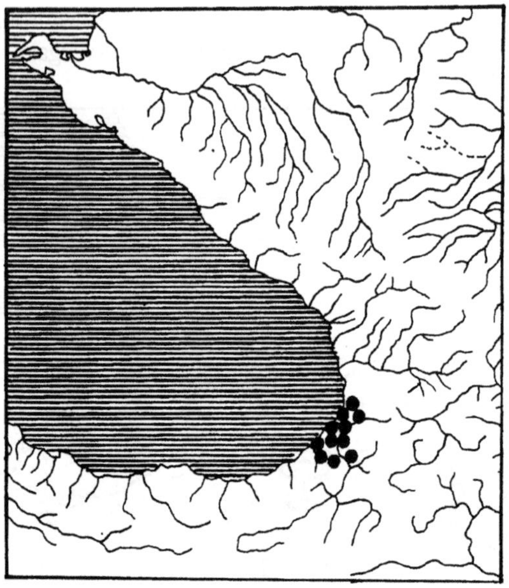

Figure 7: *Rhododendron ungernii*

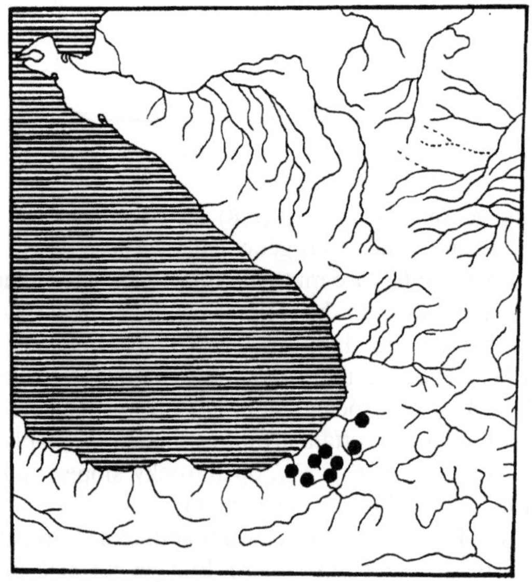

Figure 8: *Rhododendron smirnowii*

The linking of the Caucasus-Balkanian and Pyrenean populations are likely to have taken place before the Pliocene (Dolukhanov 1980).

The Europe-Mediterranean-Colchian species group is also established by *Ilex colchica*. European - *Ilex aquifolium* and Macaronesia-Andaluzia - southern Spain - *I. perado*. The Colchian *I. colchica* is vicarious to *I. spinosa* (*I. hyrcana*) of Talysh. The species of eastern Asia *I. hylonoma*, *I. centrochinensis*, *I. cornuta* (Sect. *Aquifolium*, ser. *Aquifolioides*) (Loesener 1901-1908)are again closely related to this group. Like *I. aquifolium* the group varies morphologically, especially the leaves. The Caucasian *I. colchica* forms the subsp. *imerethica* (Gagnidze 1965) (Figure 9, 13, 14, 15).

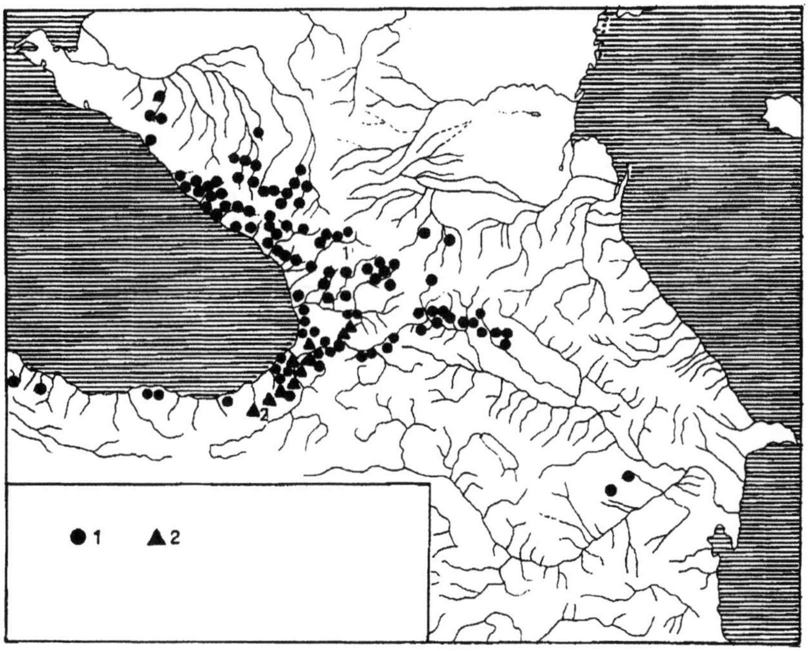

Figure 9: *Ilex colchica*: 1 - subsp. *colchica*, 2 - subsp. *imerethica*

Ruscus colchicus is a particular endemic of the Colchis (Figure 10). It is closely related to *R. hypophyllum* of Mediterranean Europe. The area of the latter is divided into numerous isolated fragments. Its most western fragment is situated in the northern part of Atlantic Morocco and in southern Spain. Both species are closely related to *R. hypoglossum* of Europe-Mediterraneis-Anatolia. The latter is more xeromorphic. The Mediterranean-Caucasian species group is establiched by *Daphne albowiana* and *Daphne pontica* (Figure 16).

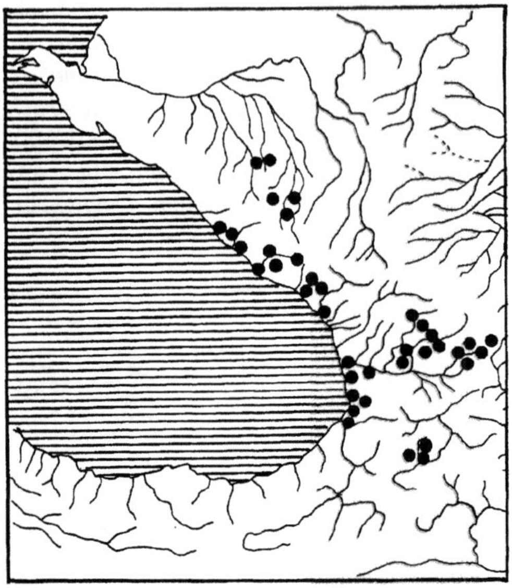

Figure 10: *Ruscus colchicus*

Species connected with the centre II

Hedera colchica and *H. pastuchowii* belong to the Caucasus–East Asia relative group. The former relates to the Chinese species *H. robusta*; *H. pastuchowii* to the species of South-Eastern Asia *H. tobberi* and *H. schensiensis*. *H. colchica* grows in the moist forests of the Colchis; *H. pastuchowii* in rather warmer areas like the Caspian Basin, the Great Caucasus and Talysh.

Species connected with the centre II and III

The model of the florogenetic group of East-Asia and North America is *Epigaea gaultherioides* - a "regressive relict" of the southern Colchis and Lazistan. The genus is oligotypic and represented by 3 species: *E. asiatica*, in Japan; *E. repens* in the eastern part of North America (Figure 11, 12) (from the Great Lakes and Newfundland up to Florida) Takhtajan 1941, 1978, 1987; Wood 1961; Hersey & van der Kloet 1976).
The southern Colchis-Lazistanian species are closer related to the Japanese ones. The North American genera *Oxydendrum* and the Himalayan *Gaultheria* are closely related to the genus *Epigaea* (Takhtajan 1941).
It is evident that the relations of the Caucasian flora with the centres of development of evergreen dendroflora are ancient. The determination of these

relations and the expansion of ancestral species are two of the most difficult and interesting issues of the Caucasian phytogeography.

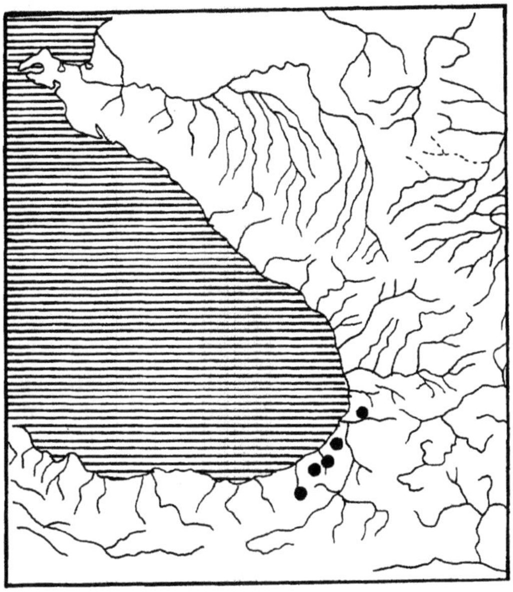

Figure 11: *Epigaea gaultherioides*

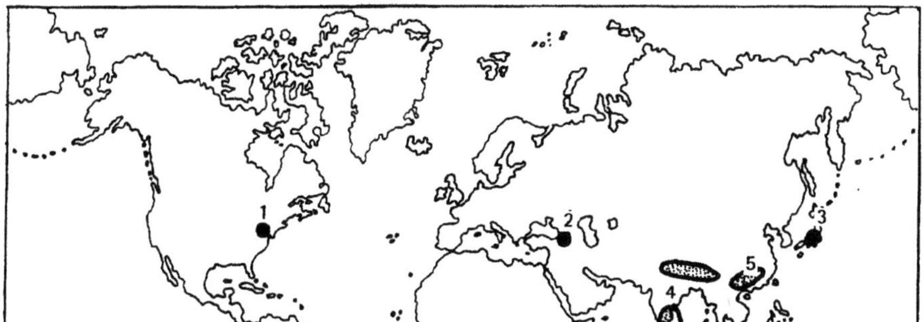

Figure 12: 1 - *Epigaea repens*, 2 - *E. gaultheroides*, 3 - *E. asiatica*, 4 - *Rhododendron arboreum*, 5 - *R. ririei*

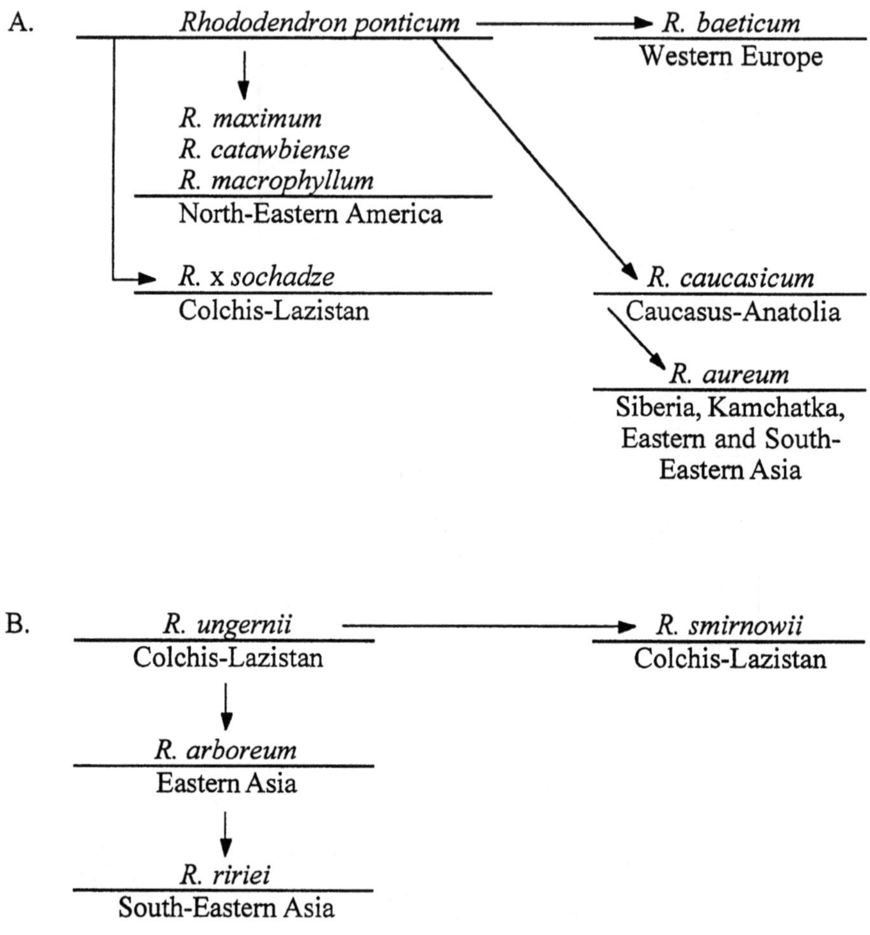

Scheme 1: The related connection of Caucasian *Rhododendrons*

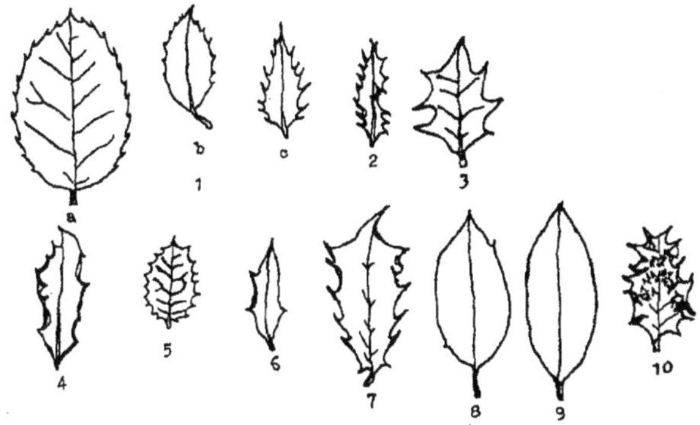

Figure 13: *Ilex aquifolium* 1. var. *ciliata*, forms: a. *platyphylla*, b. *nigricans*, c. *serrata*; 2. var. *crassifolica*; 3. var. *latispina*; 4. var. *calamistrata*; 5. var. *ovata*; 6. var. *carnosa*; 7. var. *monstrosa*; 8. var. *heterophylla*; 9. var. *laurifolia*; 10. var. *ferox*

Figure 14: *Ilex colchica* subsp. *colchica*

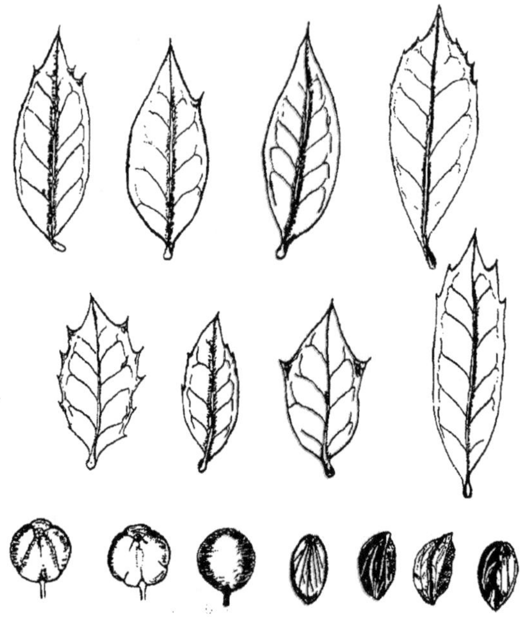

Figure 15: *Ilex colchica* subsp. *imerethica*

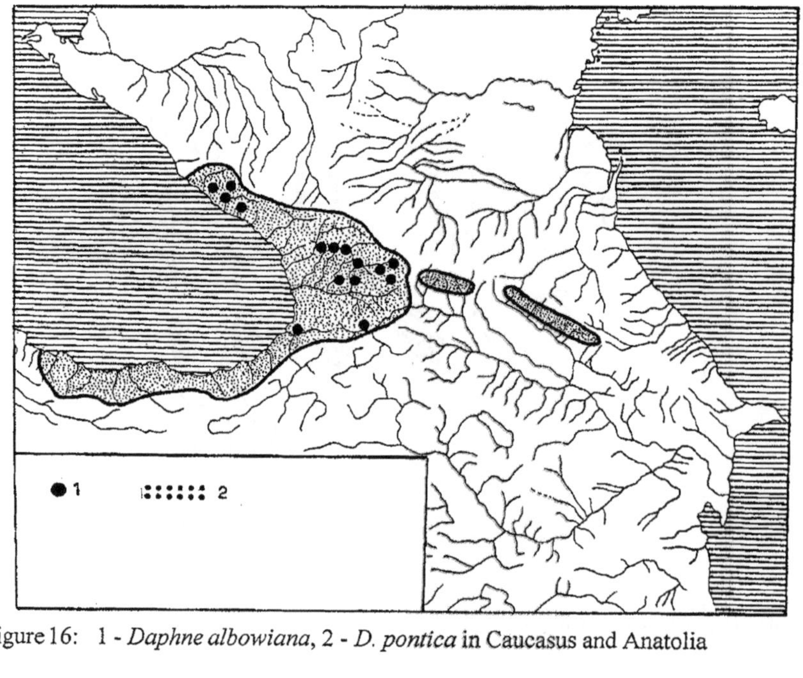

Figure 16: 1 - *Daphne albowiana*, 2 - *D. pontica* in Caucasus and Anatolia

Conclusions

On the basis of arealogical analysis the following conclusions may be drawn. One of the favourable factors of Colchian evergreen broad-leaved dendroflora's spreading is the positive balance of atmospheric moisture throughout the year. The altitudinal and latitudinal distribution of Colchian evergreen underwood depends on this factor. The same factor is significant in the altitudinal-latitudinal distribution of the closely related species components of the Colchian dendroflora.

The florogenetic groups, similar to the Colchian evergreen mesophyllous complex, in the Holarctis and the Ancient Mediterranean Kingdom are situated between 20° and 25° N latitude.

The formation of the complex of such a biocoenological flora is connected with broad-leaved and dark-coniferous highmountain ecosystems. The above mentioned complex and the synusia of semiprostrate underwood was formed in the colder and moister mountainous conditions of the Palaeogen.

The ancient mountain systems of eastern and south-eastern Asia might have been the centre of variety and diversity, where the species of this complex are concentrated (Dolukhanov 1980).

According to Dolukhanov the floristic complex of the ombro- and thermophilous evergreen woody plants was supposed to be formed in the Euxinian Basin and subcaspian mountain systems at the end of the Palaeogen. It was formed under conditions of a positive balance of atmospheric moisture with an annual mean temperature of 18-20°C at the altitude of more than 1500m a.s.l. The temperature optimum of the Atlantic period of the Holocene (Margalitadze 1995) illustrates the Colchian mesophyllous complexes of the late altitudinal expansion.

So the Colchian mesophyllous semiprostrate evergreen underwood and florogenetic elements' group were formed under the conditions of a positive balance throughout the year during the long-lasting evolution.

The species of the mesophyllous dendroflora in the Colchis have favourable conditions for the development and spreading. Outside the Colchian Phytogeographical Province in the eastern Transcaucasus, these species appear to be "regressive relicts" in florogenetic centres of southern Europe and the Mediterranean area. Though they are characterized by a high viability in the western Transcaucasia.

A synusia similar to the Colchian evergreen dendroflora also appears in dark coniferous and broad-leaved forest coenosis in the florogenetic centre of Atlantic North America, especially in the mountainous system of the Appalachian Mountains.

The strongest modern florogenetical centre where the synusia, similar to the Colchian evergreen dendroflora appeared, is the centre of eastern and south eastern Asian, from the Himalaya up to Korea and the Japanese Islands. These species are spread there at the northern latitude of 24-30°. It should be mentioned that between 20° and 25° lies the border of the Holarctic and Palaeotropical Kingdoms (Takhajan 1978).

The synusia of an evergreen dendroflora had been formed in this centre at the height of 2800-3800(4000)m a.s.l. at the highest belt of dark coniferous forest and sometimes in the subalpine belt (*Rhododendron arboreum* which is closely related to *R. ungernii* and *R. aureum*, which is itself closely related to *R. caucasicum* grows under these conditions) (Chamberlain 1982).

The atmospheric circulation of precipitaions in the centres of Asia and North America, unlike the centres of the Transcaucasus is influenced by the monsoon. The elements of the evergreen dendroflora are related to the mountain ecosystems of a monsoon climate (Klötzli 1988).

It is interesting that the amplitude of the altitudinal distribution of evergreen woody plant species is limited in the above mentioned centres, although these centres appear to be under the influence of a monsoon climate.

On the basis of climate data analysis Dolukhanov (1980) assumes that the distribution of continental atmospheric masses in winter causes deep temperature depressions even on the coastline. The evergreen dendroflora of underwood in north-eastern America is generally concentrated between the northern latitude of 37-39° (in comparison with the Colchis at about 3-4° further to the south).

The climate analysis shows that to the North of 27° to 28° latitude, where the species of evergreen dendroflora are concentrated the climate is rather rough (in the lower montane belt) (Dolukhanov 1980). It is the same in Eastern Asia.

The reasons discussed will explain the modern geographical allocation of the Colchian-like evergreen mesophyllous dendroflora and subprostrate synusia. The conclusions may also be used for the restauration of the habitats of similar distributed complexes in the historical past.

References

Aleksandrova M.S. (1975): The *Rhododendron*'s of natural flora of SSSR. Nauka, Moscow (in Russian).

Chamberlain G.F. (1982): A revision of *Rhododendron*, II. Subgenus *Hymenanthes*. Notes from the Royal Botanic Garden Edinburgh 39, 2: 202-485

Davis P.H. (1971):Distribution patterns in Anatolia with particular reference to endemism. In: Plant life of South-West Asia (ed. P.H. Davis) Edinburgh, 15-27

Dolukhanov A.G. (1980): Colchian underwood. Metsniereba, Tbilisi (in Russian).

Gagnidze R. (1965): Critical review of Caucasian species of the genera *Ilex* L. Notulae Systematicae ac Geographicae Institutum Botanici Thbilissiensis 25: 32-45 (in Russian)

Gagnidze R. (1974): Botanical and geographical analysis of the floral coenotic complex tall harbaceous vegetaion of the Caucasus. Metsniereba, Tbilisi (in Russian).

Gagnidze R. (1996): Plants geography. Publishing of the Tbilisi University, Tbilisi (in Georgian)

Gagnidze R. & Margalitadze N. (1981): Concerning the South-Eastern boundary of the Mediterranean phytogeographical region. Bulletin of the Academy of Sciences of the Georgian SSR, 102, 1: 125-128

Grossheim A.A. (1967): Flora Kavkaza, 7. Nauka, Leningrad (in Russsian).

Hersey R.E. & van der Kloet S.P. (1976): Taxonomy and distribution of *Gaultheria* in the Caribbean. Canadian Journal of Botany, 54, 21: 2465-2472.

Kharadze A. & Davlianidze M. (1969): To investigate introgressive hybridisation in genera *Rhododendron* L. in the Caucasus . Notulae Systematicae ac Geographicae Institutum Botanici Thbilissiensis 27: 82-86 (in Russian)

Kharadze A. & Gagnidze R. (1970): Review of hemixerophilic endemic elements in the flora of the Novorossiysk subprovince of the Caucasus. Notulae Systematicae ac Geographicae Institutum Botanici Thbilissiensis 28: 56-82 (in Russian)

Klötzli F. (1988): On the global position of evergreen broad-leaved (non ombrophilous) forests in the subtropical and temperate zones. Veröff. Geobotanisches Institut ETH, Stiftung Rübel, Zürich, 98: 169-196.

Kolakovsky A.A. (1960): The flora and vegetation of Colchida. The Publishing House of Moscow University, Moscow (in Russian)

Loesener T. (1901-1908): Monographia Aquifoleacearum, 1. (1901), 2 (1908). Halle

Maleev V.P. (1940): The vegetation of Black Sea coasts in the Euxinean Province of Mediterranean, its origin and correlations. Trudi Botanicheskogo Instituta Akademii Nauk SSSR, Ser. Geobotanica, 3 (4), Leningrad (in Russian)

Maleev V.P. (1941): The Tertiary relicts in the flora of Western Caucasus and general etaps of history of its flora and vegetation. Materialy po Istorii Flori i Rastitelnosti 1: 61-144 (in Russian)

Margalitadze N. (1995): Holocene history of the vegetation of Georgia. Metsniereba, Tbilisi (in Russian)

Shatalina M.S. (1964): On the distribution of some *Rhododendron* in the Caucasus. Bulletin Mosckovskogo Obshestva Ispitatelei Pryrodi, Otdel Biologicheskyi 59, 1: 135-139 (in Russian).

Stevens P.F. (1978): *Rhododendron* L. in: Davies H.P. Flora of Turkey, 6, Edinburgh University Press, 90-94.

Stevenson J.B. (1947): The species of *Rhododendron* (ed. 2). London

Stevenson J.B. (1972): *Rhododendron* L. (incl. *Azalea* L. pro parte) in Flora Europaea, 3 (ed. by Tutin T.B. et al.). Cambridge University Press, 8-9.

Takhtajan A.L. (1941): Genera *Epigaea* in Caucasus. Notulae Systematicae ac Geographicae Institutum Botanici Thbilissiensis, 10: 29-35.

Takhtajan A.L. (1978): The floristic regions of the World, Nauka, Leningrad (in Russian).

Takhtajan A.L. (1987): Systema Magnoliphytorum. Nauka, Leningrad (in Russian).

Wood C.E. (1961): The genera of Ericaceae in the Southeastern United States. Journal Arnold Arboretum, 42: 10-80.

Recent shifts in vegetation boundaries of deciduous forests,
especially due to general global warming, pp. 217–236
edited by F. KLÖTZLI and G.-R. WALTHER

Vegetation and changing environment on Mt. Etna

Emilia POLI MARCHESE
Istituto di Biologia ed Ecologia Vegetale
95128 Catania, Italy

Abstract

Some changes in the vegetation in the Mt. Etna territory, caused by environmental changes, are described. As the Etna is an active volcano, the environmental factors arising from the volcanic activity play an extremely important role.

From the data given, which refer to various periods, it can be seen that the changes in the vegetation include the disappearance of some plant communities, the reduction of the distribution range of others and the appearance of new forms of vegetation in recently formed environments.

Introduction

In an active volcanic area, besides the environmental factors affecting all areas, there are also the factors arising from the volcanic activity which are often the most important ones. These factors are the cause of numerous changes: changes in the physical and biological characteristics and, indirectly, in the climate and microclimate.

Plants growing in the area are particularly affected by these changes, which operate together with other environmental factors such as climate and man's activity.

A lot of authors, particularly in the last two centuries (Recupero 1815; Ferrara 1818; Sartorius von Waltershausen 1880; Gemmellaro 1830; and many others), have written about the damage in plants and plant communities caused by volcanic eruptions. However, most of the above-mentioned studies mainly concentrate on the volcanological and most spectacular aspects of eruptions.

It was an awareness of these limitations combined with the great deal of knowledge that we have accumulated in more than thirty years of research on the Etna vegetation that led me to carry out the work dealt with here. The objective is to show, on the basis of both the research done and existing documents, the ways in which some aspects of the vegetation on Mt. Etna have been modified by environmental changes arising from the volcanic activity. Changes in the vegetation caused by variations in the climate and man's intervention will also be mentioned.

Mt. Etna

Mt. Etna is situated in the north-east of Sicily, at 15° latitude north and 35° 50' longitude east. It is a relatively recent volcano in that it was formed at the beginning of the Quaternary period. It is the highest volcano of Europe. It rises as a solitary peak in the heart of the Mediterranean region between the much lower mountain ranges to its north and the plain of Catania to its south, and facing the Mediterranean Sea to the east.
Reaching a height of 3370m a.s.l. and with a basal area of 1.500km^2,the volcanic structure rests on a layer of Pleistocene clays (Romano 1982).
The morphology of the Etna is characterized by the presence of more than 200 secondary craters formed during lateral eruptions, and by a large depression - an ancient caldera - on its eastern slope, called "Valle del Bove". Numerous lava flows flow down the slopes of the volcano, in various directions, as far as the sea.
The composition of the volcanic substrate is the result of lava flows of different ages. The vegetation becomes established much more rapidly on loose substrates and between small blocks of lava than on the more compact pahoehoe lava where plant colonization is much slower.
The extremely permeable nature of volcanic soil fosters the absence of flowing water and accentuate the xeric character of the Mediterranean region.
The climate in the Etna territory, like that found on other high mountains in the Mediterranean region is mediterranean. From ombrothermic diagrams (Figure 1), which were constructed following Bagnouls and Gaussen's (1957) methodology, it can be seen that along the southern coast, in Catania (65m), the climate is of the eu-mediterranean type (warm xerothermic with a xerothermic index of 140). Higher up, on the eastern slopes (595m), it becomes meso-mediterranean, while in still higher areas it is oroxerothermic (1880m)becoming cold xerothermic at around 2950m a.s.l. (Poli 1965).

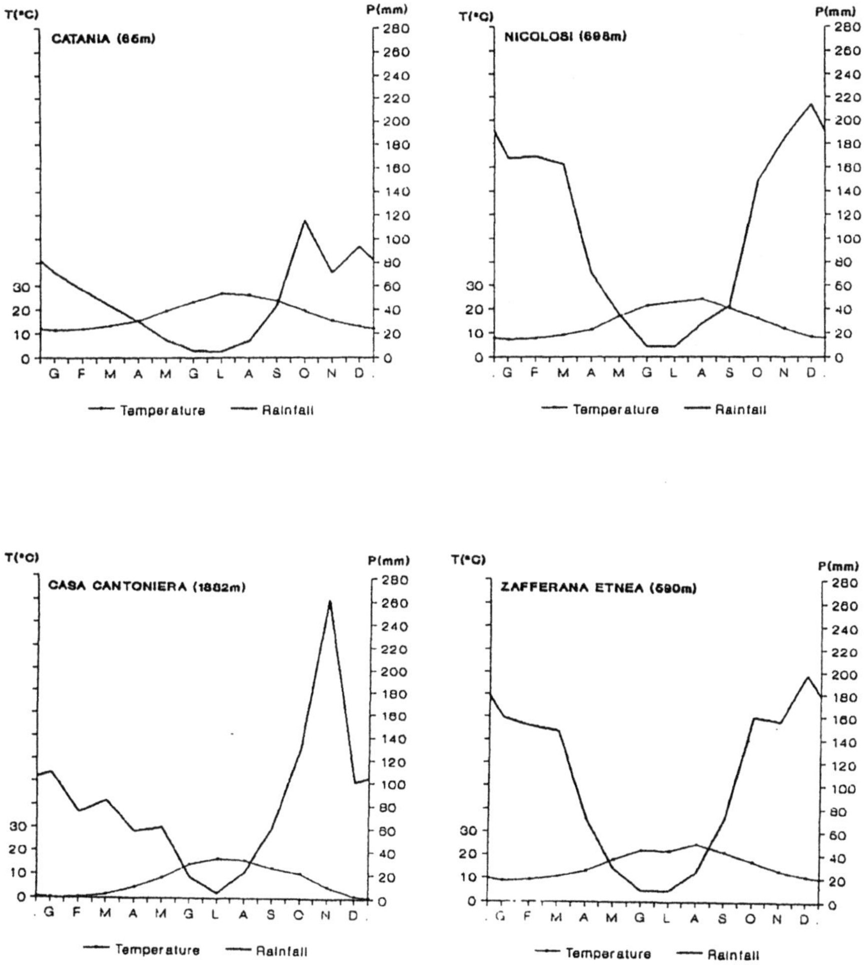

Figure 1: Ombrothermicdiagrams

The temperature variations in relation to the altitude are more or less regular on all slopes while the rainfall is heavier on the eastern slope because of its exposure to the sea. This is reflected in the vegetation.
The vegetation on Etna varies greatly. On the higher slopes there are vast areas of natural vegetation while lower down there are intense cultivations and increasing urbanization. At these lower altitudes the natural vegetation is limited to the edges of cultivated fields and to the outskirts of urban areas, to more recent lava flows and to substrates that cannot be used for agriculture. In the higher slopes the vegetation takes on a more individual identity with the presence of communities peculiar to the Etna territory.
On the basis of various data it is possible to propose the following altitudinal scheme (Figure 2) of the vegetation. The vertical distribution of vegetation reflects the essentially Mediterranean location of the volcano.

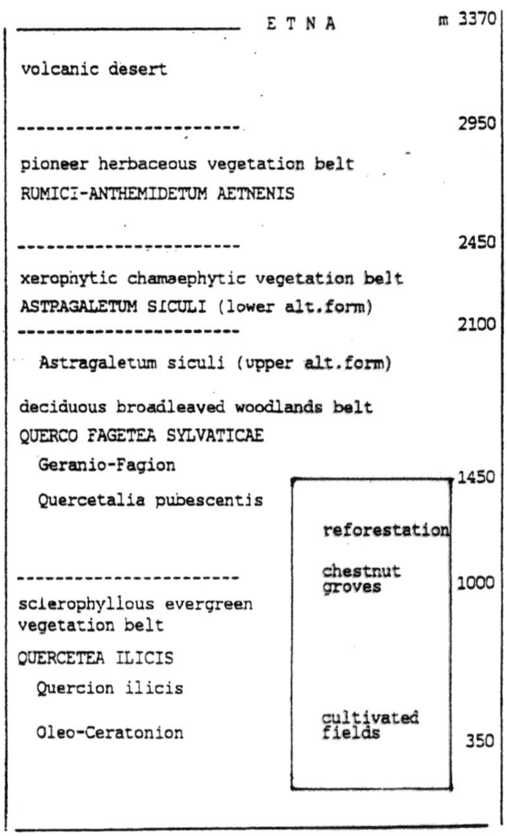

Figure 2: Vertical distribution of the vegetation

Methodology

This study is based on the knowledge acquired during long years of geobotanical research on the Mt. Etna territory. The vegetation was studied using the methodology of the Zurich-Montpellier school (Braun-Blanquet 1964).

Significant data were obtained by comparing cartographic documents, among which the vegetation maps produced by Pirola and Vecchio (1960), Poli (1963), Poli et al. (1983) were particularly useful. The geological map of the Etna (Romano et al. 1979) was of great help in providing information regarding some characteristics of the volcanic substrate. Some data were obtained by cartographic field work and with the help of photo-interpretation, while others were taken from the existing literature.

Environmental changes

Among the changes that have affected the Mt. Etna territory, those altering the plant cover are the result, as mentioned above, of factors arising from the volcanic activity, the climate and man's intervention.

The Etna is an extremely active volcano. Every eruption is characterized by explosive activity and effusion, during which an enormous quantity of material is ejected. It has been calculated (Patti, personal comm.) that in the course of 10 years (1977-1987) there have been more than 30 volcanic eruptions with the emission of an average of 540 million of cubic metres of material. This has had a great effect on the vegetation, which is constantly reduced. Some plant communities cover smaller and smaller areas, while others have completely disappeared. Extensive forests are replaced by sterile lavas and new morphologies: small volcanic cones, ridges and depressions. The continual changes in the forms of the relief and the substrates mean that the plants are constantly obliged to re-begin their slow, but continuous, colonization. Therefore new types of vegetation are found on the new lava substrates.

In some cases new environments such as fumaroles, cracks and cavities are formed. These provide a suitable environment for highly specialized forms of vegetation that previously could not live there.

The disappearance of vegetation, particularly of forests, in extensive areas not only is changing the vegetation, but also brings climatical changes: both the microclimate and the climate in general are affected. These changes further modify the vegetation, xerophilous types being priviledged. On the new lava surfaces the evapotranspiration by the forest vegetation is now lacking. It is well known that this water is significant. In some woodlands in central Europe (Ehrendorfer in Strasburger 1991) it is equivalent to 45% of the rain water. Since volcanic eruptions are frequent and each eruption affects a wide area, it is clear that the phenomenon has widespread consequences.

The climatic changes that affected the earth in ancient times, subsequent to the last ice age, also affected the vegetation on the Etna and its distribution on the slopes of the volcano. The most mesophilous vegetation types were affected the most. They were reduced in surface, with a change in their altitudinal limits (Poli et al. 1978).

It is well known that there are climatic variations depending on altitude as mentioned above. In every belt of vegetation (Figure 2), moreover, there are numerous other variations depending on the morphology and exposition of the slope and the nature of the substrate. There are also other climatic variations, in particular as regards the microclimate. These are due to factors such as the volcanic activity, as mentioned above, and to man's intervention. They are the cause of changes within the belts of vegetation and in some areas can bring about changes in altitudinal limits.

Since ancient times man has been responsible for modifications to the Etna territory which have had significant repercussions on the vegetation. As reported by Diodoro Siculo, vast forests were cut down even in Greek times to build the tyrants of Siracusa's fleet: other forests were cut down or burnt to provide pasture and arable land. There has been further deforestation since those times up to the present day.

All this has led to a reduction in the distribution of some types of vegetation and the disappearance of others. Moreover, the continuous use of the soil has led to the impoverishment and degradation of the floristic composition of various types of vegetation. In some areas there has been reforestation which has led to an increase in forest vegetation.

The overall effect of the above stated events has been a change in the vegetation and in its distribution on the slopes of the volcano.

Effects of environmental changes on vegetation

Due to the continuous environmental changes in the Etna territory, it is clearly impossible to deal fully with the countless changes which the Etna vegetation has undergone. Some examples will be given on the basis of the study carried out.

Woodland vegetation

The Etna woodlands are distributed in a fragmentary way on the slopes of the volcano at altitudes between a few hundred meters and 1900-2000metres above sea level.

On the lower slopes, up to a height of 1000-1100ma.s.l. there are Quercion ilicis Br.-Bl. (1931) 1936 communities. Various communities have been identified (Gentile 1969; Poli & Maugeri 1974; Poli et al. 1979 and 1981; Romano 1974). They include Querco-Teucrietum siculi Gentile 1969, principally distributed at higher altitudes on more mature soils; the *Quercus ilex* L. and *Fraxinus ornus*

L. community loosely belonging to Orno-Quercetum ilicis Horvatic 1958 and represented on the eastern slope of the volcano and the *Quercus ilex* and *Celtis tournefortii* Lam. community (Celtido-Quercetum ilicis Poli et al. 1981), represented on the south-western and western slopes, where the climate is driest.

In the next belt up, between 1000-1100 and 1400-1500 m a.s.l., there are communities of Quercetalia pubescentis Br.-Bl. (1931) 1932 in which it is possible to distinguish syntaxonomically similar communities, characterized by oaks of the group *Quercus robur* L. (*Q. virgiliana* (Ten.) Ten., *Q. dalechampii* Ten., *Q. congesta* Presl, *Q. pubescens* Willd.). In the higher stations of the eastern slopes, *Q. cerris* L. woodlands are well represented.

In some areas some highly impoverished woodland fragments (mixed forest) have been found (Poli et al. 1979). These can be identified as belonging to Orno-Ostryon Tomaz 1940. In the next belt higher up, between 1400 and 1800-1900m a.s.l., in the cooler sites there are fragments of beech woodland belonging to Aquifolio-Fagetum Gentile 1969 of Geranio-Fagion Gentile 1969 (Poli et al. 1979 and 1981).

In the same belt *Betula aetnensis* Raf. communities occur, often as open brushwoodland.

In the area between the deciduous oak and beech belts, laricio pine woodlands are common. On some lava surfaces these are of a pioneer character. The more mature woodlands tend to be formed of Quercetalia pubescentis lower down and Geranio-Fagion in the cooler stations at higher altitudes (Poli et al. 1981).

Regarding the changes that environmental factors and particularly volcanic eruptions have brought about in the woodland vegetation, a period of more than 170 years – from 1825 to today – is considered.

The first data on the distribution of the Etnean woodlands date back to Scuderi (1825). The same data are subsequently given by Philippi (1832) and Strobl (1880). These authors took only the woodlands located in the "woodland zone" ("nemoral zone") into account. The woodland area considered by Speranza (1960) includes the anthropic woodlands (chestnut groves).

The data are reported in Table 1, which include our data that have been obtained from the Etna vegetation map (Poli et al. 1983) and from our new unpublished cartographic data referring to 1996.

From the table it emerges that from 1825 until today, a period of more than 170 years, the total natural woodland area decreased by 12.409,94 hectares (67,46%). The beech forests decreased by 31,82%, the laricio pine forests by 65,81 % and the oak forests by 72,37 %.

Since 1993, the year in which the latest eruption took place, the total area of natural woodlands has been nearly 6.000 ha of which more than half (52%) is now occupied by oak woodlands, 1/3 (33%) by laricio pine woodlands and about 1/6 (15%) by beech woodlands.

SURFACE (hectares)					
	woodland zone	woodlands total area	beech	laricio pine	oaks
Scuderi (1825-27)	30.969,75	18.397,44	1.292,44	5.794,22	11.310,78
Philippi (1832)	30.968,88	18.396,98	1.292,24	5.794,22	11.310,52
Strobl (1880)			1.292,24	5.794,22	
Speranza (1960)		▫18.808,00			
Hofmann (1960)			724,00		
Poli et al. (1983)		6.487,50	* 906,25	° 2.093,75	◆ 3.487,50
Poli (1996, unpubl.)		5.987,50	* 881,25	° 1.981,25	◆ 3.125,00

Table 1: Surface occupied by Etnean natural woodlands (hectares)
▫ Chestnut groves included
* The areas where the beech is associated with other forms are included
° The areas with pioneer woods are not included
◆ The woodland areas located below the "woodland zone" are included

Bearing in mind that since 1825 until today there have been more than 60 volcanic eruptions (Romano & Sturiale 1982, Azzaro & Neri 1992) and that the lava flows have an average maximum length of 8 - 9 kilometres and a width varying from a few hundred metres to as much as a kilometer and a half (Patti, personal comm.), the reduction in the Etna forests does not seem at all surprising.

Among the eruptions that have caused most damage to the forests in the last one and a half centuries, it is worth remembering the eruptions of 1832 and 1843 on the western face and the eruptions of 1892, 1910 and 1983 on the southern face. Great damage to forests was caused by the lava flows of 1852, 1865, 1911, 1923, 1928, 1971 which affected the eastern, the northern and north-eastern slopes of the volcano.

It should be considered that in certain periods in particular in recent time, the amount of woodland area has also been modified by man's intervention. In areas where felling, fires and transformations of the territory have been carried out, the woodlands have clearly been reduced. On the other hand, reforestation has increased the woodland area. This increase is a phenomenon of the last forty years, following the interventions of the Regional Forestry Commission. However, notwithstanding the importance of man's role in changing the woodland area, it is by far inferior to that of the volcanic eruptions.

The vegetation of the high Mediterranean belt

On Etna the high Mediterranean belt is characterized by a discontinuous chamaephyte vegetation dominated by an endemic thorny species: *Astragalus siculus* Biv. Such vegetation (Astragaletum siculi Poli 1965), with its great cushions, makes lapilli and volcanic sands firm. It is present at altitudes up to 2400-2450m a.s.l.

It is situated, above all, on very old loose volcanic substrates. At the same elevations, on the rocky substrates of lava flows, but not very recent, there are various types of pioneer vegetation, belonging to Rumici-Anthemidetum aetnensis - unstable form, Poli 1965.

Above the climatic limit of Astragaletum siculi (2450m a.s.l.), the vegetation becomes more and more sporadic. Thus, the higher belt is characterized by a very poor pioneer community: Rumici-Anthemidetum aetnensis - stable form, belonging, like Astragaletum siculi, to the endemic alliance Rumici-Astragalion siculi Poli 1965 and to endemic order and class (Rumici-Astragaletalia and - Astragaletea siculi).

At the upper limit of this pioneer vegetation, which goes up to an altitude of 2900-2950m a.s.l., only four species manage to adapt to the difficult environmental conditions (*Hypochoeris robertia* Fiori and the endemic *Anthemis aetnensis* Schouw, *Senecio aetnensis* Jan, *Rumex aetnensis* Presl). These can grow at a height of up to 3000-3050m a.s.l. and are continually subjected to the destructive action of the volcano. Above this level the survival of any plant is impossible: it is the area of the terminal craters, called "volcanic desert" (Poli 1970).

In the high Mediterranean belt, the environmental changes that have taken place in recent times have mainly been caused by the volcanic activity. Only a few changes have come about because of man's activity.

From the existing documents (Poli 1963; Poli et al. 1983) and from recent unpublished cartographic data, it is possible to identify the changes in the plant cover caused by the volcano in an altitudinal belt where the vegetation is more characteristically that of Etna, both because of the floristic composition (there are various endemic species) and the physiognomy it assumes.

Table 2, where the available data are reported, shows the following:

In the period between 1963 and 1996, the vegetation underwent various changes. The volcanic activity, mainly confined to the summit zone, gave rise to frequent lava streams.

There were many lava flows between 1963 and 1971 affecting mainly the southern, northern and north-western slopes. In the subsequent period, extensive lava flows ploughed the high Etna zone, destroying both areas covered by Rumici-Anthemidetum aetnensis and areas where Astragaletum siculi grew. The most damage was caused by the lava flow of 1983 and 1985 on the southern face. The last lava flow, that of 1991-1993, affected a vast area in the eastern face within the Valle del Bove.

VEGETATION	SURFACE (hectares)	
	Poli 1963	Poli (unpubl.) 1996
Rumici-Anthemidetum aetnensis (*)	6.175,00	5.029,44
Astragaletum siculi	2.799,07	2.577,07

Table 2: Surfaces occupied by vegetation (Rumici-Astragalion siculi) on the highest belt * both forms: stable and unstable

The effect of this activity was: a reduction in the area occupied by the endemic vegetation (Rumici-Anthemidetum aetnensis and Astragaletum siculi) and a contemporary increase in the sterile area adjoining the volcanic desert. In other words the desert extended its territory downwards (Table 2).
Another community found exclusively on Etna was also principally destroyed by this lava flow: the *Hieracium crinitum-Luzula sieberi* community (Poli 1965). It was at the base of the southern wall of the Valle del Bove.
As well as destroying certain communities, some lava flows also created new environments: fumaroles, near the erupting craters. In the 1991-93 lava flow these were quickly colonized by primitive organisms. Only two years after the eruption the first colonies of Cyanobacteria characterized by a thermophilous species of the *Chlorogleopsis* genus (Grilli et al. 1995) were already established where the air temperature was about 60-65°C and besides these there where colonies characterized by thermophilous moss species. In both cases it is a question of new forms of vegetation that had previously not been found on Etna.
To these changes caused by the volcanic activity, those arising from anthropic activity must be added. The latter are much more limited and difficult to quantify but as important as the prior. They include the work involved in the construction of embankments to deviate the 1983 lava flow, which led to the destruction of vegetation and the movement of large quantities of earth and various other interventions.

The vegetation of the basal belt

On the lowest slopes of the volcano the main cause of vegetation changes is man. Since ancient times man has been responsible for the disappearance of the original forests and the hypothetical maquis-forest of Oleo-Ceratonion, probably represented by a "Lentiscetum". The intensification of anthropic activity and the consequent climatic changes towards greater environmental xericity, have prevented the continuation of even a fragment of any original maquis-forest. Today in the climax belt under consideration the most evolved type of vegetation would seem to be represented (Maugeri et al. 1979) by the *Euphorbia dendroides* L. maquis (Euphorbietum dendroidis). In many areas

this maquis and other strips of natural vegetation, mainly limited to the old lava flows, have been destroyed by man's intervention.
An example of this is the area occupied by the lava flow of 1381, which affected the lowest zone of the volcano: from about 400m a.s.l. to sea level.
About 25 years ago (Poli & Grillo 1974-75) the types of vegetation listed in Table 3 grew on the lava flow.

	VEGETATION	year 1996
1	*Halimione portulacoides-Suaeda fruticosa* comm. (Salicornietea)	AA
2	*Sedum coeruleum-Arenaria leptoclados* comm. (Tuberarietea guttatae)	R
3	*Cymbopogon hirtus-Stipa capensis* comm. (Tuberarietea guttatae)	R
4	*Crithmum maritimum* comm. (Brometalia rubenti-tectori)	AR
5	*Galactites tomentosa* comm. (Brometalia rubenti-tectori)	R
6	*Trifolium nigrescens-Ornithopus compressus* comm. (Brometalia rubenti-tectori)	AA
7	*Euphorbia dendroides- Calicotome infesta* comm. (Quercetalia ilicis)	* R
8	*Ceratonia siliqua-Olea europaea* v. *sylvestris* comm. (Quercetalia ilicis)	AA
9	*Olea europaea* v. *sylvestris* comm. (Quercetalia ilicis)	AA
10	*Quercus pubescens* s.l.-P*istacia terebinthus* comm. (Quercetalia ilicis)	AR
11	*Quercus ilex-Pistacia terebinthus* comm. (Quercetalia ilicis)	AR

Table 3: Plant communities on lava flow of 1381 (from Poli & Grillo 1974-75). AA: no longer to be found; A: rarely found; R: floristically impoverished, reduced in area and frequency; * disappeared from the floristic composition of *Pistacia lentiscus*

From recent observations (year 1996) it appears that due to the intense urbanization these communities have disappeared from more than 60% of the area they previously occupied. Of particular importance is the fact that the following plant communities are no longer to be found (Table 3): the herbaceous communities of the lava surfaces near the sea (1); the nitrophilous vegetation (6); the *Olea europaea* ssp. *sylvestris* maquis (8 and 9).
The halophilous *Crithmum maritimum* community (4) and the more mature vegetation types: the *Quercetalia ilicis* communities (10 and 11) have almost completely disappeared.

The floristic composition of the following vegetation is impoverished, much less frequent and covering smaller areas: the transient vegetation made up of dwarf therophytes (2); the examples of "Mediterranean steppe" (3) and some of nitrophilous communities (5); examples of maquis (7).
Some species have no longer been found in the whole area. Of these it is worth mentioning: *Cheilanthes vellea* (Aiton) F.V.Muell., a very rare plant in Sicily with only two sites on Etna, one of which was on the lava flow of 1381 (Poli & Grillo 1973); *Pistacia lentiscus* L., is another important element in the Mediterranean maquis, it is now in the process of disappearing from the entire Etna territory. On the other hand, the new adventitious species, *Nicotiana glauca* Graham, noted on this lava flow in 1973 (Poli & Grillo 1973), is now much more widespread in the coastal area.

The vegetation of Val Calanna

Val Calanna, affected by the latest eruption of Etna in 1991-93, is an example of a single morphological unit recently changed by a lava flow.
This valley, situated on the eastern face of the volcano has an unusual form - almost like an amphitheatre - having originated from the subsiding of a caldera (Romano 1982). It reaches an altitude of 1380m a.s.l., at the edge of the Valle del Bove, and drops to 980m a.s.l. in the valley floor. It stretches from west to east and has very steep walls on the orographic right: on the orographic left there are more gentle slopes. The valley floor, the slopes of the western wall and those of the orographic left have been modified by successive lava flows of 1792, 1819, 1852, 1950-51, 1978 and the most recent one in 1991-93 which are either partially or totally superimposed. The areas in the valley floor which have been spared by the lava, and whose soil is more mature have, for some time, been used for agriculture.
From a study carried out in 1960 by Pirola and Vecchio, which includes a map of the vegetation, it appears that at that time the types of vegetation listed in Table 4 were present. Their location in the valley and surface area is given for 1960 and 1996.
Table 4 shows that some plant communities are only partially present in the valley, while others still occurr in the same area, depending on their location in the valley. The 1991-93 lava flow, which is responsible for these changes, modified the valley floor, the slopes of the western wall, the slopes of the orographic left and the basal part of the wall on the orographic right. This can be seen in figure 3, in which the limits of this lava flow are shown.
As well as changing the distribution of the plant communities indicated in Table 4, the lava flow destroyed some sites of significant arboreal elements, such as *Ilex aquifolium* L. and *Erica arborea* L., species that are still present in few other sites in the Etna territory.

VEGETATION	location	surface (ha) in 1960	surface (ha) in 1996
Fagus sylvatica comm. (Geranio-Fagion)	1	16,0	16,0
Populus tremula comm. (Geranio-Fagion)	1	2,4	2,4
Quercus ilex comm. (Quercion ilicis)	1,2	23,0	21,0
Castanea sativa comm. (Geranio-Fagion / Quercion ilicis)	1,2,6	10,5	1,0
Genista aetnensis comm.	1,4,5,6	76,0	44,0
Ailanthus altissima comm.	6	0,5	0,5
pioneer vegetation on the lava flows	4,5,6	20,5	1,3
new pioneer vegetation on the lava flows		absent	23,50
cultivated fields	6	11,0	2,0
uncolonized areas	1,2,3,4,5,6	62,0	163,5

Table 4: Vegetation in Val Calanna in 1960 (from Pirola & Vecchio 1960).
1: wall on orographic right; 2: orographic left, SE face of Mt. Calanna; 3: orographic left, East face Mt. Calanna; 4: slopes on orographic left; 5: slopes on western wall; 6: valley floor.

The modifications in the vegetation in the valley caused by the lava flow have been accentuated by man's intervention. The construction of a barrier at the exit of the valley during the 1991-93 eruption caused the magma level in the valley to rise. This brought about the destruction of some woodland communities located in the basal part of the wall on the orographic right and on the orographic left, therefore including the relief of Mt. Calanna.

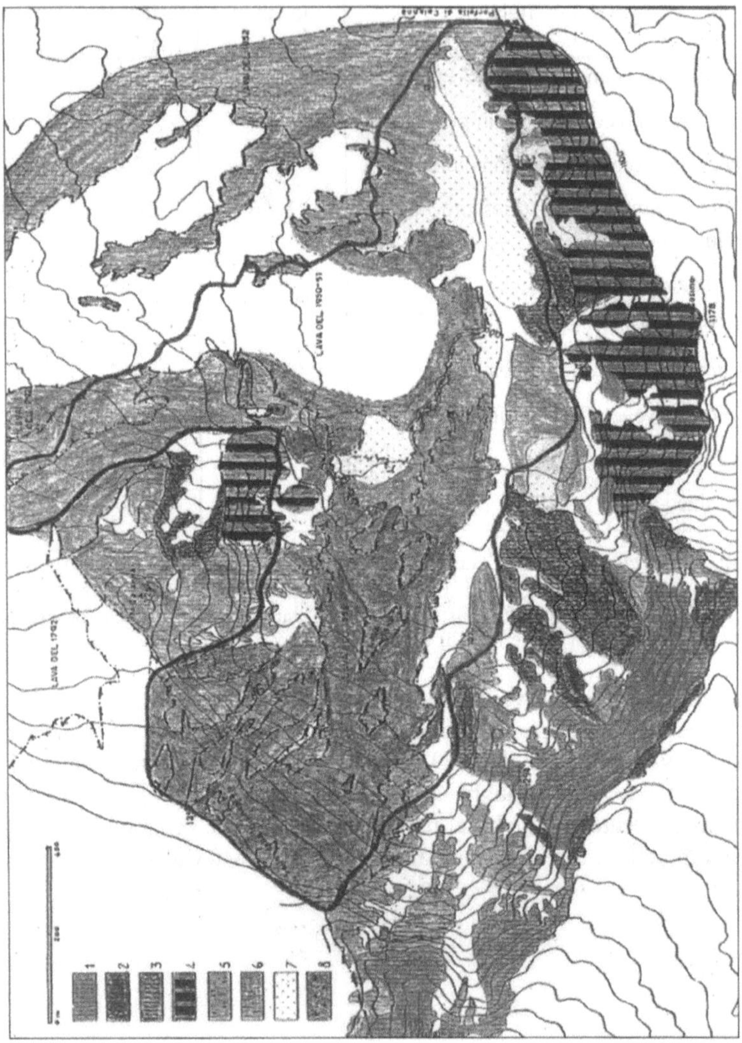

Figure 3: Vegetation map of Calanna valley on Mt. Etna (from Pirola & Vecchio 1961) 1: *Fagus sylvatica* community, 2: *Populus tremula* community, 3: *Castanea sativa* community, 4: *Quercus ilex* community, 5: pioneer vegetation on lava flows, 6: *Genista aetnensis* community, 7: cultivated fields, 8: *Ailanthus altissima* community.
Within the black line: area occupied by the 1991-93 lava flow.

Conclusion

From the results stated above it is clear that the vegetation on Etna is subject to continuous variations.
Both natural and anthropic factors affect the volcano, the different effects being superimposed on one another leading to one overall result: the Etna vegetation, with its mutability and precariousness.
Man has brought about great transformations; the climate has caused variations in the vegetation according to the altitude and with changes in time and the volcano has caused large scale destruction, with the subsequent creation of new environments and substrates.
Very significant species have disappeared from the floristic composition of some plant communities.
The areas in which various plant communities are distributed have been reduced at all altitudinal levels. Some plant communities found exclusively on Etna are only to be found in ever diminishing sites. It is sufficient to consider that in more than 170 years the natural forests have been reduced by 67,46%.
These are serious losses: in the woodland vegetation there are plant communities – the beech woodlands (Geranio-Fagion) – that are living in extremely precarious conditions. On Etna they reach their southern-most location in Europe: they can be considered relicts of a period when the climate was more favourable for them. Where they are destroyed they are unlikely to become established again.
Another serious loss is that concerning the vegetation of the high Mediterranean belt – the Rumici-Astragalion communities. As this is endemic vegetation, it can never be re-established.
The increase in the sterile surfaces is considerable. They give, however, new environments and new possibilities of plant life. The first stages of new plant communities begin to appear in recently formed fumarole environments. A vast range of pioneer communities becomes established in the great new spaces. They are gradually replaced by even more complex forms of vegetation until these are destroyed by new eruptions and new lava flows, after which plant life begins to appear again.
This continuous alternation between destruction and periods of reconstruction with the consequent rejuvenation of the relief, and of the plant cover is the main characteristic of the Etna territory. Here nothing lasts, everything is changeable. This means that rigorous measures must be taken to safeguard the plant life. The institution of the Etna Natural Park does not guarantee these sufficiently. The Park's management criteria established claim that the most suitable and practical ways of protecting the plant life must be found. The Park's land plan, which is currently being drawn up, will give directions regarding this matter.

Acknowledgement

I would like to thank Dr. R. Galesi of the Institute of Biology and Plant Ecology of Catania University and the geologist Dr. G. Patti for having calculated the areas given in the tables for the various types of vegetation.

References

Agnello A. (1861): Tavole prontuarie officinali della reciproca riduzione di misure pesi e monete del sistema metrico decimale eddel sistema metrico legale antico di Sicilia. Palermo.

Azzaro R. & Neri M. (1992): The 1971-1991eruptive activity of Mt. Etna. (First steps to relational data-base realization). C.N.R. IIV, Open file Report, 3/92, 1-10.

Bagnouls F. & Gaussen H. (1953): Saison sèche et indice xérothermique. Documents pour les cartes des productions végétales. 3, 1, 8.

Braun-Blanquet J. (1951): Pflanzensoziologische Einheiten und ihre Klassifizierung. Vegetatio 3, 1-2 (1950): 126-1333.

Braun-Blanquet J. (1964): Pflanzensoziologie. Wien.

Braun-Blanquet J., Roussine N. & Nègre R. (1952): Les groupements végétaux de la France méditerranéenne. Montpellier.

Ehrendorfer F. in Strasburger E. (1991): Lehrbuch der Botanik für Hochschulen. Stuttgart, Jena, New York

Ferrara F. (1818): Descrizione dell'Etna. Palermo

Gemmellaro C. (1830): Sulla vegetazione di alcune piante a varie altezze dal cono dell'Etna. Con prospetto topografico. Atti Accademia Gioenia di Scienze Naturali Catania, 4.

Gentile S. (1969): Remarques sur les chenaies d'yeuse de l'Apennin mèridional et de la Sicile. Vegetatio, 17, 1/6: 214-231.

Huepfer P. (1895): Die Regionen am Aetna. Wissenschaftliche Veröffentlichungen des Vereins fuer Erdkunde zu Leipzig, 2: 316-322.

Maugeri G., Lo Giudice R., Privitera M. & Tinè R. (1979): La vegetazione ad Euphorbia dendroides L. sull'Etna. Bollettino Accademia Gioenia di Scienze Naturali Catania, se. 4, 13, 10: 115-126.

Philippi R. (1832): Ueber die Vegetation am Aetna. Linnaea 7: 727-764.

Pirola A. & Vecchio S. (1960): Osservazioni sulla vegetazione della valle di Calanna (Etna). Bollettino Istituto Botanico Università Catania 2, 2: 131-142.

Poli E. (1963): Carta della vegetazione altomontana dell'Etna. C.N.R., Roma.

Poli E. (1965): La vegetazione altomontana dell' Etna. Flora et Vegetatio Italica, mem. 5. Sondrio.

Poli E. (1970): Vegetationsgrenzen in Vulkangebieten. Archivio Botanico e Biogeografico Italiano, 46, se. 4, 15, 1/2: 1-24.

Poli E. (1979): Les ètages de vègetation de l'Etna et du Teyde. C.R. Societè de Biogéographie, 483: 111-126.

Poli E. & Grillo M. (1973): Su alcune specie della flora etnea. Bollettino Accademia Gioenia Scienze Naturali Catania, se. 4, 11, 7/8: 1-9.

Poli E. & Grillo M. (1974-1975): La colonizzazione vegetale della colata lavica etnea del 1381. Atti Istituto Botanico e Laboratorio Crittogamico Università Pavia, se. 6, 10: 127-185

Poli E., Lo Giudice R. & Ferlito N. (1979): La vegetazione della Valle San Giacomo (Etna). Atti Accademia Gioenia Scienze Naturali Catania, se. 7, 10: 253-319.

Poli E. & Maugeri G. (1974): I boschi di Leccio del versante nord-occidentale dell'Etna. Bollettino Accademia Gioenia Scienze Naturali Catania, se 4, 12, 5/6: 741-759.

Poli E., Maugeri G. & Ronsisvalle G. (1981): Note illustrative della carta della vegetazione dell'Etna. C.N.R. AQ/1/131, Roma.

Poli E., Maugeri G. & Ronsisvalle G. et al. (1983): Carta della vegetazione dell'Etna. C.N.R. Roma

Recupero G. (1815): Storia naturale e generale dell'Etna. Catania

Romano E. (1993): La vegetazione naturale del piano basale dell'Etna. Tesi di Dottorato, Università di Catania.

Romano R. (1979): Presentazione della nuova carta geologica dell' Etna. Evoluzione del vulcanismo etneo. Atti seminario Sottoprogetto "Sorveglianza dei Vulcani attivi e rischio Vulcanico". P.F. Geodinamica, Catania.

Romano R. & Sturiale C. (1982): The Historical Eruptions of Mt. Etna (Volcanological data). Memorie Società Geologica Italiana, 23: 75/97.

Sartorius von Waltershausen (1880): Der Aetna. Leipzig.

Scuderi S. (1825-27): Trattato dei Boschi dell'Etna. Atti Accademia Gioenia Scienze Naturali Catania, 1, 2, 3.

Speranza F. (1960): Dei limiti altimetrici della vegetazione sull'Etna, Catania.

Strobl G. (1880): Flora des Aetna. Oesterreich. Botanische Zeitschrift.

Plenary discussion

Question: G. CARRARO

The process that you described is like an "oromediterranisation", so you are loosing *Fagus* forest in which you have also a part of the tertiary relicts as *Hedera*, *Ilex* and so on.

Answer: E. POLI

I think *Ilex aquifolium* was widespread in tertiary time, but now in Sicily it isn't because the climate is drier. *Ilex aquifolium* now has on Etna very rare refugies in few areas. And in our time it was in *Fagus* forests. *Hedera helix* principally is in *Quercus ilex* forests and in some areas in *Quercus pubescens* forests, but it isn't in *Fagus* forests.

Statement: A. BORHIDI

My impression was a thing to the vegetation belts. The situation is a little bit more similar to that of east Balkanian vegetation and Crimean vegetation where upon the mediterranean belt then comes the *Quercus pubescens* belt with the *Pinus pallasiana* belt which is practically substituted by the *Pinus laricio* belt here in the Etna and then comes the Fagion belt and this Fagion belt is too mountainous for having evergreen elements because it reaches practically to the subalpine region which forms the timberline. So it is more or less the same situation that is in the east Balkanian where practically only in the lower Fagion belt occur evergreen elements. But here this lower Fagion belt is substituted by the *Quercus pubescens* and *Pinus laricio* belt. For this reason this Fagion belt is different, not so evergreen influenced as it is for instance in the more atlantic Fagion belt. The other thing I would like to mention is that volcanic activity could influence very much the climate as well. For instance this is a very good example that during the Pleistocene the Mexican volcanos were active and for that reason in the Mexican highlands there were not glaciations. The effect is an advantageous situation to the migration of the neotropical elements northwards although it was in the Pleistocene where from the northern part of America came down the arctotertiar. And this was the situation where practically the two kinds of flora introgrediates. And it was practically conditioned by the volcanoes activity in the Mexican highlands. So this is very important in the verification of the migration.

Answer: E. POLI

The *Fagus* forest on Mt. Etna haven't evergreen elements. The vegetation in general in this volcanic region is very poor. Also in the typical floristic composition there is an impoverishment. This is the general character of the Etnean vegetation. If you look at the mediterranean climate you might think because of its dry character it is impossible for *Fagus* to live here but northexposures and maybe sometimes mist allow *Fagus* to survive.

Question: K. FUJIWARA

This is very interesting about beech forest occuring in misty areas. Beech forest is just replacing laurel forests as in the Canary Islands. That's the case because Sicily is very dry. Therefore *Quercus ilex* forest occurs instead of laurel forest. But how about *Laurus nobilis* doesn't it make pioneer scrub or something like that? Because in Miyake jima and Sakura jima islands (volcanic islands) in Japan, *Persea* (*Machilus*) forest is a late succesional and also pioneer forest, more than a *Quercus* forest: evergreen *Quercus* or oak forest occurs after *Persea* (*Machilus*) forest, after laurel forest, then it changes to evergreen *Quercus* forest. But in Sicily you don't have such species like *Cinnamomum* or *Persea*.

Answer: E. POLI

We have only sometimes mist but there is no mist belt as on the Canary Islands. *Laurus nobilis* is only in some places near human habitations, planted by humans, but wild *Laurus nobilis* is very, very rare in Sicily, for example near rivers, but always in very little areas. And *Ilex aquifolium* is very rare in Etnean territory; it is in the area of *Fagus*. In older time there were vast *Fagus* forest but now they are few and little. Many lava flows are without vegetation and this causes climate changes. And *Ilex* has now very little refugee stands, principally on very old (praehistoric) substrate.

Question: F. KLÖTZLI

I have a little bunch of questions. First of all this reminds me very much of the Sierra Nevada in the area of southern Spain with your concoction of species and environmental conditions, however, there is not *Fagus* but other deciduous species. This just as a remark. Then you mentioned several times the climate was getting drier. Now I would be interested to know, where, when and how much drier. You showed climate diagrams, can you show us on a climate diagram where the changes occurred during the season?

Then I wonder, how much of the *Fagus* forest is left, because this is of geobotanical importance.

How high are the evergreen species going? Of course, I mean now the sclerophyllous ones in your case.

Answer: E. POLI

Evergreen sclerophyllous species are: *Quercus ilex*, *Phillyrea latifolia*, *Rhamnus alaternus*, *Olea europea* var. *oleaster* and others. They are in the Quercetalia ilicis belt. *Quercus ilex* is up to 1000-1100m and on rocky soil to 1400m a.s.l. But normally on 1400m a.s.l. there is the *Fagus* belt while the *Quercus pubescens* belt is between 1000m and 1400m a.s.l.

Unfortunately there are only very few meteorological stations. We found that the climate changed from old time to today. Over all now there is less precipitation and the temperature is rising. We have a botanical garden on 1700m a.s.l. on southwest side of the volcano. And there we have a little meteorological station. There we make experiments about lava colonization. And we planted little *Pinus laricio* on lava to experiment the possibility of reforestation in other sites. Some years ago the climate destroyed *Pinus* because the dry summer period was longer.

There are two reasons why the *Fagus* forests was destroyed: lava flows and men. Lava flows are today in areas once occupied by forests and from old time man destroyed a lot of Etnean forests; in Greek time the fleet of the tyrants of syracus was built of wood from Etna.

Question: H. BRUELHEIDE

A question about the climatic diagrams of the Etna region. When I tried to make a comparison between the distribution range of *Fagus sylvatica* and climatic parameters I could not find any climatic diagrams of higher altitudes. Are there any weather stations within the *Fagus* range on Mount Etna?

Answer: E. POLI

We can not do this correlation because we have very few meteorological stations and it is impossible to compare the presence of *Fagus* and the climate conditions.

Question: H. BRUELHEIDE

Provided there is no change in climate, how many years would it take for *Fagus* to be established on fresh volcanic substrate?

Question: K. FUJIWARA

How old is the lava flow where *Fagus* is growing on?

Answer: E. POLI

Individuals of *Fagus* can grow on young lava flows, as mentioned above, but in deep lava cracks.

In the places where you can find *Fagus* forests there are very old substrates (undatable soils).

Recent shifts in vegetation boundaries of deciduous forests,
especially due to general global warming, pp. 237–251
edited by F. KLÖTZLI and G.-R. WALTHER

Laurophyllisation in Ireland - the case of *Rhododendron ponticum*

Gerry DOYLE
Department of Botany, University College Dublin,
Dublin 4, Ireland

Abstract

Since its arrival in Ireland some 200 years ago, *Rhododendron ponticum* has invaded extensive areas of native vegetation. Blechno-Quercetum oakwoodlands have been particularly affected, with the elimination of (a) the understorey, normally dominated by *Ilex aquifolium*, (b) the woodland ground flora, and (c) the epiphytic flora. Dense shade cast by *Rhododendron* ensures that regeneration of trees and other species is inhibited. Spread of *Rhododendron* into scrub woods with birch (*Betula pubescens*), heathlands dominated by *Calluna vulgaris* and *Ulex* species, and marginal areas in peatlands has also been recorded and is invariably associated with disturbance. The expansion of *Rhododendron ponticum* in Ireland predated whatever modern climate change has been demonstrated, indicating that disturbance combined with the inherent invasive abilities of the exotic species contributed significantly to the process of laurophyllisation. The Irish experience of *Rhododendron ponticum* suggests that environmental changes other than modern climatic shifts should be investigated if the process of laurophyllisation is to be properly understood. Since the eradication and control of *Rhododendron* are expensive, such measures will be confined to limited areas of particular conservation or economic value and as a consequence the continued expansion of *Rhododendron* in Ireland is likely.

Introduction

The most notable example of laurophyllisation in Ireland concerns the spread of *Rhododendron ponticum*, which was first introduced to Britain in 1763 (Elton

1958) and to Ireland probably later that century (Cross 1982). Since then, the shrub has been widely planted for ornamental purposes, as a wind break and as game cover. It has become thoroughly naturalised in suitable habitats, extending from sea level up to 530m (Cross 1975) in parts of the Kerry mountains in south-western Ireland.

Spread of *Rhododendron* in Ireland

The history of *Rhododendron* in Ireland has been closely associated with its spread into Blechno-Quercetum oakwoodlands, especially those in the vicinity of Killarney, County Kerry, undoubtedly the country's finest semi-natural deciduous woodlands.

Cross (1981) and Watts (1984) trace the history of invasion of *Rhododendron* in the Killarney region. The shrub was probably first planted there towards the end of the 18th or beginning of the 19th century, since the first written account of its occurrence was in Hall and Hall (1843) who referred to large *Rhododendron* bushes growing within the Killarney oakwoods at Derrycunihy. Planting presumably continued after that time, since growth ring analysis sets their planting date at about 1890 (Cross 1975). By 1912 *Rhododendron* had become an obvious feature of the oakwoodlands as Rübel (1912) described the luxurious growth and sub-spontaneous spread of the shrub at Killarney. He presented a fascinating description of these oakwoods, which was later quoted in Praeger's (1934) seminal work on the Irish flora and vegetation:

> 'The dominant tree is *Quercus sessiliflora* (*petraea*) which reminds us that we are in north-western Europe. This is confirmed by a few of its associates such as *Betula pubescens*, *Salix cinerea*, *Prunus spinosa*, *Corylus*, etc. But these are quite accessory, the striking feature in this oakwood being the glittering, shining effect produced by the reflection of light which takes place in the second layer of the wood. The smaller trees and the shrubs in fact have shiny leaves and you fancy yourself in the southern country of the evergreen Laurel woods. The holly, *Ilex aquifolium*, is the general dominant and specially contributes to this effect. Frequently interspersed and locally even dominant we find one of the most interesting woody plants of Ireland, the strawberry tree, *Arbutus unedo*, directing our thoughts to the West Mediterranean. This species embellishes the Killarney woods, not only as a shrub of 3-5m in height, as we are used to seeing it in Corsica, but a tree of considerable height. Another laurel-leaved plant, *Rhododendron ponticum*, catches your eye with its luxuriance. It is not native, but it spreads sub-spontaneously and evidently feels quite at home here. The many epiphytes give a nearly tropical aspect to the woods; the oaks bear

> epiphytically *Polypodium vulgare*, *Geranium robertianum*, *Ilex aquifolium* and *Saxifraga spathularis*. Even *Rhododendron ponticum* ascends the lofty habitat of a branch of *Prunus spinosa*.'

While *Rhododendron* was obviously growing luxuriously in some places, its spread was not yet so obvious, since Turner and Watt (1939) observed that seedlings of *Rhododendron* are occasionally found far from the parent bushes, but so far invasion of the Killarney Woods is mostly marginal". Tansley (1939) commented on the spread of *Rhododendron* by seed dispersal, with seedlings appearing at considerable distances from the parent plants, and through vegetative propagation by suckering from the tips of its procumbent branches. He also observed that the shrub appeared to have replaced *Ilex aquifolium* in some areas of woodland, where, owing to the greater shade cast, the field and moss layers were severely reduced, leaving the soil almost bare. Tansley considered that the invasion by *Rhododendron* was progressive and that such a replacement was a rare example of the successful invasion of an undisturbed community by an exotic species. By 1952 the progress of invasion had accelerated, as Lüdi (1952) mentioned its prodigious spread in the Killarney woods where the shrub provided a dense and almost impenetrable understorey. Another account from that time suggested that *Rhododendron* was replacing *Ilex* in the shrub layer and proving a menace to the native vegetation (Warburg 1953). Some time later the extent of the *Rhododendron* problem was confirmed as Kelly (1981) estimated that by 1969 some 40% of the Killarney oakwoods were colonised, and needed major clearance to allow for natural forest regeneration.

The spread of *Rhododendron* in Ireland was not restricted to the oakwood habitat or confined to a single geographical area. Praeger (1934) described its spread at Achill Island in County Mayo, in the north-west of Ireland, where *Rhododendron* had been planted, had naturalised and had spread widely, an occurrence that was common in many exposed peaty places elsewhere in Ireland. Praeger also described its spread onto heathland in the Mourne Mountains of County Down in the north-east of the country. The plant had been introduced into the undergrowth of a wood, which was later cut, and through the influence of disturbance spread "in thousands" over some areas of adjacent stony hillsides. Later works mention the spread of *Rhododendron* in other disturbed situations, such as in conifer plantations on Atlantic Blanket Bog in the west of Ireland (Doyle and Moore 1982).

The extent of the spread of *Rhododendron* since its introduction to Ireland was summarised by Scannell and Synnott (1972), who considered the shrub to be possibly naturalised in each of the forty Irish vice-counties, and indicated in a distribution map (Figure 1) based on the Atlas of the British Flora (Perring and Walters 1976) and modified by Cross (1982).

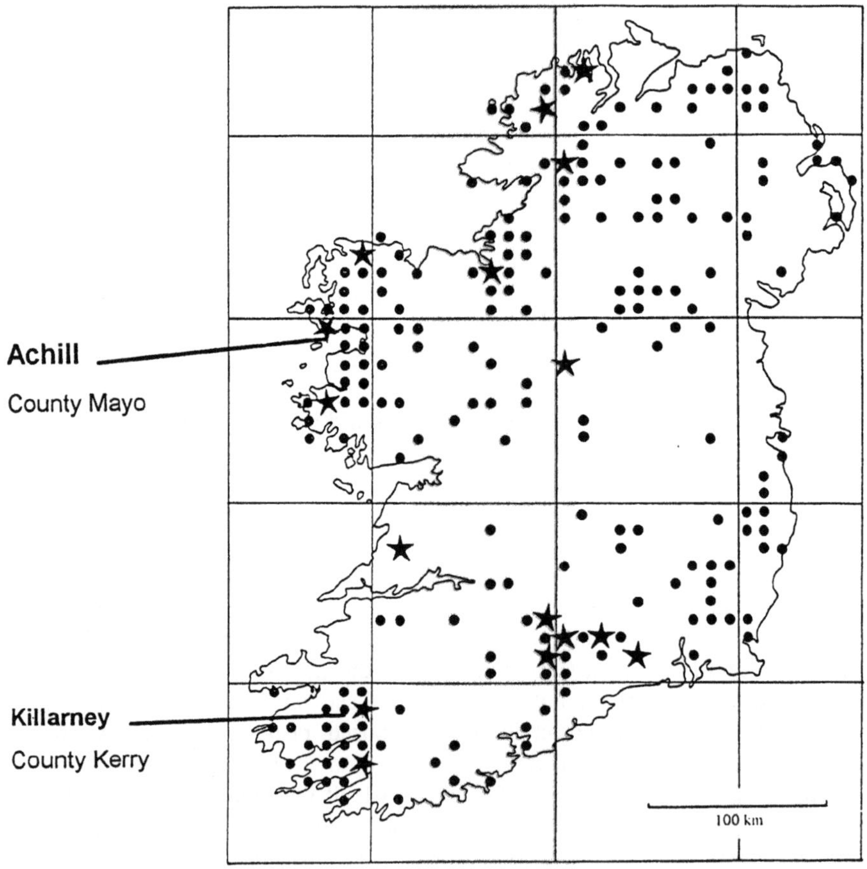

Figure 1: Distribution map showing the modern distribution of *Rhododendron ponticum* in Ireland.
The map is taken from Cross (1982), which was modified from the Atlas of the British Flora (Perring and Walters 1976). Each dot indicates one or more records in a 10km square of the Irish National Grid. Stars indicate those areas where extensive colonisation has taken place.

Features of *Rhododendron* habitats in Ireland

In the Irish situation the climate is hyper-atlantic, with winter temperatures around 6°C and summer temperatures around 16°C. Raindays in the extreme west may exceed 200, annual rainfall may exceed 1600mm, and relative humidity is constantly high. The plant is most vigorous in the constantly humid oceanic lowlands. The plant extends on deep podsolised sands and well-drained humus soils, but does grow on peat, especially in habitats on bog margins and along drainage channels where peat is not constantly waterlogged. The pH of soils supporting *Rhododendron* in Ireland varies from 3.3. to 6.4 with an average value of around 4.2 (Cross 1975). The plant invades where there are disturbances generated by man and is encouraged in its spread by grazing which reduces competition from other species (Cross 1975, Neff 1975). Litter of *Rhododendron* forms a thick, dark, friable mor humus, which when exposed by clearance produces a spongy crust that becomes baked on the surface and inhibits re-colonisation by other species (Cross 1975). In Ireland the plant attains heights of 7m to 8m in woodlands where its branches are supported by trees, but reaches just 4m in open areas and only 2m in exposed sites (Cross 1975).

Native habitats of *Rhododendron* in the Caucasus Region

The native locations for *R. ponticum* ssp. *ponticum* are in Bulgaria, Turkey, the Caucasus and Lebanon (cf. Cross 1975), in areas with distinctly continental climates.

Zazanashvili (1998) presents an excellent account of the habitats and communities in which *R. ponticum* is found in Colkheti (West Transcaucasus), the most humid region of the Caucasus in the catchment area of the Black Sea. The shrub forms part of the tertiary relict evergreen flora of the Caucasus region (*Rhododendron ponticum*, *R. ungernii*, *R. smirnowii*, *R. caucasicum*, *Laurocerasus officinalis*, *Ilex colchica* and *Ruscus colchicus*) and is concentrated in a number of woodland habitats. In this region *R. ponticum* is generally confined to northern slopes of humid mountain ranges and especially to valleys subjected to humid air masses. In Colkheti, *R. ponticum* avoids calcareous substrates, and occurs on relatively well aerated soils. It is confined to areas with high levels of precipitation spread rather evenly throughout the year. It occurs where precipitation ranges from 1300mm and up to 4000mmin some mountain situations: it is not found where precipitation is less than 1000mm. Prolonged snow cover, lasting from 90 to 196 days, is a prerequisite for the survival of the evergreen relict species in the cold mountainous areas. In lowland situations the shrub occurs where annual air temperatures range from

12°C to 14°C, while in its mountain habitats annual air temperatures may be about 4°C.

According to Zazanashvili (op. cit.) *Rhododendron ponticum* occurs in woodlands at altitudes ranging from 0m to 2300m. It is chiefly associated with beech, chestnut, and hornbeam-beech-chestnut forests, occasionally occurring in other woodland types. The typical woodland canopy is relatively open, ranging from 30% to 60% where the climate is humid, while in drier sites the shrub is found in more shaded situations with canopy cover ranging from 50% to 60%. In the lowlands and up to about 500m in the Caucasus foothills where human impact has not been severe, *R. ponticum* is found in the "Colkhic mixed broad-leaved forests", forming a shrub layer with many of the other relict evergreens. The forest dominants include *Castanea sativa*, *Quercus hartwissianna*, *Fagus orientalis*, *Carpinus caucasica* and *Alnus barbata*, together with *Ulmus elliptica*, *Tilia caucasica*, *Fraxinus excelsior*, *Acer campestre* and *Diospyros lotus*. In such situations and where there is a relatively warm winter temperature regime *R. ponticum* sometimes forms a straight low tree. The shrub also occurs in the lower mountain belt (500m to 1000m) in communities of chestnut, beech-chestnut, and beech-hornbeam-chestnut forests, and in the beech (*Fagus orientalis*) forest that dominates in the 1000m to 1500m altitudinal belt, where some 50% of the area supports a well-developed evergreen underwood. The shrub occurs less frequently in the dark coniferous forest (fir, spruce-fir, and beech-fir) at altitudes ranging from 1400m to timberlines at 2250m.

It is of interest to note that forest tree regeneration is inhibited in native habitats in some mild and damp situations where *R. ponticum* forms in places a treeless scrub (Zazanashvili 1998).

Irish plant communities invaded by *Rhododendron*

Woodland communities

According to Cross (1982) *Rhododendron* occurs mainly in oakwoods assigned to the Blechno-Quercetum and the subassociation scapanietosum (Kelly and Moore 1975, Kelly 1981, White and Doyle 1982) which is typical of western Ireland. These woods are dominated by *Quercus petraea*. Some *Betula pubescens* and *Sorbus aucuparia* occur in the canopy, while *Ilex aquifolium* is predominant in the understorey. The ground flora contains *Luzula sylvatica*, *Vaccinium myrtillus*, *Blechnum spicant*, *Dryopteris aemula* with some *Calluna vulgaris* and *Pteridium aquilinum* found in places. The community is characterised by an abundance of *Hymenophyllum* (*H. tunbrigense* and *H. wilsonii*) and a rich and well-developed bryophyte flora (Richards 1938, Kelly 1981). At one time, the Killarney fern, *Trichomanes speciosum*, was common in these woodlands, but owing to its filmy appearance and relatively large size was almost eradicated through collection during the 19th century. In 1987 a new station for the fern was discovered in the oakwoodland habitat at Killarney, in an area that had been cleared of *Rhododendron* (Doyle 1987). Perhaps other such occurrences have been obscured and thereby protected by the invasive shrub. Epiphytes are a characteristic feature of these woods, with bryophytes and lichens covering trunks and branches to the uppermost canopy levels. Ferns, such as *Polypodium* species, herbaceous species including *Saxifraga spathularis* and *Geranium robertianum*, together with woody species that include *Ilex aquifolium* and *Rhododendron* all adopt an epiphytic habit growing in axes between trunks and branches of oak in these highly humid woodland situations.

In Killarney, *Rhododendron* is also found in the Blechno-Quercetum coryletosum (which has affinities to ashwoods belonging to the Corylo-Fraxinetum), wherever the soil pH in that community is below 5.5 (Cross 1982).

While these oakwoods represent important conservation areas in the relatively treeless Irish landscape, they cannot be viewed as undisturbed communities as suggested by Tansley (1939). These oakwoods have been strongly influenced over the recent centuries by native mammals such as red deer (*Cervus elaphus*), by introduced sika deer (*Cervus nippon*), by clearance and alteration by man, and through grazing by domestic stock (Watts 1984). As a result they are best described as semi-natural. In such conditions of constant disturbance the spread of *Rhododendron* has been promoted with the exposure of an abundance of safe sites for germination (Cross 1981). The natural cycle of moss carpet development, where the carpets fragment and slip down slopes, also exposes bare soils for colonisation. Neff (1975) and Cross (1975) observe that heavy grazing does not affect *Rhododendron* owing to the toxic properties of its leafy material. In heavily grazed situations the plant is favoured in comparison to more palatable species such as *Ilex aquifolium*. The twigs and leaves of *Ilex* are palatable to many animals including deer, cattle and sheep, its bark may be gnawed by both rabbits and deer in winter; in situations of heavy browsing young *Ilex* individuals may be checked indefinitely (Peterken and Lloyd 1967). Equally, *Rhododendron* gains an advantage in heavily shaded situations, where the shade intolerant *Betula* is inhibited. Once *Rhododendron* overtops its competitors their regeneration is effectively curtailed. These problems are of greater significance in the western woods than in the those found close to the east coast of Ireland where the relatively drier climate appears to ameliorate the impact of *Rhododendron* (Neff 1975).

Scrub woods

In marginal areas on acid soils, and where birch (*Betula pubescens*) forms scrub with *Pteridium aquilinum*, *Calluna vulgaris* and *Erica cinerea*, invasion of *Rhododendron* has taken place, again reflecting the disturbance that is a feature of such places.

Heathlands

Invasion of heathland follows fires and disturbance, and has occurred extensively in some areas, on Achill Island for example. Not unexpectedly, heather (*Calluna vulgaris*) dominated communities tend to be readily invaded from adjacent woodland areas, since heathlands are dependent on constant disturbance in the form of burning and grazing for their maintenance. In addition, mature heather communities in the degenerate phase of their life cycle provide bare areas in the centre of bushes that are readily colonised by *Rhododendron*. The shrub is particularly suited to heathland habitats since it is resistant to burning and readily regenerates from its root stock, while seeds from surrounding areas allow rapid colonisation following fire episodes (Cross 1975).

In north-west Mayo *Rhododendron* has been found invading *Erica erigena* heath, around Lough Carromore, for example. Such invasion is particularly significant in this case since this heather is one of Ireland's rarer plants, reaching its most northerly location in Europe in this area.

Peatlands

The plant does not colonise intact, waterlogged ombrotrophic bog surfaces, but does invade relatively well drained peats along drainage channels in atlantic blanket bogs, on dried-out turf banks and cutaway bogs and has spread extensively in such habitats in many parts of Ireland.

Conifer plantations on peat

In several places, *Rhododendron* has been planted as a shelter belt around peatland areas that have been planted with *Picea sitchensis* and *Pinus contorta*. The habitat perturbation that is inherent in the afforestation process in such areas, which includes the insertion of surface drains, the up-turning of bare peat sods, and the supply of phosphate fertiliser, combine to present conditions favourable for *Rhododendron* colonisation. Doyle and Moore (1982) traced the floristic changes in such plantations during the first twelve years of development and found that native bog species (such as *Schoenus nigricans*) were first eliminated and that a discrete suite of invaders, including *Rhododendron*, had appeared in twelve year old plantations.

Impact of *Rhododendron* on invaded communities

Cross (1982) outlined the impacts on those communities invaded by *Rhododendron*. Elimination of shade intolerant species results since only 2% of radiation is transmitted through a *Rhododendron* canopy - such plants include many woodland bryophytes, epiphytes of the lower branches and trunks of native trees, woodland herbs and dwarf shrubs. The richness of habitats suitable for epiphytes is not substituted by *Rhododendron* since it casts such heavy shade and continuously exfoliates its bark. Some ground layer species are eliminated through the smothering effects of deep litter accumulation and development of a deep humus layer.

According to Cross (1975) there is a sparse insect fauna associated with *R. ponticum*, there are no significant insect predators, and no serious fungal pathogens associated with the shrub in its Irish habitats. The negative impact of *Rhododendron* on regeneration of shrubs and trees has been discussed by Cross (1975, 1981, 1982) and Neff (1975). Regeneration is almost entirely eliminated, particularly where *Rhododendron* invasion is associated with the impacts of heavy grazing. Since it is not browsed, *Rhododendron* out-competes the native species. Clearance of *Rhododendron* and reduction in numbers of grazing animals are required to ensure proper regeneration in Irish oakwoods. Clearance of *Rhododendron* is technically difficult and costly since detailed attention must be given to its complete clearance even from rock habitats (Neff 1975). The current practice involves clearing of the shrubs by cutting, followed by herbicide application to the regrowth. Emerging seedlings are manually removed as they appear. Such site manipulation is costly and is generally only implemented in oakwoodlands of major conservation importance that are in state ownership.

Discussion

The spread of *R. ponticum* in Ireland preceded recent climatic changes, but was clearly facilitated in its expansion by disturbance caused directly by man through woodland destruction and replanting, peatland drainage and peat harvesting, by the influence of domestic animals including cattle and sheep, and the influence of native red deer and introduced sika deer. The impact of these processes is to reduce the competitive ability of native species and to provide safe sites for the regeneration and spread of the evergreen *Rhododendron*.

The widespread expansion of *Rhododendron* in Ireland may provide an insight into the processes that are currently influencing the expansion of evergreen species in the European ecotonal areas, either as a direct result of climate change or through the invasive process mediated by altered forest management. The forest systems, in southern Switzerland for instance, are no longer such important sources of wood for either energy production or for wood-based products. These forest systems are now primarily of value as amenity and protection woods, and are no longer managed for production. With the withdrawal of disruptive management, and in the absence of heavy grazing, there would appear to have been an alteration in the composition of trees contributing to canopy formation and an expansion of native understorey trees and shrubs. In these circumstances, and taking the Irish situation into

consideration, it might be suggested that disturbance mediated spread of laurophyllous species might be reduced though the competitive impact of the developing native understorey. The process of laurophyllisation in such places is obviously complicated since climate change may allow expansion of the exotic evergreens and certainly is evidenced through lianisation, with native species such as *Hedera helix* assuming a climbing habit that was previously considered only typical of the mild oceanic parts of western Europe.
It is clear that in Ireland, and under the current conditions of land-use and management, *Rhododendron* will have to kept in check by man if it is not to replace significant elements in natural and semi-natural communities. The severity of this problem is partly related to the scarcity of woodland cover in Ireland, so that invasion of woodlands and reduction of tree regeneration in the surviving forest remnants presents an important conservation problem. It is likely, given the cost of physical removal and treatment with herbicides, that eradication or control will be confined to important conservation areas where the economic input may be justified. Otherwise, it would seem that *Rhododendron* will remain as an aggressive invader in Ireland.

Bibliography

Cross J.R. (1975): Biological flora of the British Isles: *Rhododendron ponticum* L. Journal of Ecology 63: 345-364.

Cross J.R. (1981): The establishment of *Rhododendron ponticum* in the Killarney oakwoods, S.W. Ireland. Journal of Ecology 69: 807-824.

Cross J.R. (1982): The invasion and impact of *Rhododendron ponticum* in native Irish vegetation. Journal of Life Sciences Royal Dublin Society 3: 209-220.

Doyle G.J. (1987): A new station for the Killarney fern (*Trichomanes speciosum* Willd.) in the Killarney Oakwoods (Blechno-Quercetum). Irish Naturalists' Journal 22: 353-356.

Doyle G.J. & Moore J.J. (1982): Floristic changes in developing conifer plantations growing on blanket peat in the west of Ireland. Struktur und Dynamik von Wäldern (Ed. H. Dierschke), pp 699-716. Cramer, Vaduz.

Elton C.S. (1958): The ecology of invasions of animals and plants. Methuen, London.

Hall S.C. & Hall S.C. (1843): A week at Killarney. Jeremiah How, London.

Kelly D.L. (1981): The native forest vegetation of Killarney, South-West Ireland: an ecological account. Journal of Ecology 69: 437-472.

Kelly D.L. & Moore J.J. (1975): A preliminary sketch of the Irish acidophilous oakwoods. Colloques Phytosociologique III. La Végétation des Forêts Caducifoliées Acidiphiles. (Ed. J.-M. Géhu), pp 375-385. J. Cramer, Vaduz.

Lüdi W. (1952):Fragmente zu Waldstudien in Irland. Die Pflanzenwelt Irlands (Ed. W. Lüdi), pp 214-223. Hans Huber, Bern.

NeffM.J. (1975): Woodland conservation in the Republic of Ireland. Colloques Phytosociologique III. La Végétation des Forêts Caducifoliées Acidiphiles. (Ed. J.-M. Géhu), pp 273-285.

Perring F.H. & Walters S.M. (1976):Atlas of the British Flora. E.P. Publishing, Wakefield.

Peterken G.F. & Lloyd P.S. (1967): Biological Flora of the British Isles: *Ilex aquifolium*. Journal of Ecology 55: 841-58.

Praeger R. Ll. (1934): The Botanist in Ireland. Hodges Figgis & Co., Dublin.

Richards P.W. (1938): The bryophyte communities of a Killarney Oakwood. Annales Bryologici 11: 108-130.

Rübel E.A. (1912): International Phytogeographical Excursion to the British Isles. V. The Killarney Oakwoods. New Phytologist 11: 54-57.

Scannell M.J.P. & Synnott D.M. (1972): Census catalogue of the Flora of Ireland. Stationery Office, Dublin.

Tansley A.G. (1939): The British Islands and their vegetation. Cambridge University Press.

Turner J.S. & Watt A.S. (1939): The oakwoods (Quercetum sessiliflorae) of Killarney, Ireland. Journal of Ecology 27: 202-233.

Warburg E.F. (1953): A changing flora as shown by the status of our trees and shrubs. The Changing Flora of Britain. (Ed. J.E. Lousley), pp 171-180. Botanical Society of the British Isles.

Watts W.A. (1984): Contemporary accounts of the Killarney Woods 1580 - 1870. Irish Geography 17: 1-13.

White J. & Doyle G.J. (1982):The vegetation of Ireland: a catalogue raisonné. Journal of Life Sciences Royal Dublin Society 3: 289-368.

Zazanashvili N. (1998):On the Colkhic Vegetation. Recent shifts in vegetation boundaries of deciduous forests, especially due to general global warming. (Eds. F. Klötzli F. & G.-R. Walther), pp. 181-197. Birkhäuser, Basel.

Plenary discussion

Statement: G.-R. WALTHER

I agree that you see *Rhododendron* in Ireland as a problem. I would like to distinguish different factors between the situation in Ireland and the one we have here in southern Switzerland: I think that in Ireland the climatic conditions have always been such that *Rhododendron* could grow. You have clearly explained the central role of disturbance in promoting the invasion of *Rhododendron* in the Irish situation. The situation in Switzerland is quite different. There is the hypothesis here that climate has changed, and we have in this area the situation where disturbance has decreased and not increased. As a consequence, in the Swiss situation, laurophyllous species occupy a newly opened but limited ecological niche. There is no reason to suggest that the laurophyllous species will spread extensively since the niche suitable for laurophyllous species is very limited, and confined to particular places, where the climatic conditions are suitable for these species. As a result we must view laurophyllisation in Ireland and in Switzerland in different perspectives.

Answer: G. DOYLE

I recognise that the problems evident in Ireland and Switzerland are quite different. Comparisons between the two situations are, however, useful in highlighting underlying mechanisms that influence the spread of exotic species. The reduction in disturbance and grazing in the Swiss forests is obvious, as evidenced by the prominence of developing tree saplings observed during our field excursions. Contrary to your assertion, it may be that these reduced impacts are the trigger for the spread of exotics and that grazing levels in the past checked the spread of these species.

Remark: F. KLÖTZLI

You have argued that the spread of *Rhododendron* in Ireland predated climate change, a contention that cannot be faulted. What would happen, however, if the temperature in Ireland were elevated somewhat, while humidity remained at its current levels? Would you see another invasion by garden escapes of New Zealand plants? Would you then have a *Hebe* problem as well as a *Rhododendron* problem?

Answer: G. DOYLE

We have planted some New Zealand plants: *Phormium tenax* has been used as a potential windbreak in the West of Ireland. This plant has not spread from its original locations.

Question: F. KLÖTZLI

Is there any evidence of recent temperature or humidity changes in the Killarney Valley, where the spread of *Rhododendron* is so obvious?

Answer: G. DOYLE

I do not have any evidence for such changes in that area. I did not specifically address this topic since expansion in *Rhododendron* took place before there was any supposed climatic change.

Question: F. KLÖTZLI

I wonder why *Rhododenron baeticum* is not in Ireland. If it were, it would be even better adapted to the climatic conditions there. It is interesting to speculate that if *R. baeticum* were found in Ireland it would be accompanied by its natural parasites owing to the geographical proximity of that subspecies' natural distribution, and that there would not be an invasive problem in Ireland. I think the Irish *R. ponticum* problem arises since there is a general absence of *R. ponticum* pathogens in your country. Do you know of any pathogen on *Rhododendron ponticum*?

Answer: G. DOYLE

The natural predators and pathogens appear to be missing in Ireland.

Question: F. KLÖTZLI

And a final question: Is there any chance that *Rhododendron ponticum* would invade the Fraxinetum in Ireland? Do you think it is really confined to the Podsol and the moment you come to Cambisol, to Pseudogley or Luvisol then there is no problem?

Answer: G. DOYLE

There is no evidence that would suggest that *Rhododendron ponticum* is likely to invade Corylo-Fraxinetum habitats, which are generally limestone-rich areas in Ireland. The shrub does invade areas of oakwood belonging to the Blechno-Quercetum coryletosum, a subassociation with affinities to the Corylo-Fraxinetum. That subassociation is, however, confined to areas with soil pH below 5.5.

Question: A. BORHIDI

What about the possibility of using limestone treatment against *Rhododendron*?

Answer: G. DOYLE

Such treatment might radically alter the soil conditions in treated areas, producing unwanted results. For example, it might promote the expansion of *Fagus sylvatica*, a tree species not native in Ireland but expanding in some oakwood areas.

Question: N. ZAZANASHVILI

In Ireland it appears that *Rhododendron ponticum* is spreading in a variety of habitats and in different plant communities. Is *Rhododendron* excluded from particular habitats in your country?

Answer: G. DOYLE

Yes! *Rhododendron* is excluded from the entire range of habitats and plant communities associated with calcareous, limestone-rich soils.

Question: N. ZAZANASHVILI

What is the altidudinal limit for *Rhododendron ponticum* in Ireland?

Answer: G. DOYLE

The shrub reaches its altitudinal limit in Ireland at 530m a.s.l. in the Kerry mountains, in the south-west of the country.

Question: H. BRUELHEIDE

Have any other colchic species been introduced to Ireland?

Answer: G. DOYLE

Prunus laurocerasus has been planted and has moved into some woodland areas, but generally remains in a peripheral position and does not cause the ecological difficulty presented by *Rhododendron.*

Statement: A. BORHIDI

Prunus laurocerasus is not so much acidophilous such as *Rhododendron.*

Question: H. BRUELHEIDE

That means it is very difficult to predict which introduced species will spread in the future and which will remain in the gardens.

Answer: G. DOYLE

Species would appear to have different lag times. It seems that *Rhododendron* was in residence for just thirty years after its introduction around 1800, before its expansion phase commenced.

Question: H. BRUELHEIDE

Are there special features of dispersial in *Rhododendron* that promote its invasive ability?

Answer: G. DOYLE

The species has very light seeds, that can be dispersed widely. So where disturbance occurs, such as denudation of peatland surfaces following fire, the shrubs can gain a foothold and spread sub-spontaneously afterwards.

Question: G. CARRARO

Arbutus unedo and many different *Rhododendron* species (though perhaps not *R. ponticum*) occur scattered in gardens. Up till now I have seen just one specimen of *Arbutus unedo* and not a single specimen of *Rhododendron* in a natural habitat, despite the prevalence in the area of acid soil. Perhaps the occasional droughts that are a feature of the area prevent the expansion of these species.

Answer: G. DOYLE

The climatic conditions around Ascona are very different to those typical of the *Rhododendron* areas in Ireland, which have mild winters, cool summers and constant high humidity. A comparison of the climatic conditions at Ascona with those that prevail in the native locations of *Rhododendron ponticum* might best shed light on this discussion.

Statement: N. ZAZANASHVILI

In the Caucasus *Rhododendron ponticum* is protected by snow in winter, since the northern slopes remain covered for around 100 days. The protection from frost is considered one of the main ecological reasons why *Rhododendron* has such a wide distribution in the mountains of the west Caucasus (Colchis).

Answer: G. DOYLE

But I imagine the climate here is closer to the climate in the Colchis than to the climate in Ireland.

Statement: G. CARRARO

It is possible that this winter drought is not so good for *Rhododendron* regeneration.

Statement: F. KLÖTZLI

In the Ascona situation, the soil conditions would be different to those that prevail in Ireland. In southern Switzerland there are no podsols, since the climate is too warm in summer. We have instead what were once referred to as crypto-podsols, which show a certain degree of podsolisation, but have a very weakly developed humus layer. If we look to gardening practice to illustrate this point, it is not feasible in Zurich, for instance, to grow *Rhododendron ponticum* in anything but a peat bed. Based on my own personal observation, such planted shrubs remain confined to that peat bed. I agree with the speaker that radical alteration of the soil conditions in places invaded by *Rhododendron ponticum* in Ireland would be unwarranted and would lead to the elimination of the Blechno-Quercetum in those areas.

Recent shifts in vegetation boundaries of deciduous forests,
especially due to general global warming, pp. 253–272
edited by F. KLÖTZLI and G.-R. WALTHER

The historical shift of the evergreen broad-leaved forest in East China

Yongchang SONG
Institute of Environmental Science, East China Normal University
Shanghai 200062, China

Introduction

Since the Tertiary Period underlies the region of east China the influence of strong monsoons with weather that is warm and humid enough to support the growth and development of different types of forest. The southern section of that region was very little affected by the glaciers in the Quaternary Period. The widely distributed evergreen broad-leaved forests (EBLF) found there at the present time, are direct descendants of the Tertiary flora and they are also the major component of the modern EBLF in East Asia.

Both the Yellow River and the Yangtze River basins are regarded as the origin of the ancient Chinese civilization. In former times they were the habitats of EBLF in China, and they are still heavily distributed over the Yangtze River basin nowadays. Coincidentally the Chinese civilization is a type of EBLF civilization.

Among the major environmental issues that we have to face these days, the global warming is critical. However, there are two different concepts to approach this intriguing topic. Some claim that the pollution of the environment caused by the industrialization is the culpit of global climatic changes, while others state that the climate itself is changing and therefore the climatic change is only the result of successive variations of nature rather than a result of the human activities.

The geological, palaeontological and palaeometeorological studies have shown that our planet has incessantly been in a process of evolvement since its formation six billion years ago, but the changes could only be perceived by thousands, thousands upon thousands or by million years. However, after the emerge of human beings, with their increasing population accompanied by technological advances, the global changing process accelerated. The change of nature nowadays may be regarded as the result of both nature per se and human activities.

The earliest human activities in east China could be traced back to the Xia dynasty, about 21th century BC i.e. almost 5,000 years ago. The approach to the shift of the EBLF in east China under the impact of human activities as well as the change of nature during that period, especially during the last 500 years, could be significant to reveal the relationship between vegetation and climatic change, and to predict the influence of global change on vegetation or vice versa.

The distribution of evergreen broad-leaved forest in east China

The vegetation of a given area is the result of the complex natural conditions, such as climate, soil, geology, topography as well as living organisms and activities of human beings in that area. The vegetation of today is the continuation of the vegetation of the past. To recognise the distribution of EBLF today even only roughly would be very helpful for searching through the historical shift of EBLF in east China.

There are different explanations for the term "evergreen broad-leaved forest" internationally. Two categories can be distinguished: one is referred to as temperate broad-leaved rain forest, temperate broad-leaved sclerophyllous forest and temperate evergreen seasonal forest (Ovington 1983), others even include all the moist evergreen broad-leaved forest of the nontropical region such as subtropical rain forest, temperate rain forest, laurineous forest without sclerophyllous forest into that category (Schröder 1983; Hübl 1988). The other category of evergreen broad-leaved forest is referred to as a special type of biome which has established under the monsoon climate and is distributed only over warm-temperate and subtropical areas. It is different from temperate rain forest, subtropical rain forest and sclerophyllous forest (Klötzli 1988). Even laurineous forest that developped under the winter-rain climate is specified as monsoon EBLF, or summer-rain EBLF (Song 1994, 1995). There are a lot of common characteristics among subtropical rain forest, temperate rain forest, sclerophyllous forest and laurineous forest, such as a canopy which is composed of evergreen broad-leaved trees; the distribution area of these forests lies beyond the tropical regions, all the characteristics from tropical rain forest are missing and an intolerability to seasonal coldness and drought prevails. All types mentioned above could be categorized as EBLF for global or wide-scale range vegetation studies conveniently, but they are clearly distinguished in the physiognomy, structure of communities, condition of habitats, and geographical distribution. Each of the forest types will be a special vegetation type, therefore, it is better to nominate the different types of vegetation separately when they are studied in details. The strict sense of EBLF is applied in this paper.

The fundamental characteristics of EBLF in China have been discussed before (Song 1988, 1995). It will not be repeated in this study. Figure 1 shows the distribution of EBLF over China.

EBLF is regarded as a kind of zonal vegetation, or a zonal biome. Its northern border is situated at the Hangzhou Bay in the east, extends westward along the south bank of Lake Taihu, across Xuancheng, the southern slope of the Dabie Mountain, the southern section of Dashengnongjia, the southern foot of Dabashan Mountain eventually reaching the vicinity of Pingwu. It is located at a latitude of 31°-32°N. The northern border is known as Daba-Dabie-Yangtze line (line A in Figure 1). The eastern part of that line coincides with the isotherm of the annual mean temperature (AT)=16°C, the warmth index (WT)=135°C, the coldness index (CI)=-1.5°C, and the potential evapotranspiration (PET)=950mm. The southern border of EBLF is located at a latitude of about 24° N. It borders on Fuzhou in the east, extends westward along the Daiyunshan Mountain, Fuping Ridge, crosses over the middle part of Jiulianshan Mountain, then extends through Dayaoshan Mountain to Liuzhou, along the Longjiang River and Jinchengjiang River it reaches the northern bank of Nanpanjiang River, across the mountains and plateaus of mid-Yunnan, and finally the Sinoburmese national boundaries (line B in Figure 1). The eastern part of the southern border matches roughly the isotherm of AT=20°C, WI=180°C and PET=1200mm. Regarding EBLF as a type of community on the vegetation classification, its individual stands such as small patches of artificial woodland, may occur at the southern slope of the Qinling Mountain and the Yangtze-Huaihe River basin. From the point of view of zonal vegetation those stands show the transition from evergreen broad-leaved to deciduous broad-leaved forest, i.e. an ecotone of evergreen deciduous broad-leaved mixed forest. That border could be called Qinling-Huaihe line (line C in Figure 1). It is located at nearly 33°-34°N latitude, and roughly matches the isotherm of AT=15°C, WI=125°C, CI=-5°C and PET=900mm (Table 1).

Some individual species of evergreen broad-leaved trees may be distributed further north or south. Ni & Song (1997) have chosen 112 popular species from the constituents of EBLF (including the south subtropical transitional evergreen broad-leaved forest and the evergreen broad-leaved moss forest in the mountainous region), and analyzed the relationship between their location and climate. The trees were categorized into three different types according to the optimal WI and CI for distribution, which is low-temperature type, medium-temperature type and high-temperature type (Table 2).

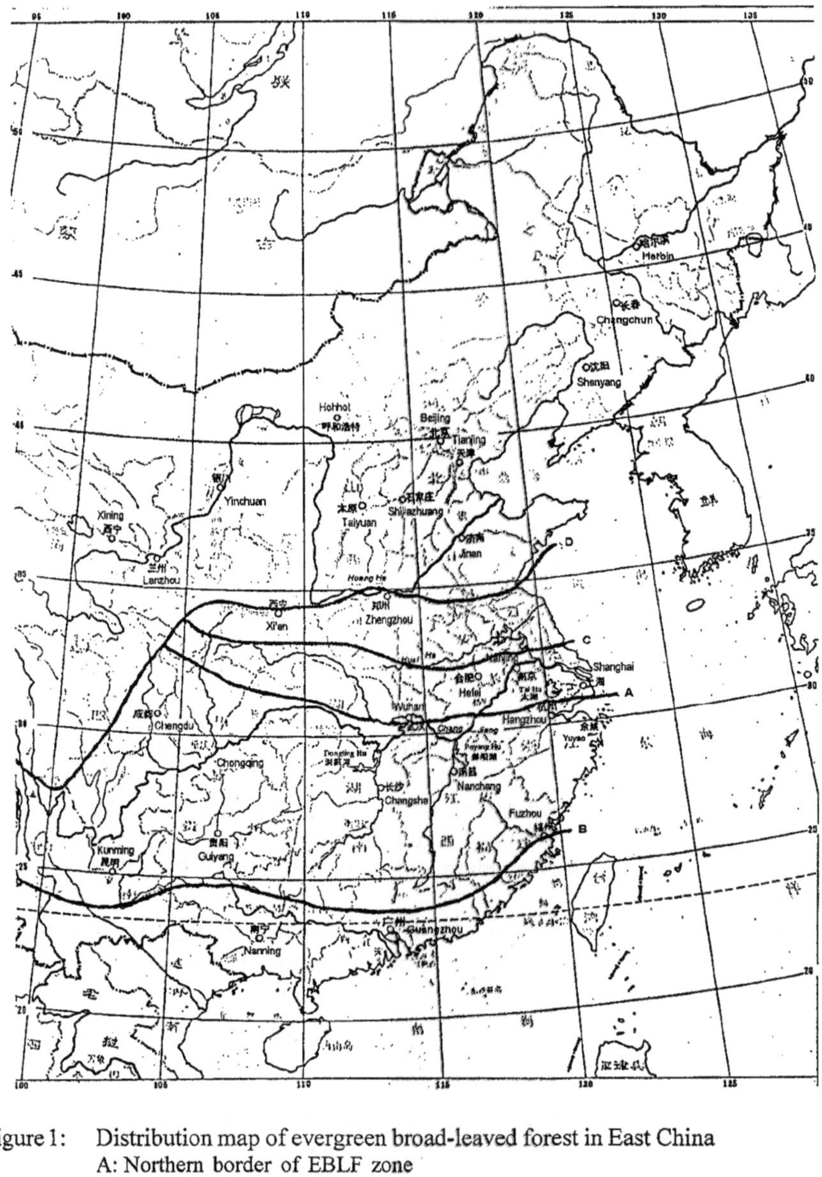

Figure 1: Distribution map of evergreen broad-leaved forest in East China
A: Northern border of EBLF zone
B: Southern border of EBLF zone
C: Northern border of stands of EBLF community
D: Northern border of evergreen trees

Border line	Annual mean tempe-rature	January mean tempe-rature	July mean tempe-rature	Winter (Nov. Dec. Jan.) mean temperature	≥10°C accumulated temperature	WI	CI	PET
Northern border of evergreen trees	13.0	-1	25.0	2.5	3500	120	-10	800
Northern border of stands of EBLF community	15.0	1.5	28.0	5.0	4000	125	-5	900
Northern border of EBLF zone	16.0	3.5	29.0	7.0	4500	135	-1.5	950
Southern border of EBLF zone	20.0	10.0	29.0	13.0	7500	180	0	1200

Table 1: Climatic index of the distribution of evergreen broad-leaved forest in East China

1. Low-temperature type (WI=100~140°C; CI=-12~-2°C): *Castanopsis delavayi, Cast. ceracantha, Cast. orthacantha, Lithocarpus dealbatus, Lith. cleistocarpus, Cyclobalanopsis delavayi, Cy. glaucoides,* etc.
2. Medium-temperature type (WI=140~180°C; CI=-2~0°C): *Castanopsis sclerophylla, Cast. carlesii, Cast. carlesii var. spinulosa, Cast. fargesii, Cast. tibetana, Cast. eyrei, Cast. fordii, Cast. jucunda, Cast. fabri, Cyclobalanopsis glauca, Cy. multinervis, Cy. oxyodon, Cy. nubium, Cy. myrsinaefolia, Cy. gracilis, Cy, gilva, Cy. genseniana, Cy. chungii, Lithocarpus hancei, Lith. polystachyus, Lith. harlandii, Litsea coreana var. sinensis, Lindera megaphylla, Cinnamomum japonicum, C. camphora, C. subavenium, Machilus thunbergii, M. leptophylla, Phoebe sheareri, Ph. bournei, Cleyera japonica, Adinandra millettii, Schima superba, Distylium myricoides, Daphniphyllum oldhami, D. macropodum, Michelia maudine, Ilex latifolia, I. suaveolens, Engelhardtia fenzelii, E. roxburghiana, Manglietia fordiana, Altingia chinensis, A. gracilis, Elaeocarpus decipiens, Sloanea sinensis, Styrax suberifolius,* etc.
3. High-temperature type (WI=180~210°C; CI=0°C): *Castanopsis hystrix, Cast. fissa, Cast. kawakami, Manglietia chingii, Schefflera octophylla, Cryptocarya chinensis, Cry. densiflora, Cry. concinna,* etc.

Table 2: The temperature types of major species in the evergreen broad-leaved forest over China

Most of the 13 species of the low-temperature type are the evergreen broad-leaved trees distributed over the Yunnan Plateaus. All 10 species of the high-temperature type are south subtropical evergreen broad-leaved trees, while 87 species of medium-temperature type are the main components of typical EBLF vegetation in east China. To categorize the evergreen broad-leaved constituents, the optimal temperature index was adjusted, although the whole range of the temperature for their distribution would be much wider. For some widely distributed species like *Cyclobalanopsis* (*Quercus*) *glauca*, its optimal WI will be 110.5-185.5°C, while its whole range extends from 35.6°C to 225.1°C, and the mean value is 148°C (Ni & Song 1997). Geographically the northern border of the major evergreen broad-leaved species is near the southern bank of the Yellow River in the east and next to the northern slope of the Qinling Mountain in the west (line D in Figure 1). The latitude is about 35° N; it roughly matches the isotherm of AT=13°C, WI=120°C, CI=-10°C, PET=800mm.

The shift of the evergreen broad-leaved forest in east China in different historical periods

By the data of spore-pollen analysis, the results from archaeological excavation and historical literature, the shift of evergreen broad-leaved forests over east China during the last 5000 years could be divided into the following phases:

The Yangshao warm phase (ca. 5000 years B.P. before)

According to the archaeological studies from the Songze site in Qingpu county of Shanghai (31°10' N, 121°26' E), the sporo-pollen assemblage of the accumulates from the bottom of a low cultural layer (determined with C14 dating, its age was 5360±105 years B.P.) was dominated by *Cyclobalanopsis glauca* and *Castanopsis* (see Figure 2), followed by *Ilex* and some deciduous trees like *Quercus acutissima*, *Ulmus* and *Morus*, etc. (Wang et al. 1980). The amount of pollen from evergreen woody plants (W) almost accounted for 40~50% of the total amount from all the woody plants (E) i.e. the value of E/W was 40~50%. Studies of the pollen in the superficial layer of soil of the natural woodland both from EBLF and evergreen deciduous broad-leaved mixed forest showed that the E/W ratio for distinguishing those two types of forest was 25% (Yang & Song 1992). Since the composition of spore and pollen from Songze was dominated by the pollen from evergreen woody plants, it would be unequivocally justified that the zonal vegetation then in Shanghai was the EBLF composed of *Cyclobalanopsis* and *Castanopsis*.

The cultural layer from Hemudu of Yuyao county in the Zhejiang province was also determined by C14 dating. Its age is 6800~5600 years B.P.; results of the spore-pollen analysis show that the composition from Hemudu was dominated by *Cyclobalanopsis*, *Castanopsis*, *Altingia*, *Quercus*, and *Liquidambar*

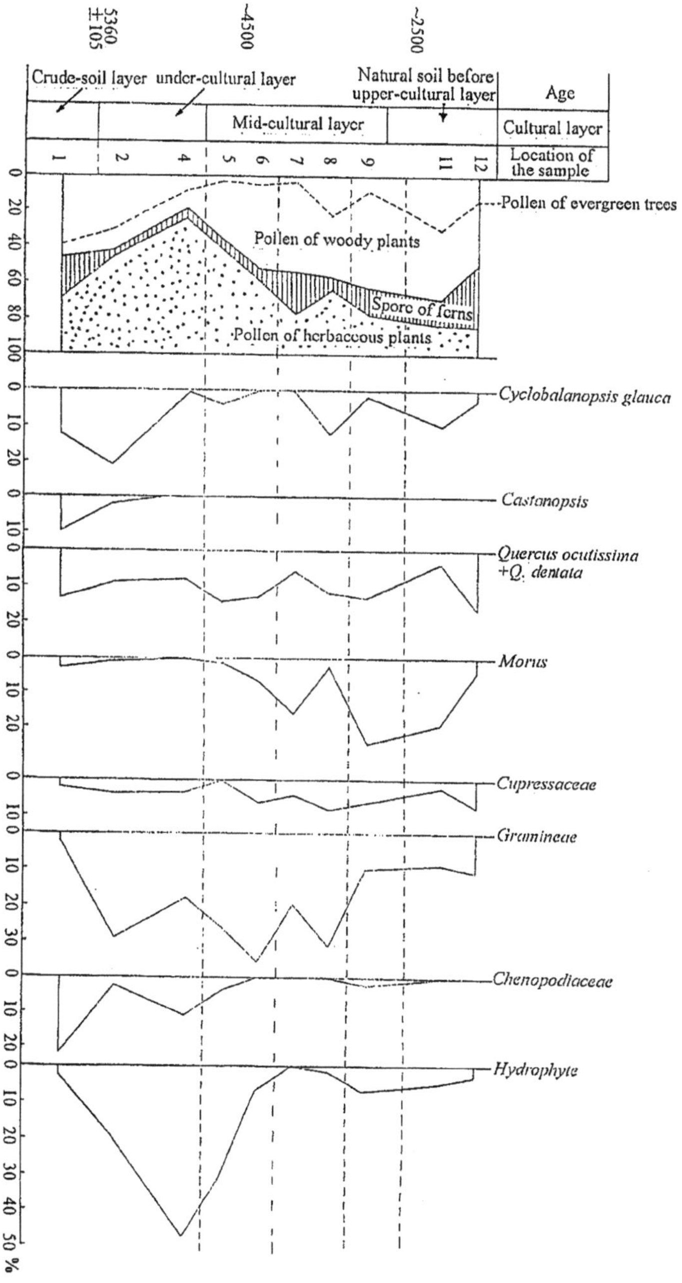

Figure 2: Pollen diagram from ruins in Songze (Redrawn from Wang 1980)

formosana, pollen from *Mallotus paniculata*, *Styrax*, *Lauraceae* and *Oleaceae* had been detected too. The spores found were from *Lygodium polystachum*, *L. microstachyum*, *L. salicifolium*, *Ophioderma pendula*, *Pyrrosia martini*, *Lemmaphyllum carnosum*, *Microsorium superficiale*, *Osmunda*, *Humata* etc. There were uncarbonized leaves and fruits from *Phoebe sheareri*, *Cinnamomum japonicum*, *Lindera glauca*, *Cyclobalanopsis myrsinaefolia*, *Cyclobalanopsis gilva*, *Castanopsis tibetana*, *Murraya exotica*, *Choerospondias axillaris*, *Magnolia coco*(?), *Camptotheca acuminata*, *Chloranthus fortunei* etc. (Sun et al 1981). Obviously, EBLF occur in the zonal vegetation. Since the EBLF dominated by *Altingia chinensis* and *Castanopsis tibetana* are distributed over Fujian province and the southern area in Zhejiang province today, and *Lygodium microstachyum*, *L. salicifolium*, *Ophioderma pendula*, *Pyrrosia martinii*, *Lemmaphyllum carnosum* are distributed over the regions of Guangdong, Guangxi, Taiwan and Hainan provinces. It would be reasonably inferred that the EBLF back then might have been similar to the types that spread from south Zhejiang province through mid-Fujian province at the present time. Besides, there were remains of *Elephas maximus*, *Dicermocerus sumatrensis*, *Rhinoceros sondaicus* and others among the dead animals (Museum of Zhejiang province 1978). All those evidences clearly show that the vegetation 5000 years ago must have been similar to the EBLF of the southern subtropical regions. Pollen from some evergreen trees, like *Cyclobalanopsis glauca*, *Castanopsis*, *Elaeagnus* etc. were found almost at the same time of this phase inside the subclay layer of 5100±300 years B.P. from Huangkou (34°25'N, 116°42' E) in Xiao county of Anhui province. In the layer of 5245±90 years B.P. from Yenchen county (34°48' N, 118°26' E) of Shandong province a lot of *Ceratopteris* and *Zygnema* were discovered. In the Longshan cultural site (35°25' N,119°25'E) in Liangchenzhen of Lichen county, Shandong province, pieces of bamboo were found and inside the site of 5950±120 and 5225±130 years B.P. from Yuanzhou (35°33' N,116°36' E) of Shandong province *Pteris vittata* and *Lygodium japonicum* originating from the subtropical region were found (Tang et al. 1992). In addition, wild elephant and rhinoceros were discovered wandering over the Yellow River basin area (He 1982). All those evidences clearly show that the northern border of the EBLF might have located at 35° N (Figure 3). That period was supposed to be the first warm phase of east China, dated 5000 years B.P. An inference could be drawn that the annual mean temperature was 2°C higher then the temperature of present. The inference is somewhat identical with the conclusion about the climate change in late 5000 years in China by Zhu (1972).

Figure 3: Shift of the northern border of EBLF in East China through different historical periods

1. The Yangshao warm phase (ca. 5000 a. B.P. before)
2. The Liangzhu cold phase (ca. 4500-4000 a. B.P. before)
3. The warm phase of Xia-Shang dynasties (ca. 3900-3500 a. B.P.)
4. The cold phase of West-Zhou dynasties (ca. 3000 a. B.P.)
5. The warm phase of Chun-Qiu and Chin-Han dynasties (ca. 2700-1800 a. B.P.)
6. The cold phase of Wie, Jin, Southern and Northern Dynasties (ca. 1800-1400 a. B.P.)
7. The warm phase of Sui-Tang dynasties (ca. 1400-900 a. B.P.)
8. The cold phase of Song-Yuan dynasties (ca. 900-600 a. B.P.)

The Liangzhu cold phase (ca. 4500~4000 year B.P. before)

The observation of the composition of sporo-pollen from the mid-cultural layer of Songze remnants (4500 years B.P.) show that the amount of *Cyclobalanopsis glauca* had been reduced, that of *Castanopsis* even dropped down to zero, that all the varieties and amounts of evergreen broad-leaved trees were reduced, the E/W ratio was brought down to 13%. The deciduous species discovered were mostly *Quercus acutissima*, *Ulmus*, *Celtis*, *Albizia*, *Rhus* and *Liquidambar*. The composition was characterized as a type of evergreen-deciduous broad-leaved mixed forest, similar to the composition from the vegetation on the hilly land in Nanjing (32° N,118°48' E) nowadays. The bottom-section of the Liangzhu-cultural layer (3855±95 years B.P.) from Tingling in Ginshan county of Shanghai was dominated by pollen from herbaceous plants, and partly from some deciduous trees and conifers, however, none were dominated from evergreen trees. Late in that period the amount of pollen grains from woody plants were rising, as were those from subtropical deciduous trees like *Liquidambar formosana*, *Pterocarya stenoptera*, *Magnolia*, etc, and also from gymnosperms like *Ketelearia*. That period stands for the first cold phase through 5000 years B.P. in east China. The annual mean temperature was 1.5°C lower than the temperature is now. The northern border of EBLF moved southward 4~5° latitude and reached the area located at latitude 30° N (see Figure 3).

The warm phase of the Xia-Shang dynasties (ca. 3900-3500 years B.P.)

The top layer of the mid-Songze cultural layer is a good representative of that period. The proportion of pollen from *Cyclobalanopsis glauca* was high, especially in the assemblage of sporo-pollen. The E/W value was as high as 55%. In the Shanghai area the vegetation was the typical EBLF. The annual mean temperature was about 1-2°C higher, the annual precipitation about 200~300mmhigher (Wang 1978) than in present time. The northern margin of the EBLF shifted to the Huaihe basin, about latitude 33° N.

The cold phase of the West-Zhou dynasties (ca. 3000 years B.P.)

The bottom section of the upper Songze cultural layer is a good example of that period. The amount of *Cyclobalanopsis glauca* is relatively sparse in the composition of sporo-pollen. The E/W value dropped to 14.5%. The northern border of the evergreen broad-leaved forest moved southward to the present location (31° N). However, the cold climate in the early Zhou dynasty did not last long. After centennials the weather turned warm again.

The warm phase of the Chun-Qiu (spring and autumn period) and the Chin-Han dynasties (ca. 2700-1800 years B.P.)

That period stands for an interval from 700 BC to 200 AD. Based on the study of a profile of the accumulations from the bottom section of the peat layer of Chengdong of Qingpu county formed 2720±70years B.P. and the top section

1530±140 years B.P.; the profile of Sijin of Songjiang county in Shanghai were formed 2920±105years B.P. (Wang et al. 1988). In their assemblages of sporo-pollen *Cyclobalanopsis*, *Castanopsis* and the deciduous *Quercus*, *Liquidambar*, *Castanea*, *Ulmus* and *Rhus* were dominant. The E/W ratio was 35~65%. In addition, *Cyclobalanopsis*, *Morus*, *Albizia* were the dominants in the accumulates from the Songze site (2500 years B.P.), where *Cunninghamia* was a big part too, and the E/W ratio was 30%. Inferred by the above evidences the zonal vegetation of the Shanghai area could be a type of EBLF, but its composition was not as diverse as the vegetation of 5000 years B.P.

In addition, as the ancient book <Yue Jue Shu> told, there was a large area of woods distributed over the hilly land in the region of Shaoxing in Zhejiang province between 500 and 300 BC. The forest was composed of *Phoebe*, *Cinnamomum camphora*, *Trachycarpus fortunei*, *Sassafras tzumu*, *Dalbergia*, *Catalpa ovata*, *Catalpa bungei*, *Vernicia*, as well as some species of gymnosperms like *Pinus*, *Cupressus*, *Juniperus* and *Torreya*. The ancient book <Wu Yue Chun Qiu> had narrated that the circumference of the trunks was twenty arm spans and the height of the trees were 50 Xuns (an ancient measure of length, equal to about 2.7m). The ancient graves with wooden verandas in Shaoxing area (475-221 BC) were made of huge wood from *Phoebe*. We assume that there was an abundance of big trees.

At that time the Huaihe basin was the northern margin of the habitat of elephants (He 1993), Bamboo was flourishing over the Shandong province as well as in the Xianyang region in the Shanxi province. The optimal annual mean temperature for the growth of bamboo is 15.6°C. However, the annual mean temperature in Xian City (34°18' N, 108°56' E) near the Xianyang region is nowadays 13.3°C. The annual mean temperature at that time must have been 1-2°C higher. That period was supposed to be the third warm phase in 5000 years in east China. During that period, the northern border of EBLF had moved to the north to the valley of Huaihe (Figure 3).

The cold phase of Wei-Jin, Southern and Northern Dynasties (ca. 1800-1400 years B.P.)

That period is dated 150~500 AD. The II zone of the profile from Chendong of Qingpu county, and the B zone of the profile from Sijin of Songjiang county in Shanghai are perfect examples of that period. The evergreen trees such as *Cyclobalanopsis* and *Castanopsis* decreased, while the deciduous trees such as *Quercus acutissima*, *Carpinus*, *Pterocarya*, *Betula*, *Ulmus* increased (Wang et al. 1988). The evidence specified that the type of zonal vegetation was evergreen-deciduous broad-leaved mixed forest. Archaeological studies showed that in the graves of the Han dynasty in the Changsha area of Hunan province dug out whole pieces of *Cunninghamia lenceolata* during that period. They were used to elaborate the mid-galley of the coffin. That was another valuable evidence for the distribution of EBLF in Hunan province.

Moreover the look «Shui Jin Zhu» written by Li Dao Yuan (466-527) in the 5th century had depicted the distribution of vegetation across China. The deciduous broad-leaved forest was composed of *Ulmus*, *Salix*, *Corylus*, *Quercus*, the bushes and the coniferous forest composed of *Pinus* and *Juniperus* were widely distributed over the basin of the Yellow River, while to the south in the middle and lower reaches of the Yangtze River valley, the big EBLF composed of evergreen *Quercus* and different species of trees occurred. Further south existed dense and humid forest composed of bamboo, palm trees and camphor trees. In short, the whole array of vegetation zones then was rather similar to that of the present time.

The warm phase of the Sui-Tang dynasties (ca. 1400-900 years B.P.)

That period is dated 600-1100AD. The composition of spore-pollen observed from the III zone of the profile from Chendong in Qingpu county and the C zone of the profile from Sijin district in Songjiang county was the typical representative in the Shanghai area then. The E/W ratio was 30%. Dominant species were *Cyclobalanopsis* and *Castanopsis.* The whole composition was similar to the one of the I zone of that profile (Wang et al. 1988). The zonal vegetation was characterized as EBLF.

During this warm phase the population was increasing rapidly in China. Consequently, not only all the plain area was exploited and farmed, but also the lakes, the sea, and the mountains were turned into farmland and their state deteriorated. The forest was widely overharvested. In the Tang dynasty (618-907 AD) the cultivation of tea was rapidly propagating. A vast area of EBLF was substituted by tea plantations. The herd of wild elephants was found in the north of the Yangtge delta again. The historical record about the activities of wild elephants in Nanjing and south Huaihe River had been set in 507~552 AD (Chang 1982). The plum was blossoming in Xian, and citrus were grown in Changan. All the above evidences show that the annual mean temperature was 1-2°C higher than now. The arrangement of the vegetation zones in east China was almost identical with that in the warm phase of the Qin-Han dynasties. The northern border of EBLF moved to the Huaihe-Yangtze River valley.

The cold phase of the Song-Yuan dynasties (ca. 900-600 years B.P.)

That period is dated 1100 to 1400 AD. The composition of spore-pollen from the IV zone of the profile from Chengdong, Qingpu County, is the perfect representation for the Shanghai area at that time. Pollen from evergreen trees like *Cyclobalanopsis glauca* and *Castanopsis* were found and pollen from deciduous trees: *Quercus acutissima*, *Betula*, and *Liquidambar formosana*. In addition, pollen from *Pinus* were discovered. The combination is typical for evergreen-deciduous broad-leaved mixed forest. Plum trees did not occur in north China during that period, except the planted ones in the royal garden. Lake Taihu was frozen in 1111 AD, and the ice was firm enough for the passage of carriages. The *Litchi chinensis* of south Fujian were all killed due to the cold climate. The annual mean temperature at that time was 1-2°C lower than it is now. The northern border of the EBLF in east China moved southward through the middle part of Zhejiang, Jiangxi and Hunan province, which is located at about 29° N latitude (Figure 3). However, although the climate has warmed up once during 1200 AD and 1300 AD, the vegetation zone remained unvaried.

The shift of evergreen broad-leaved forests in east China during the last 500 years

Historically the climate change has been the leading force for the shift of vegetation in east China either northward or southward. During the last 500 years, since the end of the fifteenth century, east China was in the so called "Little ice age". The climatic change was not as drastic as before, but the fluctuations of temperature and humidity still had a considerable impact on the growth and distribution of the EBLF, as well as on the cultivation of crops. Chinese historical literature gives detailed reports about the frozen winter which affected the agricultural production as well as the cultivation of economic plants and their phenology. Those documents are very helpful for the study of the relationship between the change of vegetation and climate .

Interestingly the area distribution of EBLF in east China coincides with the cultivation of *Citrus*, *Camellia sinensis*, *Vernicia fordii*, *Toxicodendron vernicifluum* and bamboo. Especially the distribution of *Citrus* and *Cunninghamia lanceolata* almost overlapped with the occurrence of EBLF (Figure 4). Therefore, the shift of the southernmost line along which the *Citrus* had been killed due to extreme weather, in combination with the historical records of extremely cold temperature ever since 1471 AD, both the climatic change and the shift of the EBLF distribution line could be addressed.

In the fifteenth and sixteenth century (1470~1530 AD) the first cold phase occurred in east China in five hundred years. At that time not only Huaihe River, but also Taihu Lake, Boyanghu Lake, Huangpujiang River, and Qiantangjiang River were frozen every other year. During winter, all citrus trees of the region of Shanghai, Suzhou and Taihu Lake were killed by the frigid temperature.

In the seventeenth century the second and coldest cold phase occurred (1617-1725 AD). In 50 years (1650~1700 AD) the Huaihe River, Hansui River and Taihu Lake were frozen four times, Dongtinghu Lake three times. Even Lake Boyanghu was frozen. In 1670 AD the Yangtze River was almost frozen and all citrus trees in the Jiangxi province were killed. Only few people still grew citrus after this event. The trees of *Dimocarpus longan* and *Litchi chinensis* no longer occurred in southern Fuji. The eastern part of the southern line of deadly frozen citrus lay in the Huangyan (28°38' N, 121°25' E), Quxian (28°58' N,

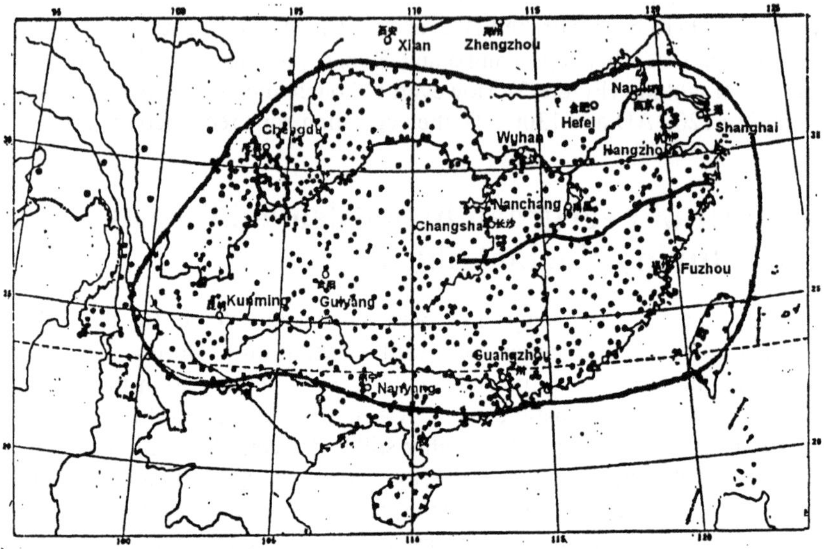

Figure 4: Distribution of *Citrus* and *Cunninghamia lanceolata*
Black dot = cultivating place of *Citrus*
Black dotted line = southern border of *Citrus* frozen deadly line
Black cycle line = Distribution area of *Cunninghamia lanceolata*

118°52' E), Nanfeng, Anfeng to Hengyang(26°56' N, 112°30' E) area. Geographically the eastern coastal region of the line of deadly frozen citrus remained at 29° N latitude, while the western part in the area of Hunan province, approximately lay at a latitude of 27° N. Compared with the cultivated northern border at the present time, the eastern part was situated at a latitude of 3° further south, and the western part differed even by 6° (Figure 4). The northern border of the EBLF was almost the same as that of the cold phase in Song-Yuan dynasties, located at latitude 29° N.

In the sixteenth century, the first warm phase that had emerged in east China, was that from 1531~1616 AD. During those eighty five years, warm winters were frequent (1517, 1604 and 1604 AD). Plum trees were blossoming in October and November in the Huaihe River valley , and litchis were ripening in October in south Fujian. The northern border of the EBLF had moved up further north. However, Lake Taihu had been frozen three times, in 1568, 1578, and 1580 AD. It is assumed that the temperature in winter then was lower than nowadays, and that the northern border of the EBLF was still further south than now.

The eighteenth century was the second warm phase in east China in 500 years. Lake Boyanghu had not been frozen, and almost no snow or frosty weather occurred in the coastal area of south Fujian. Comparing the spring phenology in the late eighteenth century and early nineteenth century with its counterpart now (Hangzhou, Suzhou, Beijing) (Gong et al. 1983), the results show that most of the phenological phenomena were later than they are nowadays. On an average they were 6 days later, however, a few years later the phenological phenomena were earlier (Table 3).

A.D.	Location	Phenological phase	Ancient data	Present data	Difference days	Source
1778	Hangzhou	Flourishing blooming	9 Apr.	25/3	-15	«Diary of Zhuting»
1780	Hangzhou	Flourishing blooming	14 Apr.	25/3	-20	«Diary of Mr. Wu Tuchuang»
1787	Hangzhou	Flourishing blooming	29 Mar.	25/3	-4	«Diary of Mr. Wu Tuchuang»
1790	Wuxi	Flourishing blooming	8 Apr.	29/3	-10	«Daily table of Mr. Wu Tuchuang»
1791	Hangzhou	Flourishing blooming	3 Apr.	25/3	-9	«Daily table of Mr. Wu Tuchuang»
1800	Hangzhou	Flourishing blooming	24 Mar.	25/3	+1	«Daily table of Mr. Wu Tuchuang»
1808	Hangzhou	Flourishing blooming	19 Mar.	25/3	+6	«Diary of Mr. Wu Tuchuang»

Table 3: Blooming dates of *Amygdalus persica*

The empirical formula of the correlationship between the dates of the beginning of the blooming of *Amygdalus persica* with the longitude, latitude and the altitude is as follows:

$$Y = 68.16 + 3.98\,(\Phi-30) + 0.71\,(\lambda-110) + 1.364h$$

Where Y= the date of the beginning of the blooming of *Amygdalus persica*
Φ = the latitude there
λ = the longitude there
h = the above sea level there

The blooming date of *Amygdalus persica* was delayed for 4 days when its location moved northward in latitude. The assumption can be made that the beginning of spring in the late eighteenth century was later than nowadays, and the difference was 6 days about latitude 1.5°. It is generally recognized that the temperature of the warmest phase within the last 500 years was lower than at present. The comparison of winter temperatures of the second warm phase in the last 500 years with todays vaslues shows that they have risen 1.0-1.5°C. However, the northern border of EBLF lay approximately at the same latitude as in the sixteenth century.

The weather of east China in the nineteenth century had been cooler again. During the seventy years from 1831 to 1900 AD there was a freezing winter every other year. Lake Taihu had been frozen twice and the frozen period lasted for more than one month. It was shown from the record of phenological observation that there was also a cold phase between 1840 and 1890 AD.

After the modern meteorological observation station in Shanghai had been established in 1873 AD, the meteorological observations were carried out regularly. Therefore, an array of continuing temperature records in 120 years was obtained. Those data have been applied in studying the change of temperature in China during that period, and a curve of the change of annual mean temperature as well as a sliding curve in five years have been derived from them (Xu 1993). In addition, it has been shown that the weather as a whole tended to get warmer most of the time. The changing rate of the temperature calculated was 0.044°C/10 years, i.e. the temperature increased by 0.52°C in the last 20 years.

The rising of the temperature had led to a movement towards the north of some evergreen broad-leaved trees, especially the evergreen economic plants. *Machilus thunbergii* from the coastal area was observed to be naturally distributed over Laoshan mountain in Qingdao as far as 36°09' northern latitude. *Trachycarpus fortunei*, *Ligustrum lucidum* as well as *Photinia serrulata* were able to grow well in the area of northwest China, like in Xian and Lingtong. *Cinnamomum camphora* and *Trachycarpus fortunei* were widely cultivated in the Shanghai area. They were growing uncultivated and even natural communities were formed. The cultivation of *Citrus*, *Cunninghamia lanceolata*, *Camellia sinensis*, bamboo and *Vernicia fordii* was expanded further to the north. The area of Shanghai became another base for citrus production. Recently the cultivation of citrus emerged at Jiashan county at the bank of the Huaihe River. Of course, the movement of the cultivation border northwards would be attributed to the advanced technique and improved seed quality too, but the rising temperature was indispensable to the widespread distribution of citrus.

Summary and comments

The EBLF in China is virtually a part of the EBLF in east Asia. It is a special type of vegetation developped under the influence of the monsoon climate. The northern border of the EBLF distribution over China is at 31° N, latitude approximately, some stands of the community are distributed up to 33° N. The climatic indices of the northern border are as follows: WI=135°C, CI=-1.5°C, AT=16°C,CMT= coldest month mean temperature =3.5°C, and WI=125°C, CI=-5°C, AT=15°C, CMT=1.5°C. In Japan the northern border of EBLF is at a latitude of 39° N (the Pacific coast) and 40° 35'N (Japan Sea coast) (Fujiwara 1981), the climatic indices are: WI=92°C, CI=-11°C,AT=11.8°C, CMT=0.9°C (Hattori et al. 1985). They are almost identical with the climatic indices of the northern border of evergreen broad-leaved trees in China. However, the northern border of the EBLF in China is in a more southern area. Since the topography in east China tends to be relatively low and flat, the Siberian cold front is able to come straight upon the south region. In Japan on the other hand, the barrier formed by the Japan Sea weakens the influence of the cold front and as a result the distribution limit moves up north a little bit. It seems that the low temperature in the winter is the critical factor for the distribution of the EBLF. On the other hand, the "evergreen broad-leaved forest" as it is called in China refers to a forest with a tree layer composed primarily of evergreen broad-leaved trees; in case deciduous broad-leaved trees or conifers are mingling, the proportion should not be over 1/4, the EBLF should have a physiognomy showing the dominance of evergreen broad-leaved species. The EBLF should exclude the evergreen-deciduous broad-leaved mixed forest, therefore, the distribution limit could be kept at a relatively low latitude. A more detailed comparison and further studies are necessary in the future.

Because of the changes in temperature in the last 5000 years, the border of the EBLF in east China has been moving northwards and southwards as well as backwards and forwards. The EBLF had been distributed in the basin of the Yellow River in the first warmest phase (Yangshao warm phase ca. 5000 years B.P.), at a latitude of 35° north. The northern border of the EBLF moved southward when the warm phase was shorter and the temperature decreased with time, i.e. the northern border of the EBLF moved to the valley of the Huaihe River, which lay approximately at a latitude of 33° in the second warm phase (Xia-Shang dynasties, ca. 3900~3500 years B.P.), and in the third warm phase (Chun-Qiu warm phase ca. 2700-1800 years B.P.). In the fourth warm phase (Sui-Tang dynasties, ca. 1400-900 years B.P.) the border of EBLF only reached the basin of the Huaihe and Yangtze River. On the other hand the cold phase appeared to be lasting longer and the extreme temperatures were much lower in the first cold phase (Liangzhu cold phase, ca. 4500~4000 years B.P.). The EBLF moved to the hilly region of the south of theYangtze River, which lay at 30° northern latitude, in the second (West Chou dynasties ca. 3000 years B.P.) and in the third cold phase (Wei-Jin dynasties, ca. 1800~1400 years B.P.), the northern border of the EBLF moved to a latitude of 31° N, while in the fourth coldest phase (Song-Yuan dynasties, ca. 900~600 years B.P.) the northern border of EBLF moved to the middle section of Zhejiang, Jiangxi, Hunan province, which lay about 29° north latitude.

Corresponding to the modern "Little ice age" in Europe, the weather in China was subject to different cold phases in the last 500 years. As a whole it seemed to be colder than the weather in the Sui-Tang dynasties (600-1100AD), but the temperature in 500 years tended to rise with time, and led to movement towards north of the distribution line of the EBLF and the economical crops.

What are the crucial factors of the dynamics of the climate in east China? It could be the continuing change of the global environment itself; the increasing global greenhouse gases produced by human actions could also play a significant role in the change. The study of the climatic changes of the last 500 years shows that the alternations of cold and warm phases occurred in 60 to 100 years. However, the later warm phase has lasted for 95 years and until now no evidence from meteorological records exists that shows that the climate will turn into a cold phase.

No doubt, the vegetation change would be corresponding to the climatic change, but the shift of natural vegetation should be progressive. It is always lagging behind the climatic change. An extremly cold winter would kill many plant species, but the successive warm temperature could help some plants to recover. The shift of the border of the EBLF either northward or southward cannot be accomplished in a few years. The vegetation shift can only be realized under the continuing influence of climatic shift, moreover, it will be strongly interfered by human action. Therefore, long-term tracking observation is very necessary.

References

Chang Tien-lin, (1982):A preliminary research of historical climatic change of the Yangtze-Delta. Jour. of East China Normal University (Nat.Sci.Ed.) 1982(4):101–120

Fujiwara K. (1982): Phytosociological investigation of the evergreen broad-leaved forest of Japan (I). Bull. Inst. Env. Sci. Tech. Yokohama Nat. Univ, (7): 67–133.

Gong Gaofa, Zhang Peiyuan, Zhang Jindong (1983):A study on the climate of the 18th century of the lower Changjiang Valley in China. Geographical Research, 2(2) : 20–32.

Hattori T. & Nakanishi S., (1985): On the distributional limits of the lucidophyllous forest in the Japanese Archipelago. Bot. Mag. Tokyo 98: 371–333.

He Yeheng (1993): The historical changes of rare beasts in China. Hunan science technique press.

Hübl E. (1988): Lorbeerwälder und Hartlaubwälder (Ostasien, Mediterraneis und Makaronesien). Düsseldorfer Geobotan. Kolloq. 5: 3–26.

Ovington J. D. (ed.) (1983): Temperate Broad-Leaved Evergreen Forests. Ecosystems of the World, vol. 10. Elsevier, Amsterdam.

Klötzli F. (1988): On the global position of the evergreen broad-leaved (non-ombrophilous) forest in the subtropical and temperate zones. Veröff. Geobot. Inst. ETH, Stiftung Ruebel (Zürich) 98: 169–196.

Ni Jian & Song Yongchang, (1997): Relationship between climate and distribution of main species of subtropical evergreen broad-leaved forest in China, Acta Photoecologica sinica, 21(2): 115–129.

Ni Jian & Song Yongchang (1997): Relationships between geographical distribution of *Cyclobalanopsis glauca* and climate in China. Acta Botanica Sinica, 39(5): 451–460.

Schroeder F. G. (1983): Die thermischen Zonen der Erde. Ein Beitrag zur Präzisierung der geobotanischen Terminologie, mit einer Vegetationskarte. Tuexenia 3: 31–46.

Song Yongchong, (1983): Die räumliche Ordnung der Vegetation Chinas. Tuexenia 3: 131–157.

Song Yongchang, (1988): The essential characteristics and main types of the evergreen broad-leaved forest in China. Phytocoenologia 16(1): 105–123.

Song Yongchang, (1994): On the scientific term of evergreen broad-leaved forest and their typology. In Jiang Shu et al. (eds.): Study of vegetation ecology, Scientific Press.

Song Yongchang (1995): On the global position of the evergreen broad-leaved forests of China. In: Box et al. (eds.): Vegetation Science in Forestry, 69–84.

Sun Xiang-jun, Du Nai-qiu & Chen Ming-hong (1981): The Palaeovegetation and Palaeoclimate during time of Homudu people. Acta Botanica Sinica 23(2): 146–151.

Tang Linyu, Sheng Caiming (1992): The vegetation and climate of big warm period of Holocene in the north Jiangsu In: Shi Yafeng & Kong Zhaohuan (eds.): Climate and environment of big warm period of Holocene in China, Ocean Press.

Wang Kaifa, Zhang Yulan, Ye Zhihua, Jiang Hui (1978): The climatic change in late six thousand years in Shanghai area deducted by the pollon-spore analysis. Scientia Atmospherica Sinica, 139–144.

Wang Kaifa et al. (1980): Palynological study of Songze in Qingpu county, Shanghai. Jour. Archaeology, 1980(1): 59–66.

Wang Kaifa ,Sheng Caiming, Lv Houyuan (1988): The vegetation and climatic change in the last three thousand years in the eastern part of Shanghai deducted by pollen-spore assemblages. Historical geography, Section IV: 13–20.

Xu Jialiang (1993): Some features of temperature variations in Shanghai for the 118 years periode. Acta Geog. Sinica (1): 26-32.

Yang Jiaowen & Song Yongchang (1992): The correlation between the species-composition and spore-pollen assemblage of soil surface in Tiantong National forest park, Zhejiang province (not published)

Zhejiang province museum nature section (1978): Identification on animal and plant residual of Hemudu site. Jour. Archaeology (1): 95–105.

Zhu Kezhen, (1972): A preliminary study of climatic change in the last five thousand years in China. Jour. of Archaeology, 1972(1): 168–189.

Recent shifts in vegetation boundaries of deciduous forests,
especially due to general global warming, pp. 273–300
edited by F. KLÖTZLI and G.-R. WALTHER

Evergreen Broad-Leaved Forests in Japan and Eastern North America: Vegetation Shift under Climatic Warming

Kazue FUJIWARA[1] and Elgene O. BOX[2]
[1] Institute of Environmental Science, Yokohama National University
Yokohama 240, Japan
[2] Geography Department, University of Georgia
Athens, Georgia, USA

Abstract

Evergreen broad-leaved forests include "laurel" forests, evergreen Fagaceae forests, tropical and temperate rainforests, and some non-Fagaceae sclerophyll forests. Laurel forests are especially well developed in humid warm-temperate to subtropical areas of East Asia and occur topogenically (moist/wet depressions) in the sandy, drier coastal plain of the southeastern USA. Evergreen Fagaceae forests also occur in both regions, but with more mesomorphic, laurophyll character in East Asia and more semi-sclerophyll character in the southeastern USA. Both warm-temperate regions have secondary forests which are deciduous and more closely related to the adjacent typical-temperate deciduous forests. Under global warming (even without drying), weedy and other highly mobile species from warmer areas can be expected to invade northward. Species of the warm-temperate evergreen broad-leaved forests, however, may also expand northward, invading the adjacent deciduous forest regions. In Japan, *Quercus myrsinaefolia* has recently expanded inland, but a relationship to climatic warming has not yet been verified. In the southeastern USA, no such northward or inland expansion has been documented, but this could be related to a substrate discontinuity and/or to the more erratic nature of the North American climatic limit for evergreens.

Introduction

Evergreen broad-leaved forests include mesomorphic, laurophyll "laurel" forests, mesomorphic to somewhat xeromorphic evergreen Fagaceae forests, tropical and temperate rainforests (generally broad-leaved but with conifers important in some areas), and some non-Fagaceae sclerophyll (e.g. eucalypt) forests. Comparative descriptions of the various evergreen broad-leaved forest regions in the world have been offered, among others, by Suzuki (1952, 1953); Miyawaki (1967); Ovington (1983); Klötzli (1984, 1988) and Hübl (1988); and more recently by Song (1995); Tawaga (1997, cf. Tagawa & Hotta 1997); Hara and Yonebayashi (1997a, 1997b); Box (1997) and Box et al. (1998).

The term "laurel forest" was suggested by the Laurisilvae of such places as the Canary Islands, which appear to be limited to relatively perhumid regions with only mild winter frost (Rübel 1930; cf. Box 1995c). These warm-temperate/humid-subtropical evergreen broad-leaved forests are especially well developed in East Asia (southern Japan, eastern China, southernmost Korea) and also occur in comparably perhumid warm-temperate/subtropical areas of southern Brazil, eastern Australia, and coastal southeastern South Africa, as shown in Figure 1. Such forests also occur in climatically similar montane belts of many tropical mountains as well as topogenically wetter areas in drier warm-temperate regions (e.g. coastal southeastern USA, eastern Australia). Cool-temperate analogs occur in perhumid areas of the Southern Hemisphere, such as southern South America, New Zealand and Tasmania, as well as tropical upper-montane belts.

Evergreen Fagaceae forests also occur in East Asia, with a mesomorphic, laurophyll physiognomy very similar to that of the laurel forests. In East Asia this leaf type is often called 'lucidophyll', after the typically shining appearance of the leaves (Kira 1969, 1977). Evergreen Fagaceae forests also occur, however, in drier areas such as:

1. the southeastern USA (with more semi-sclerophyll physiognomy);
2. behind more maritime cool-temperate laurel forests in the Southern Hemisphere (*Nothofagus*); and
3. in mediterranean and other subhumid warm-temperate/subtropical climates (with more sclerophyll character), including the southwestern USA.

In this report, we consider only the laurel forests and evergreen Fagaceae forests of East Asia and the southeastern USA.

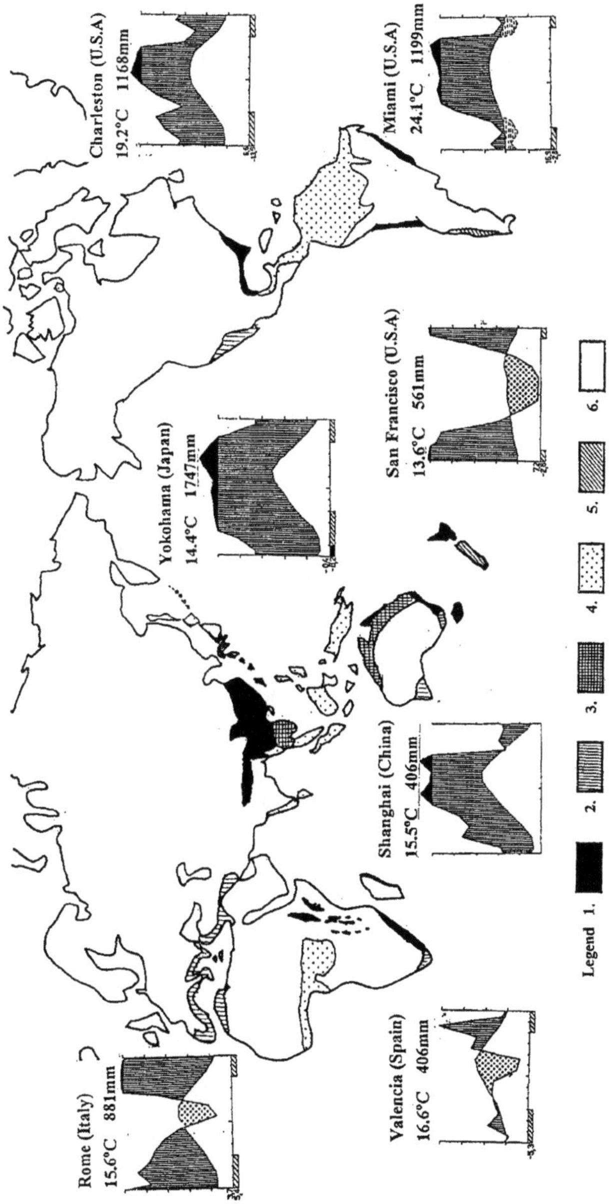

Figure 1: Global distribution of evergreen broad-leaved forests (map)
1. subtropical and warm-temperate EBLF, 2. Mediterranenan-type sclerophyll forest, 3. dry tropical evergreen forest, 4. tropical rainforest, 5. cool-temperate EBLF, 6. other land areas

Evergreen Broad-Leaved Forests near their Northern/Inland Limit in East Asia

The original distribution of evergreen broad-leaved forests in East Asia is shown in Figure 2.

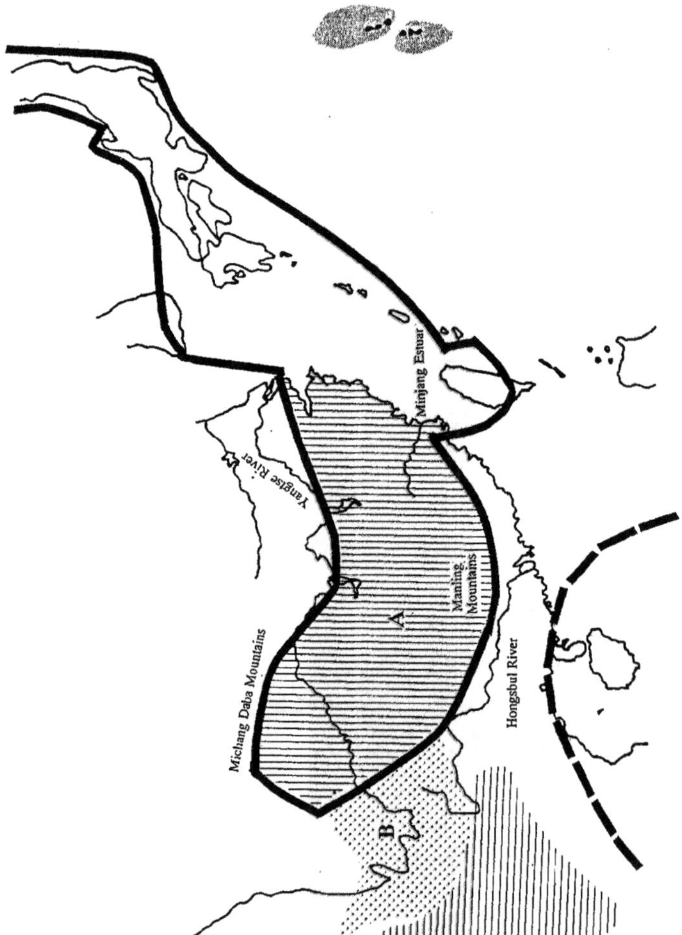

Figrue 2: Distribution of evergreen broad-leaved forests in Asia (map)
A: moist EBLF region in eastern China
B: semi-moist EBLF region in western China and adjacent

The northern limit of the Polysticho-Perseetum thunbergii type of evergreen forest (see Table 1) runs up the coasts of Japan to northern Honshu. A Polystichopso-Castanopsietum sieboldii type cuts across the southern tip of the Korean peninsula. Similar types of evergreen forests extend across eastern to central China from somewhere in the East China Plain north of Shanghai (cf. Table 2) through the central mountains to Sichuan in the west. Evergreen broad-leaved forests on the Boso Peninsula of Pacific Japan, near their northern limit, are summarized in Table 1. These evergreen forests and those of Korea show essentially the same communities, with coastal forests dominated by Lauraceae and inland forests dominated by evergreen Fagaceae (Fujiwara 1981-86; Song et al. 1990; Yim et al. 1996). The evergreen broad-leaved forest types of China are somewhat different, with distinct differences between the forest dominants and community composition in eastern and western China (cf. Wang 1961; Song 1986, 1995).

Evergreen Fagaceae are largely if not completely missing from coastal areas of northern Japan (both sides of southern Tohoku), and the evergreen broad-leaved forests are true "laurel" forests dominated by *Persea thunbergii* (Polysticho-Perseetum thunbergii). Inland, often on the first ranges of coastal hills, forests dominated by *Persea thunbergii* largely disappear and evergreen Fagaceae become dominant, especially *Castanopsis* but also evergreen *Quercus*, along with *Abies firma*, *Illicium*, *Pieris*, etc. Some laurophyll species are common throughout the evergreen broad-leaved forest region of Japan and Korea, including *Persea thunbergii*, *Camellia japonica*, *Neolitsea sericea*, *Eurya japonica*, *Aucuba japonica*, *Hedera rhombea*, *Ophiopogon caespitosus*, etc. Certain other laurophyll species are common throughout the evergreen broad-leaved forest region (coastal and more inland) but do not extend to the northern and inland limits, except where moisture is abundant (such as the Toyama and Yamagata alluvial areas): *Neolitsea sericea*, *Cinnamomum japonicum*, *Ilex integra*, *Liriope jaburan*, *Elaeagnus pungens* and *E. glabra*, and *Dryopteris erythrosora*, etc.

The region from Kanto (around Tokyo Bay, on the Pacific side) to Chubu (inland central Japan) and across to the Japan Sea side of central Honshu has laurel forests, evergreen Fagaceae forests, and mixed Lauraceae-Fagaceae forests, with some subtropical species such as *Elaeocarpus*, *Maesa*, *Chloranthus*, *Polystichopsis* (*Rhumora*) (e.g. Fujiwara 1986). The evergreen broad-leaved species reaching farthest north in Japan are *Camellia japonica* (reaching northernmost Honshu), *Persea thunbergii* (along coastal areas only), and *Quercus myrsinaefolia* (planted as hedges in urban areas within the deciduous, i.e. *Fagus-Quercus* forest region). *Quercus glauca* occurs in the driest inland areas. *Quercus acuta* and *Qu. salicina* reach the highest elevations within the evergreen broad-leaved forest region and occur on coastal hills. *Qu. acuta* occurs mainly in transitional areas, in transitions from coastal laurel forest to inland Fagaceae forest and in the transition from evergreen to summergreen broad-leaved forest. *Qu. myrsinaefolia* occurs as a

Table 1. Evergreen Broad-Leaved Forests near their Northern Limit in Japan (Boso Peninsula).

Community type	1	2	3	4	5	6	7	8	9	10	11	12	13	14	15	16	17	18	19	20	21	22	23	24
Number of relevés	5	7	12	7	7	12	11	34	29	15	12	4	12	14	5	7	6	6	8	5	3	8	13	11
Mean number of species	25	27	25	35	38	30	46	40	43	35	32	34	30	26	26	32	48	39	62	75	40	54	53	50
Characteristic species of association																								
Polystichum polyblepharum	V	III	III	III	.	.	.	I	+	.	.	.	.	.	.	II	II	.	.	.	.	I	I	.
Differential species of subassns. and variants																								
Cyrtomium fortunei	V	.	.	.	.	.	.	.	.	.	.	.	.	+	II	II	I	II	.	.	.	.	.	.
Ardisia crispa	IV	I	.	I	.	.	.	.	.	.	+	1	.	.	.	.	I	.	.	.	.	I	.	.
Dryopteris uniformis	III	I	.	.	.	.	.	.	.	.	.	1	.	+	.	I	II	III	.	.	.	.	.	.
Farfugium japonicum	.	IV	+	.	II	+	I	r	r	.	+	.	.	.	.	.	.	.	.	.	.	.	.	.
Euonymus fortunei var. *radicans*	.	I	.	III	.	.	.	.	.	.	+	.	.	II	I	I	.	I	.	.	.	.	.	.
Prunus lannesiana var. *speciosa*	.	.	.	III	.	.	.	.	.	.	.	.	.	.	.	.	.	.	.	.	.	.	.	.
Characteristic species of association																								
Arachniodes aristata	.	.	.	.	III	IV	II	.	.	.	.	.	.	.	.	.	.	.	.	.	.	II	IV	I
Damnacanthus indicus	.	.	+	.	III	III	V	+	I	II	I	.	.	.	.	.	.	.	I	.	.	II	.	II
Arachniodes sporadosora	.	.	.	.	III	II	IV	.	.	.	.	.	.	.	.	.	.	.	I	.	.	II	V	V
Maesa japonica	.	.	.	.	III	III	III	.	.	.	.	.	.	.	.	.	.	.	.	.	.	IV	IV	V
Myrsine sequinii	.	.	.	.	I	II	II	.	.	.	.	.	.	.	.	.	.	.	.	.	.	II	III	III
Differential species of subassn.																								
Elaeocarpus sylvestris var *ellipt.*	.	.	I	.	I	I	III	.	.	.	.	.	.	.	.	.	.	.	.	.	.	.	.	.
Desmodium laxum	.	.	.	.	.	+	III	r	r	.	.	.	.	.	.	.	.	.	I	.	1	.	I	+
Characteristic species of alliance and order																								
Piper kadzura	.	IV	V	I	IV	III	II	+	r	I	.	.	.	.	.	.	.	.	.	.	.	.	.	.
Elaeagnus macrophylla	.	III	I	IV	II	I	+	II	I	II	.	.	.	.	.	.	.	.	.	.	.	.	.	.
Podocarpus macrophyllus	II	II	IV	II	V	II	III	II	II	III	.	.	.	.	II	.	.	.	II	.	.	.	II	+
Cyrtomium falcatum	I	IV	II	I	II	.	.	+	r	+	.	.	.	.	.	.	.	.	.	.	.	.	.	.
Daphniphyllum teijsmannii	.	.	II	.	III	I	II	I	r	+	+	.	.	.	.	.	.	.	.	.	1	.	II	I
Marsdenia tomentosa	.	.	+	.	.	I	I	.	.	I	.	.	.	.	.	.	.	.		.	.	I	.	I
Characteristic and differential species of assn.																								
Quercus myrsinaefolia	.	.	.	.	.	+	.	r	II	.	+	1	III	II	IV	V	V	V	II	II	2	II	.	I
Thea sinensis	II	.	.	.	.	.	.	r	I	.	.	.	.	I	I	II	V	I	.	.	.	.	.	.

Differential species of variant and subvariant																								
Magnolia kobus	.	.	.	.	.	.	.	.	.	.	.	.	.	.	.	II	III	V	.	.	.	.	.	.
Prunus grayana	.	.	.	.	.	.	.	.	.	.	.	1	I	.	.	II	IV	II	.	.	.	.	.	.
Aphananthe aspera	.	.	II	.	.	.	.	.	.	II	+	3	II	I	I	.	V	V	.	.	.	.	.	.
Athyrium niponicum	.	.	.	.	.	.	.	.	.	+	+	1	.	.	I	.	III	V	.	.	.	I	.	.
Celtis sinensis var. *japonica*	.	.	I	.	.	.	.	.	.	II	.	1	I	II	.	.	III	III	.	.	.	.	.	.
Gynostemma pentaphyllum	I	.	.	II	.	.	.	.	.	.	.	1	.	.	.	.	II	IV	.	.	.	.	.	.
Characteristic and differential species of assn.																								
Abies firma	.	.	.	.	.	.	.	.	.	I	.	.	.	.	.	.	.	.	II	V	1	II	IV	IV
Illicium religiosum	.	.	.	.	.	.	II	.	I	.	.	.	.	+	.	.	I	.	II	IV	1	V	II	V
Pieris japonica	.	.	.	.	.	.	.	.	I	.	.	.	.	.	.	.	.	.	II	III	1	I	III	III
Skimmia japonica f. *rugosa*	.	.	.	.	.	+	II	.	.	II	I	.	.	.	.	.	.	.	II	I	.	.	+	II
Callicarpa mollis	.	.	.	.	.	.	.	.	.	.	.	.	.	.	.	.	.	.	IV	V	1	IV	II	III
Mitchella undulata	.	.	.	.	.	.	.	.	.	+	+	.	.	.	.	.	.	.	IV	III	1	I	II	IV
Dryopteris nipponensis	.	.	.	.	.	.	.	.	.	+	.	.	I	.	.	.	.	I	IV	V	2	.	III	IV
Stauntonia hexaphylla	.	I	.	.	.	I	.	.	.	II	.	1	.	.	.	.	.	.	II	V	1	IV	III	.
Cleyera japonica	.	.	.	.	II	+	III	.	r	+	I	.	.	.	.	.	.	.	V	III	2	IV	V	V
Asarum kooyanum	.	.	.	.	.	+	II	.	II	+	I	.	.	.	.	.	.	.	III	IV	3	IV	IV	V
Differential species of subassn. and variant																								
Cornus kousa	.	.	.	.	.	.	.	.	.	.	+	.	I	.	.	.	.	.	III	III	.	I	.	.
Magnolia obovata	.	.	.	.	.	.	.	.	.	.	.	.	.	.	.	.	.	I	II	III	.	.	.	.
Carex dolichost. v. *glaberrima*	.	.	.	.	.	.	.	.	.	.	.	.	.	.	.	.	.	.	V	.	.	.	II	+
Rhus trichocarpa	.	.	+	.	.	.	.	.	.	.	+	.	.	.	.	.	.	.	IV	I	.	I	I	+
Carpinus laxiflora	.	.	.	.	.	.	.	.	.	.	I	.	.	.	.	I	.	.	I	III	.	I	.	.
Deutzia scabra	.	.	+	.	I	+	II	.	.	.	+	.	.	.	.	.	.	.	.	I	.	II	III	II
Skimmia japonica	.	.	.	.	.	+	.	.	I	+	+	.	.	.	.	.	.	.	.	.	.	I	IV	III
Tsuga sieboldii	.	.	.	.	.	.	.	.	.	.	.	.	.	.	.	.	.	.	.	I	.	II	.	V
Daphniphyllum macropodum	.	.	.	.	.	+	.	r	I	+	+	1	+	.	.	.	.	.	I	.	.	.	+	IV
Euchresta japonica	.	.	.	.	.	.	.	.	.	.	.	.	.	.	.	.	.	.	.	.	.	.	.	III
Actinodaphne longifolia	.	.	.	.	.	.	.	.	.	.	.	.	.	.	.	.	.	.	.	.	.	.	.	II
Symplocos prunifolia	.	.	.	.	.	.	.	.	.	.	.	.	.	.	.	.	.	.	.	.	.	.	+	II
Rhododendron kiyosumense	.	.	.	.	.	.	.	.	.	.	.	.	.	.	.	.	.	.	.	.	.	.	.	II
Characteristic and differential species of alliance and order																								
Quercus salicina	.	.	.	.	.	+	III	r	III	I	I	.	I	+	.	.	II	I	V	III	2	V	V	V
Osmanthus heterophyllus	.	.	.	.	.	I	III	r	III	II	II	1	+	+	.	I	.	.	V	IV	1	V	V	V
Quercus acuta	II	I	.	.	.	II	III	IV	III	II	III	.	IV	III	IV	.	III	.	IV	III	2	III	III	V
Ainslaea apiculata	.	.	.	.	.	I	.	II	II	II	.	.	.	+	II	II	I	.	IV	V	2	III	III	III
Cephalotaxus harringtonia	IV	.	+	.	I	II	III	I	III	II	+	1	I	+	.	II	III	I	III	IV	1	III	II	III
Cymbidium goeringii	.	.	.	.	I	I	IV	I	IV	I	II	1	V	III	.	III	I	II	V	II	1	IV	V	III
Torreya nucifera	.	.	.	.	.	.	.	.	III	II	.	4	+	+	.	.	.	.	III	II	2	IV	III	IV
Rohdea japonica	.	.	+	.	.	+	I	r	II	I	+	.	+	+	.	.	II	.	I	.	.	II	+	.
Elaeagnus glabra	.	.	II	.	.	I	III	II	III	III	III	.	.	+	.	.	.	.	V	V	3	IV	IV	IV
Dryopteris fuscipes	.	.	.	.	.	.	.	+	+	.	.	.	.	.	.	.	.	.	I	I	I	.	II	+

Characteristic and differential species of class

Castanopsis cuspidata v. *sieboldii*	III	V	III	IV	V	V	V	V	V	V	V	4	V	V	IV	III	IV	V	V	IV	3	IV	V	V
Ardisia japonica	IV	III	II	I	III	IV	IV	V	IV	V	V	3	IV	III	V	IV	III	III	IV	IV	3	V	IV	V
Eurya japonica	III	V	IV	IV	III	V	IV	V	V	V	IV	2	V	IV	II	V	V	III	V	III	2	V	V	V
Hedera rhombea	V	III	IV	V	III	V	IV	V	V	V	V	4	V	V	IV	IV	V	V	V	III	1	IV	IV	III
Camellia japonica	IV	V	IV	V	IV	III	III	III	IV	III	III	3	IV	IV	IV	II	III	V	IV	III	2	V	IV	IV
Ophiopogon japonicus	IV	IV	III	III	III	III	V	V	V	IV	IV	3	IV	V	IV	II	III	V	IV	III	2	V	IV	IV
Neolitsea sericea	V	V	IV	V	V	III	IV	IV	V	III	III	4	V	V	V	IV	V	V	II	III	2	V	II	IV
Persea thunbergii	V	V	III	V	V	IV	V	V	V	IV	II	3	IV	III	V	III	III	I	V	I	2	IV	V	III
Ilex integra	.	V	II	IV	III	III	V	V	V	V	IV	3	V	IV	II	II	II	III	V	IV	3	IV	IV	V
Aucuba japonica	V	V	V	V	V	V	.	V	II	V	III	4	V	V	V	V	V	V	V	IV	2	IV	V	V
Trachelospermum asiat. v *interm.*	IV	V	V	II	V	V	V	IV	V	V	V	3	V	.	III	.	I	II	V	V	3	V	V	V
Cinnamomum japonicum	IV	V	V	V	V	IV	V	V	V	IV	IV	.	.	.	V	I	.	III	V	V	3	.	IV	V
Dryopteris erythrosora	IV	III	.	IV	V	III	V	V	V	III	II	4	IV	III	I	III	I	IV	V	IV	2	IV	IV	IV
Liriope platyphylla	IV	III	IV	III	III	II	II	III	III	IV	III	3	III	II	III	III	V	IV	IV	I	2	II	II	+
Trachycarpus fortunei	II	III	I	III	III	II	III	III	III	II	III	3	V	III	IV	III	IV	V	IV	I	1	I	III	.
Ardisia crenata	IV	IV	II	II	IV	V	II	IV	III	III	II	2	V	III	.	I	I	.	V	III	3	III	IV	V
Kadsura japonica	V	III	+	II	II	I	III	II	III	III	II	3	II	+	V	.	II	II	IV	II	1	IV	III	+
Fatsia japonica	III	IV	IV	III	IV	IV	IV	IV	III	III	II	2	+	III	I	I	.	I	II	I	.	II	II	I
Ligustrum japonicum	II	I	.	II	IV	II	III	II	II	II	II	3	IV	III	.	III	I	I	III	.	1	I	III	+
Dendropanax trifidus	.	III	I	IV	V	III	V	V	IV	III	III	.	.	IV	I	II	.	I	V	III	2	IV	V	V
Dryopteris bissetiana	.	III	.	III	III	.	I	V	IV	II	I	2	V	IV	III	IV	.	I	V	II	1	II	II	II
Dryopteris pacifica	.	III	.	II	IV	+	II	:V	III	II	+	2	IV	III	II	.	II	I	IV	.	.	.	III	II
Reineckea carnea	III	II	1	.	III	II	IV	II	II	.	.	3	.	.	.	.	I	III	.	.	1	1	II	.
Pteris cretica	V	.	+	.	I	+	II	+	I	.	.	1	+	+	II	.	I	I	.	I	.	.	II	III
Cinnamomum camphora	.	.	.	.	I	II	I	.	I	II	.	.	.	II	III	.	.	III	.	I	.	.	+	+
Nandina domestica	.	.	.	.	.	.	.	.	I	I	I	2	II	II	.	I	III	II	.	.	.	.	.	.
Pittosporum tobira	.	III	IV	V	III	III	IV	V	III	II	II	1	II	II	.	.	.	.	II	.	1	I	IV	III
Ficus nipponica	.	V	.	.	III	+	IV	II	II	II	I	.	.	.	.	.	.	.	I	I	2	III	IV	V
Ophiopogon planiscapus	II	I	.	I	I	+	II	+	II	I	II	.	.	+	.	I	.	.	IV	III	.	II	II	I
Ilex rotunda	.	.	+	.	II	I	IV	II	II	II	+	.	.	.	.	.	.	.	IV	.	1	II	III	I
Rubus buergeri	III	.	.	.	I	I	III	r	II	+	+	.	.	.	.	.	.	.	II	I	1	IV	I	II
Actinodaphne lancifolia	I	I	.	.	I	+	II	I	+	I	+	.	.	.	.	.	.	.	II	II	1	III	+	II
Alpinia japonica	.	.	+	.	.	+	II	.	I	+	.	.	.	.	.	.	.	.	.	I	.	.	II	+
Liparis nervosa	.	I	.	I	.	I	II	.	.	+	.	.	.	.	.	.	.	.	.	.	.	I	II	.
Myrica rubra	.	.	.	.	.	I	II	I	I	+	.	.	.	.	.	.	.	.	I	.	.	.	+	.
Elaeagnus pungens	.	II	.	II	.	.	.	II	I	.	.	.	.	.	.	.	.	.	.	I	1	.	.	.
Lemmaphyllum microphyllum	.	.	.	.	.	I	.	.	I	+	.	.	.	+	.	.	.	.	II	I	.	II	I	II
Vaccinium bracteatum	.	.	.	.	.	I	.	.	.	.	.	.	.	.	.	.	.	.	II	I	.	I	.	II
Damnacanthus major	.	.	.	.	I	.	I	r	+	.	.	.	.	.	.	.	.	.	.	.	1	.	II	I
Diplazium subsinuatum	.	.	.	.	.	II	I	.	.	.	.	.	.	.	.	.	.	.	.	.	.	I	.	I
Prunus spinulosa	.	.	.	.	II	.	+	.	.	.	.	.	.	.	.	.	.	.	.	.	.	II	.	.
Pyrrosia lingua	.	.	.	.	.	I	.	.	.	.	.	.	.	.	.	.	.	.	I	.	.	.	+	.
Anodendron affine	.	.	.	.	.	.	.	.	.	.	.	.	.	.	.	.	.	.	.	.	1	.	.	I

For list of companion species and references for relevé locations, see larger table in Miyawaki and Fujiwara (1983).

Relevé No.	7	4	5	1	2	3	6
Number of species	70	60	53	53	47	30	39
Differential species of communities							
Choerospondias axillaris	4.3	.	.	.	.	.	.
Acer oliveranum	2.2	.	.	.	.	.	.
Aleurites fordii	2.3	.	.	.	.	.	.
Illicium lanceolatum	1.2	.	.	.	.	.	.
Lindera rubronervia	+.2	.	.	.	.	.	.
Celtis biondii	+	.	.	.	.	.	.
Schima superba	.	1.1	1.2	1.1	1.1	3.2	3.2
Symplocos anomale	.	1.2	.	.	+	1.2	.
Quercus gracilis	1.2	.	+	.	.	.	3.2
Loropetalum chinensis	.	+	.	.	+	.	+
Differential species of sub-units							
Symplocos prunifolia	.	1.2	2.2	.	.	.	.
Vaccinium sprengellii	.	+.2	+	.	.	.	.
Rubus buergeri	+.2	1.2	+	.	.	.	.
Oplismenus undulatifolius	1.2	+	+	.	.	.	.
Daphniphyllum oldhamii	.	.	.	+.2	+	1.1	.
Ilex wilsonii	.	.	.	+	1.1	.	.
Syzygium buxifolium	.	.	.	+	1.1	+	+
Rhododendron ovatum	.	.	.	.	.	2.2	2.2
Symplocos botryantha	.	.	.	.	2.2	+	.
Species of higher units							
Castanopsis fargesii	1.2	4.3	3.2	4.3	3.2	2.2	+
Castanopsis carlesii	+	2.2	3.2	1.1	1.1	+	3.2
Lithocarpus henryii	2.3	2.2	+.2	2.2	.	1.2	2.2
Litsea elongata	2.2	3.3	+.2	+	1.1	.	1.1
Cinnamomum subavenium	1.1	2.2	2.2	.	.	1.2	2.2
Neolitsea aurata v. *chekiang.*	.	2.3	+.2	1.2	2.2	2.3	+
Eurya rubiginosa v. *attenuata*	.	.	2.2	+	+	.	1.1
Trachelospermum jasminoides	3.3	2.2	1.2	1.2	+.2	+.2	.
Quercus rubina	.	.	+	.	.	.	3.2
Species common to Camellietia japonicae							
Quercus myrsinaefolia	1.1	.	.	+	1.1	.	1.1
Quercus glauca	1.1	1.2	+	+	.	.	+
Lithocarpus glabra	.	.	+	.	+	+	.
Persea thunbergii	1.2	1.2	1.2	.	.	.	1.1
Ardisia crenata	1.1	1.1	+	+	+	.	+
Maesa japonica	.	1.2	+.2	+.2	.	.	.
Symplocos lancifolia	+.2	1.2	1.2	1.2	1.2	.	.
Symplocos stellaris	.	2.2	1.2	+	+	.	1.2
Meliosma rigida	.	1.1	.	1.2	1.2	.	.
Ilex purpurea	.	+	1.2	.	.	+	.
Cleyera japonica	.	.	+	+	1.2	1.1	1.1
Myrica rubra	.	.	+	+	1.1	.	.
Photinia serrulata	.	.	+	.	.	+.2	+
Woodwardia japonica	1.2	3.3	.	2.3	2.3	2.2	3.3
Dryopteris erythrosora	.	1.2	+.2	+	+	.	.
Dryopteris pacifica	.	2.3	2.2	.	.	.	2.2
Companion species							
Cunninghamia lanceolata	.	1.2	.	.	1.1	.	+
Lophatherum gracile	.	.	1.2	1.2	+	.	+
Dichronopteris dichotoma	.	+	.	.	2.2	2.2	2.3

Table 2: Evergreen broad-leaved and related forests at Tiantong Reserve, near Shanghai

late-successional species on geologically young substrates such as alluvial areas and volcanic ash plains.

The mixed Lauraceae-Fagaceae forest near its northern limit in Japan is similar to the evergreen broad-leaved forests of the southern tip of Korea, including islands between Korea and Japan. In both regions, the transition from the coastal laurel forest to the evergreen *Quercus* forests inland is characterized by a band of relatively simple forest dominated by *Quercus acuta*. On the other hand, volcanic Cheju-do, the largest island between Korea and Japan, does not have *Quercus acuta*-dominated forest but rather a mixed *Castanopsis-Quercus gilva* forest (cf. Song et al. 1990).

Drier areas in inland central Japan, such as Yamanashi, Gunma and Nagano Prefectures (still wetter than Korea and China), have evergreen forests of *Quercus glauca* on the driest sites and *Qu. myrsinaefolia* on more mesic sites, such as alluvial areas and low-lying plains.

In China, the inland areas do not have forests dominated by Lauraceae, even in eastern China. Evergreen Fagaceae forests, dominated by species of *Castanopsis* as well as by evergreen *Quercus* (still often called *Cyclobalanopsis* in Chinese literature), often also with *Lithocarpus* (*Pasania*), once occurred from eastern China south of roughly Shanghai to Sichuan and Yunnan in the southwest (Wu 1980, Hou 1983). Few Fagaceae species, however, occurred in both the more humid eastern and the drier western subregions (Wang 1961). Coastal areas of eastern China have some of the same species as in Japan, such as *Persea thunbergii*, *Quercus glauca*, *Qu. myrsinaefolia*, and *Qu. gilva*. Forests in the Tiantong sanctuary, near and representative of the northern limit of the eastern evergreen forests, are summarized in Table 2 (see more details in Box et al. 1991).

Geologically young areas, such as volcanic and subtropical limestone islands (e.g. Ogasawara Islands and the Iwo-Jima group), have Lauraceae forests, as also in the Canary Islands (Wildpret de la Torre and Osorio 1997). In the small subtropical islands of Japan, Lauraceae, Fagaceae and other subtropical taxa such as Elaeocarpaceae form mixed forests where soils are sufficiently deep.

Secondary forests in the warm-temperate region of Japan are mainly deciduous, with deciduous Fagaceae such as *Quercus serrata*, *Qu. acutissima*, and *Castanea crenata*, as characteristic of the warmer, drier parts of the deciduous forest region. Along the coastline, (sub)tropical deciduous species such as *Mallotus japonicus* and *Fagara altissima* penetrate to northern Honshu. Forests formed by these *Mallotus* and *Fagara* species correspond to the areas of potential Lauraceae forest (see Figure 3; cf. Fujiwara 1981-86).

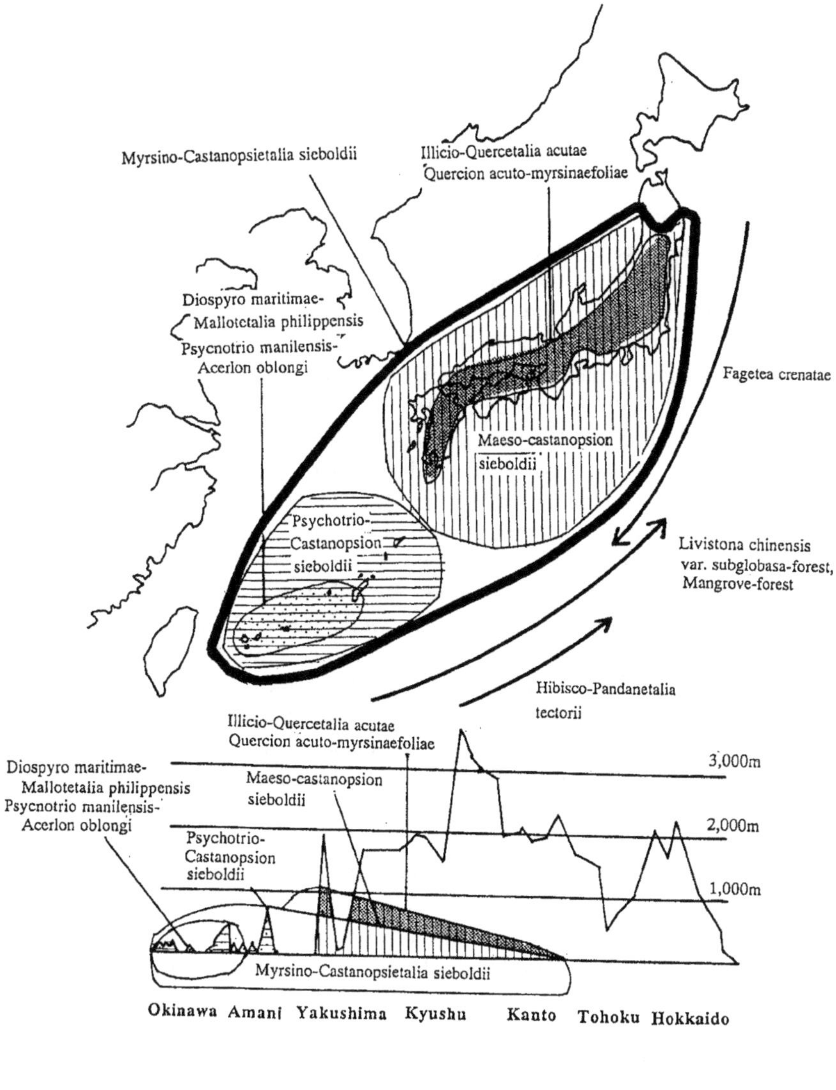

Figure 3: Distribution of temperate to subtropical forests in Japan (map)

Evergreen Broad-Leaved Forests near the Northern/Inland Limit in the Eastern USA

The warm-temperate region of the southeastern USA (the outer coastal plain from southeasternmost Virginia to the dryness limit in southern Texas) represents a region in which evergreen broad-leaved forests are climatically possible (e.g. Greller 1980, 1989; cf. Box and Fujiwara 1988, Box 1994, 1995b). This region has sometimes been portrayed as having essentially no evergreen broad-leaved forests (e.g. Olson 1983), and indeed broad-leaved forest cover of all kinds is extremely patchy, due to the low-lying topography, azonal sand substrate, and prevalence of fire (e.g. Christensen 1988, Wells 1942). Even the "original" vegetation (as described by the first European explorers, cf. Bartram 1792) was a mosaic of topogenic wetlands and forest patches (called hammocks in Florida and south Georgia) in a matrix of pine savannas and woodlands which maintained themselves by frequent ground fires (e.g. Myers and Ewel 1990).

A preliminary summary of the main evergreen broad-leaved forest types of the warm-temperate region of the southeastern USA (i.e. excluding subtropical south Florida) is shown in Table 3. Throughout the warm-temperate region, almost all evergreen broad-leaved species are from genera which also occur in East Asia, usually with more species (Box and Fujiwara 1988). True laurophyll forests (types 17-18 in Table 3) are generally called "bay" forests (after the trees whose aromatic "bay leaves" were used in place of *Laurus nobilis* by early European settlers) and occur only in moist depressions within the otherwise droughty sand substrate. Three main laurophyll taxa are involved, all with close, important laurophyll relatives in East Asia: *Persea borbonia* (red bay, plus less common and almost indistinguishable *P. palustris*); *Magnolia virginiana* (sweet bay, less sclerophyllous than the well known Southern magnolia, *M. grandiflora*); and *Gordonia lasianthus* (loblolly bay, Theaceae; *Gordonia* does not occur in Japan but has about 15 species in China). *Magnolia grandiflora* and *Ilex opaca*, both relatively slow-growing sclero-laurophylls, are not characteristic of bay forests but rather occur more widely, in somewhat more xeric forests, often in association with *Quercus virginiana* (type 3 in Table 3; cf. Daubenmire 1990, Fujiwara and Box 1994).

The potentially more extensive evergreen Fagaceae forests are generally dominated by *Quercus virginiana* (live oak) – the character tree of the American South – a spreading, semi-sclerophyllous, long-living tree which extends from maritime forests as far north as southeastern Virginia (cf. type 1) to shorter, xeric forests and woodlands (type 5) in central Texas (cf. Box and Fujiwara 1988). Other evergreen broad-leaved canopy tree species include *Qu. hemisphaerica*, a widespread dominant of secondary forests especially in Florida and southern Georgia (types 6-8); *Qu. laurifolia*, a floodplain species which is semi-evergreen toward its northern limit (types 9-12); and *Qu. geminata*, a shorter tree similar to *Qu. virginiana* but very sclerophyllous and dominating only on more extreme, deep sterile sand, in dense to open woodlands (types 13-15).

Near the northern limit of the evergreen broad-leaved forest region, along the coast of North Carolina and southeasternmost Virginia, *Qu. virginiana* forms almost monospecific canopies in older maritime forests (e.g. Bourdeau and Oosting 1959). Both *Qu. virginiana* and *Qu. hemisphaerica* occur in younger dune forests. Laurophyll *Magnolia virginiana* and *Persea borbonia* (and/or *P. palustris*) may occur in lower, wetter areas. Inland, *Magnolia virginiana* continues farthest northward, occurring in floodplains somewhat into the southern piedmont of southeastern North America and up the Mississippi River floodplain. Of the true laurophyll species, *Magnolia virginiana* also extends the farthest west, into eastern Texas. The more sclero-laurophyllous *Ilex opaca*, however, occurs both farther north (into southern New Jersey and well into the Southern piedmont) and slightly farther west (in Texas), albeit only as an understorey tree. At the continental limit, in drier central Texas, the last evergreen broad-leaved tree is *Quercus virginiana* (*s.l.*), which forms xeric forests with *Juniperus ashei* and grades into the very similar *Qu. fusiformis* and eventually other, xeromorphic evergreen *Quercus* species in interior parts of Mexico.

Similarities between the vegetation of East Asia, especially Japan, and that of eastern North America were recognized quite early (Gray 1846). Over very large inland areas of what may appear to be part of the warm-temperate climatic zone, i.e. over most of the southern piedmont, the mature forests are actually deciduous rather than evergreen. Although winter temperatures here are generally not extreme (mean temperatures 3-6°C higher than at the same latitude in Japan), temperatures occasionally fall to well below -15°C and to about -20°C along the piedmont-coastal plain boundary. These infrequent but extreme events may control the forest type (Box 1988, 1995a). This situation may be similar to that of the 'warm-temperate deciduous forests' described by Kira (1977, 1991) in inland parts of central-western Honshu. In addition, as in Japan, the secondary forests in the warm-temperate region with evergreen broad-leaved forest potential are also deciduous, mainly dominated by combinations of deciduous *Quercus* species, plus *Carya* and other species from the adjacent deciduous forest zone. These forests persist, especially with sporadic moderate disturbance, and have been recognized by some as the potential natural forest type (e.g. Quarterman and Keever 1962).

Vegetation Development in Experimental Restoration Plots in Japan

Experimental forest restoration projects have been carried out in Japan since 1972, mainly in the warm-temperate evergreen broad-leaved forest region but also in the deciduous north and in subtropical Okinawa (e.g. Miyawaki and

Table 3. Evergreen Broad-Leaved Forest Types (warm-temperate) of the Southeastern USA.

Community type	1	2	3	4	5	6	7	8	9	10	11	12	13	14	15	16	17	18
Number of relevés	10	4	5	4	4	10	7	15	10	4	3	3	5	9	4	3	7	5
Mean number of species	16	34	26	12	27	22	24	27	31	32	74	33	17	20	14	25	19	17
Differential species of communities																		
Polypodium polypodioides	III	4	III	.	.	II	.	II	.	.	.	.	.	.	.	.	.	.
Pinus glabra	.	4	.	.	.	.	.	+	.	.	3	1	.	.	.	.	.	.
Quercus michauxii	.	3	.	.	.	.	.	+	.	.	1	1	.	.	.	.	.	.
Chasmanthium laxum	.	3	.	.	.	+	.	I	.	2	.	.	.	.	.	.	.	.
Sabal palmetto	+	.	V	.	.	II	.	I	V	.	.	.	.	.	.	.	.	.
Oplismenus setarius	.	.	III	.	.	+	.	.	.	.	.	.	.	.	.	.	.	.
Tillandsia recurvata	.	.	III	.	1	IV	.	.	I	.	.	.	IV	.	1	.	.	.
Tillandsia setacea	.	1	II	.	.	+	.	.	+	.	.	.	.	.	.	.	.	.
Nephrolepis sp.	.	.	II	.	.	.	.	.	.	.	.	.	.	.	.	.	.	.
Pinus taeda	.	.	.	4	.	.	V	II	.	.	.	.	II	.	1	.	.	.
Juniperus ashei	.	.	.	.	4	.	.	.	.	.	.	.	.	.	.	.	.	.
Mahonia trifoliata	.	.	.	.	4	.	.	.	.	.	.	.	.	.	.	.	.	.
Yucca sinuata	.	.	.	.	3	.	.	.	.	.	.	.	.	.	.	.	.	.
Quercus sinuata var. *brev.*	.	.	.	.	4	.	.	.	.	.	.	.	.	.	.	.	.	.
Quercus texana	.	.	.	.	3	.	.	.	.	.	.	.	.	.	.	.	.	.
Ulmus crassifolia	.	.	.	.	3	.	.	.	.	.	.	.	.	.	.	.	.	.
Quercus hemisphaerica	.	2	II	.	.	V	V	V	.	.	.	.	.	.	.	.	.	.
Magnolia grandiflora	+	4	.	.	.	V	III	III	.	.	.	.	.	.	.	.	.	.
Anisostichus capreolata	.	.	.	.	.	III	I	I	.	.	.	.	.	.	.	.	.	.
Juniperus silicicola	.	.	.	.	.	I	V	.	.	.	.	.	.	.	.	.	.	.
Quercus laurifolia	.	.	.	.	.	.	.	.	V	4	3	3	.	V	.	.	.	.
Ulmus americana	.	1	I	.	.	.	.	.	III	.	2	.	.	.	.	.	.	.
Psilotum nudum	.	.	.	.	.	.	.	.	.	4	3	.	.	.	.	.	.	.
Sabal minor	.	1	.	.	.	.	.	.	.	2	.	.	.	.	.	.	.	.
Taxodium distichum	.	.	.	.	.	.	.	.	.	.	.	2	.	.	.	.	.	.
Quercus geminata	.	.	.	.	.	.	.	.	.	.	.	.	V	V	4	.	.	.
Quercus myrtifolia	.	.	.	.	.	.	.	.	.	.	.	.	III	III	1	.	.	.
Quercus chapmanii	.	.	.	.	.	.	.	.	.	.	.	.	IV	.	1	.	.	.
Pinus clausa	.	.	.	.	.	.	.	.	.	.	.	.	III	.	.	.	.	.
Sabal etonia	.	.	.	.	.	.	.	.	.	.	.	.	III	.	.	.	.	.
Pinus palustris	.	.	.	.	.	.	.	.	.	.	.	.	.	.	4	.	.	.
Quercus minima	.	.	.	.	.	.	.	.	.	.	.	.	.	.	.	3	.	.
Vaccinium darrowii	.	.	.	.	.	.	.	.	.	.	.	.	.	.	.	3	.	.
Schizachyrium scoparium	.	.	.	.	.	.	.	.	.	.	.	.	.	.	.	3	.	.
Aristida stricta	.	.	.	.	.	.	.	.	.	.	.	.	.	.	.	3	.	.

Common evergreen species in evergreen *Quercus* forests																		
Quercus virginiana	V	4	V	4	4	IV	IV	III	.	1	.	.	.	I	1	.	.	.
Tillandsia usneoides	V	4	III	.	.	V	I	IV	V	.	.	.	IV	III	1	.	.	.
Persea borbonia	IV	2	I	1	.	IV	V	II	IV	.	.	.	I	V	1	3	.	.
Serenoa repens	V	2	.	.	.	V	.	.	.	.	.	.	V	IV	1	3	.	.
Ilex opaca	III	4	I	1	.	I	V	III	.	3	2	3	.	II	.	.	.	.
Ilex vomitoria	I	.	.	1	2	V	III	I	.	.	.	1	.	.	.	.	.	.
Differential species of bay forests																		
Persea palustris	.	.	.	.	.	.	.	.	+	.	.	.	.	.	.	.	IV	.
Nyssa biflora	.	2	.	.	.	I	.	II	.	.	2	.	.	.	.	.	III	I
Magnolia virginiana	.	.	.	.	.	.	II	+	I	3	1	2	.	II	.	+	II	I
Gordonia lasianthus	.	1	.	1	.	.	.	.	.	.	.	.	.	.	.	.	V	V
Pinus serotina	.	.	.	.	.	.	.	.	.	.	.	.	.	.	.	.	.	V
Ilex coriacea	.	.	.	.	.	.	.	.	.	.	.	.	.	.	.	.	I	IV
Cyrilla racemiflora	.	.	.	.	.	.	.	.	I	.	.	.	.	.	.	.	.	IV
Lyonia mariana	.	.	.	.	.	.	.	.	.	.	.	.	.	.	.	.	.	IV
Ilex glabra	.	1	.	.	.	.	.	.	.	1	.	.	.	I	3	.	.	III
Companions																		
Carya glabra	II	3	III	.	.	III	II	II	.	.	2	.	.	.	.	.	.	.
Quercus nigra	I	4	I	.	.	+	.	II	.	.	.	2	.	I	.	.	.	.
Liquidambar styraciflua	I	4	I	1	.	+	.	II	+	2	3	3	.	.	.	.	.	IV
Celtis laevigata	II	.	.	.	.	.	.	II	I	1	2	.	.	.	.	.	.	.
Acer rubrum	I	2	.	3	.	.	.	.	IV	3	2	3	.	.	.	.	IV	IV
Vaccinium arboreum	IV	4	.	.	.	III	II	III	.	.	.	.	.	II	.	.	.	.
Vaccinium corymbosum	.	.	.	4	.	.	.	.	.	.	.	1	.	.	.	.	.	.
Lyonia ferruginea	.	.	.	.	.	.	.	.	.	.	.	.	III	.	.	2	.	.
Lyonia lucida	+	1	.	.	.	.	I	.	.	.	1	.	II	II	.	2	IV	III
Myrica cerifera	.	.	.	2	.	.	.	.	.	.	.	.	.	II	.	.	.	.
Callicarpa americana	I	.	III	.	.	I	I	IV	.	.	.	1	.	.	.	.	.	.
Vitis rotundifolia	III	4	III	.	.	V	V	IV	II	1	2	.	IV	I	.	.	.	.
Parthenocissus quinquefolia	+	2	I	.	.	II	II	II	II	.	.	.	III	I	.	.	.	.
Gelsemium sempervirens	.	2	II	3	.	II	III	IV	.	.	.	1	.	II	.	.	.	.
Mitchella repens	.	4	.	.	.	I	IV	III	+	2	2	.	.	.	.	.	.	.

Community Types:

1-3 *Quercus virginiana* forests (mature)
4 *Qu. virginiana* - *Pinus taeda* forests (secondary)
5 *Qu. virginiana* - *Juniperus ashei* woodland (mature)
6-8 *Quercus hemisphaerica* forests
9-12 *Quercus laurifolia* forests
13-15 *Quercus geminata* woods
16 *Qu. minima* - *Serenoa repens* scrub
17-18 Laurophyll "bay" forests

Fujiwara 1988; Miyawaki 1993; Miyawaki et al. 1993). In the evergreen region, growth rates and post-planting invasion histories have been surveyed at several sites. Weed species invade these plantations only during the first three years. Thereafter, native shrub and other woody species from nearby gardens begin to invade, such as *Trachycarpus fortunei* (Palmae), *Rhaphiolepis umbellata*, *Pittosporum tobira*, and *Ligustrum japonicum*. After about five years, understorey and canopy tree species, often with fleshy fruits, also begin to reinvade, including *Aphananthe aspera*, *Celtis sinensis* var. *japonica*, *Neolitsea sericea*, *Cinnamomum japonica*, and a potential dominant *Persea thunbergii*. After 10 years, evergreen *Quercus* species begin to appear, including *Qu. myrsinaefolia* and *Qu. glauca*. The composition of some experimental evergreen broad-leaved forest plots in Japan is shown in Table 4.
Based on these results, it appears that species with fleshy fruits, which can be dispersed by birds, are the most vagile and can invade the new forests earliest. In general, near the northern limit of the evergreen broad-leaved forest region in Japan, evergreen broad-leaved forest species cannot grow under deciduous forest but can survive in evergreen conifer plantations (*Cryptomeria*, *Pinus densiflora*) (Fujiwara 1981-86). The appearance of some evergreen broad-leaved species in the understoreys of deciduous forests and their survival in dense tree plantations of evergreen seedlings thus suggests two possibilities: either these sites are in fact in the evergreen broad-leaved forest region or the climate has warmed.

Climate Change and Vegetation Shift

Comparison of any adequate maps of potential natural vegetation and actual land cover, globally or locally, will show that the warm-temperate and subtropical evergreen broad-leaved forests of the world are among the most decimated of all major vegetation types (e.g. Suzuki 1982). Even without climate change, it has been seen in various areas that herbaceous weed and some other vagile secondary species from tropical and subtropical zones can survive in the warm-temperate zone after destruction of the zonal forest. On the other hand, the dominants of secondary forests in the warm-temperate forest zone are generally temperate deciduous *Quercus* species. Arborescent secondary species from the tropical and subtropical zones seem to penetrate the warm-temperate zone only sporadically.
Some aspects of potential vegetation changes can be studied using simple envelope models based on the climatic limits of species ranges. Even in flat areas with largely azonal substrate, such as Florida, it has been shown that a simple climatic-envelope model could capture the main geographic features of the ranges of native woody species and may serve as an adequate basis for suggesting range shifts under climatic warming (Box et al. 1993). This model suggested that the climatic range of *Quercus virginiana* could retreat almost

completely from South Florida (where it is currently important even in the subtropical forests) with warming of only 1-2°C and without climatic drying (see Figure 4, from Box and Crumpacker 1993). Although envelope models do not include compensatory changes in species metabolism, such models are easy to apply and can be used to predict potential range shifts (which can be monitored) as well as potentials for the breakup of present-day natural communities as the climatic ranges of their main structural species diverge.
An idea of possible vegetation shifts in East Asia under climatic warming is suggested by the zonation scheme shown in Table 5 plus general considerations of the relative vagility of secondary and weedy species versus species characteristic of mature communities. The following general patterns might be expected:

1) Natural forest types cannot migrate quickly but weedy and secondary species can, so weed and secondary forest communities can be expected to expand northward; in the evergreen broad-leaved forest areas this means a particular danger of invasion by weedy tropical vegetation;

2) If the warming is not compensated by increased precipitation, then the climate becomes effectively drier, fire frequency and extent may increase significantly, and pine forests may expand;

3) Mangroves may also expand poleward, as seawater temperatures rise;

4) Those laurel-forest species which can be dispersed readily by birds can be expected to expand northward, at least to mesic sites with deep soil;

5) Evergreen *Quercus* and other laurophyll-forest species (with larger, heavier seeds) may also expand northward, but this will happen much later (and only on mesic and drier sites).

In terms of the communities shown in the phytosociological tables, the most probable northward expansions may include:

1) communities 1-10 of Table 1, which involve dominants with berries or which are wind-dispersed;

2) the *Choerospondias-Acer* (both deciduous) community of Table 2; and

3) the *Quercus hemisphaerica* communities (types 6-8) of Table 3.

Fagaceae species and communities dominated by Fagaceae may move most slowly.

Table 4. Composition of Environmental Protection Forest Plots in Japan.

Relevé number	1	2	3	4	5	6	7	8	9	10	11	12	13	14	15	16	17	18	19	20	21	22	23	24	25	26	27	28	29	30	31	32	33	34	35
Year planted	72	73	76	76	76	76	77	74	71	73	73	77	78	78	73	78	74	77	79	76	77	82	81	82	72	74	80	82	80	79	82	81	81	83	75
Canopy height	8	9	9	7	5	7	6	7	12	9	5	6	5	5	5	6	6	7	7	4	4	2	3	1	5	4	4	1	2	2	1	3	2	1	1
Species planted																																			
Persea thunbergii	.	.	.	.	.	2	4	+	2	1	1	3	3	1	(1)	.	1	.	1	4	5	2	.	.	.	.	2	.	+	1	2	4	.	2	+
Cinnamomum camphora	5	5	3	5	3	2	2	5	3	5	2	3	2	2	.	.	3	.	1	1	1	3	5	4	.	.	+	.	1	1	.	.	.	2	3
Quercus glauca	1	2	5	3	2	3	.	1	(+)	1	+	3	4	3	.	2	1	.	3	.	1	2	.	.	.	.	4	.	.	.	.	.	3	.	.
Quercus myrsinaefolia	1	.	.	.	.	+	.	(+)	(+)	1	(+)	.	.	.	.	2	5	3	1	1	.	.	1	+	+	.	3	1	1	.	.	.	.	.	.
Pasania edulis	.	.	.	.	4	.	.	+	3	+	+	.	.	+	(+)	2	.	.	.	3	.	.	.	.	.	3	1	1	2	2	.	.	.	.	.
Castanopsis cuspidata v. *sieb.*	(1)	.	.	.	.	.	.	+	(+)	(+)	(+)	3	2	2	.	.	.	4	4	3	.	.	.	.	.	.	.	.	.	1	2	.	3	1	+
Elaeocarpus sylvestris v. *ellip.*	.	.	.	1	.	.	3	2	.	.	4	.	.	.	1	.	.	.	.	.	.	.	.	.	+	.	.	1	.	.	.	.	.	.	.
Myrica rubra	.	.	.	.	.	.	1	.	.	.	(+)	2	2	3	.	2	.	.	.	.	.	2	.	.	.	.	.	.	.	2	2	.	.	1	.
Ilex rotunda	.	.	.	.	.	.	.	.	.	+	(+)	1	1	2	2	2	.	.	.	.	.	2	.	.	.	.	.	.	.	.	.	.	.	1	.
Quercus gilva	.	.	.	.	.	.	.	.	.	.	.	.	.	.	(1)	.	.	.	.	.	.	.	.	.	.	.	.	.	.	.	.	4	.	.	.
Differential Species of Sub-units																																			
Aphananthe aspera	+	+	+	(+)	.	(+)	(+)	+	.	.	.	+	.	(+)	.	.	.	.	.	.	.	.	.	.	.	.	.	.	.	.	.	.	.	.	.
Celtis sinensis var. *japonica*	+	+	+	1	(+)	(+)	.	1	.	.	.	(+)	.	.	(+)	.	.	.	.	.	.	.	+	+	.	.	.	.	.	.	.	.	.	.	.
Aucuba japonica	.	(+)	+	.	.	+	.	.	+	.	.	.	.	.	.	.	.	.	.	.	.	.	.	.	.	.	.	.	.	.	.	.	.	.	.
Quercus phillyraeoides	.	.	.	.	.	.	.	.	.	.	.	.	.	.	.	.	.	.	.	.	.	.	.	.	.	.	.	.	.	.	.	.	.	.	.
Fatsia japonica	+	.	+	(+)	.	.	.	(+)	.	.	.	.	.	.	.	.	.	.	.	.	.	.	.	.	.	.	.	.	.	.	.	.	.	.	.
Cinnamomum japonicum	+	.	+	.	.	.	.	.	.	(+)	.	.	.	.	.	.	.	.	.	1	.	.	.	.	.	.	.	.	1	.	.	.	.	.	.
Trachycarpus fortunei	.	.	+	.	.	(+)	.	.	.	.	+	.	.	.	.	.	.	.	.	.	.	.	.	.	.	.	.	.	.	.	.	.	.	.	.
Neolitsea sericea	.	.	+	.	.	.	.	.	+	+	.	.	.	.	.	.	.	+	.	.	.	.	.	.	.	.	.	.	.	.	.	.	.	.	.
Eurya japonica	+	.	.	.	.	.	.	.	.	.	.	.	+	.	+	+	.	.	.	.	.	.	.	.	.	.	.	.	.	.	.	.	+	.	.
Osmanthus heterophylla	.	.	.	.	.	.	.	.	.	.	.	.	.	.	.	.	.	.	.	.	.	.	.	.	.	.	.	.	.	.	.	.	.	.	.
Hedera rhombea	+	+	.	.	.	.	.	.	.	.	.	.	.	.	.	.	.	.	.	.	.	.	.	.	.	.	.	.	.	.	.	.	.	.	.
Cucculus orbiculatus	+	+	.	.	.	.	.	.	.	.	.	.	.	.	.	.	.	.	.	.	.	.	.	.	.	.	.	.	.	.	.	.	.	.	.
Ampelopsis brevipedunculata	+	+	.	.	.	.	.	.	.	.	.	.	.	.	.	.	.	.	.	.	.	.	.	.	.	.	.	.	.	.	.	.	.	.	.
Elaeagnus macrophylla	.	+	.	.	.	.	.	.	.	.	.	.	.	.	.	.	.	.	.	.	.	.	.	.	.	.	.	.	.	.	.	.	.	.	.
Mahonia japonica	.	+	.	.	.	.	.	.	.	.	.	.	.	.	.	.	.	.	.	.	.	.	.	.	.	.	.	.	.	.	.	.	.	.	.
Distylium racemosum	.	.	.	.	.	.	.	2	2	4	3	.	.	.	.	.	.	.	.	.	.	.	.	.	.	.	.	.	.	.	.	.	.	.	.
Quercus variabilis	.	.	.	.	.	.	.	.	.	1	1	.	.	.	.	.	.	.	.	.	.	.	.	.	.	.	.	.	.	.	.	.	.	.	.
Pueraria lobata	.	.	.	.	.	.	.	.	.	.	.	+	+	.	.	.	.	.	.	.	.	.	.	.	.	.	.	.	.	.	.	.	.	.	.
Rhus trichocarpa	.	.	.	.	.	.	.	.	.	.	.	+	+	.	.	.	.	.	.	.	.	.	.	.	.	.	.	.	.	.	.	.	.	.	.
Potinia glabra	.	.	.	.	.	.	.	.	.	.	.	.	+	.	.	.	.	.	.	.	.	.	.	.	.	.	.	.	.	.	.	.	.	.	.
Osmanthus aurantiacus	.	.	.	.	.	.	.	.	.	.	.	.	+	.	.	.	.	.	.	.	.	.	.	.	.	.	.	.	.	.	.	.	.	.	.
Podocarpus macrophylla	.	.	.	.	.	.	.	.	.	.	.	1	.	.	.	.	.	.	.	.	.	.	.	.	.	.	.	.	.	.	.	.	.	.	.
Miscanthus sinensis	.	.	.	.	.	.	.	.	.	.	.	.	.	.	.	.	.	.	.	.	+	.	.	.	.	.	.	.	.	.	+	.	+	+	3
Erigeron sumatrensis	.	.	.	.	.	.	.	.	.	.	.	.	.	.	.	.	.	.	.	.	.	.	.	.	.	.	.	.	.	.	+	.	+	1	.
Acalypha australis	.	.	.	.	.	.	.	.	.	.	.	.	.	.	.	.	.	.	.	.	.	.	.	.	.	.	.	.	.	.	.	+	+	+	.
Paspalum thunbergii	.	.	.	.	.	.	.	.	.	.	.	.	.	.	.	.	.	.	.	.	.	.	.	.	.	.	.	.	.	.	+	.	.	.	+
Lapsana humilis	.	.	.	.	.	.	.	.	.	.	.	.	.	.	.	.	.	.	.	.	.	.	.	.	.	.	.	.	.	.	1	.	+	.	.

Digitaria adscendens	.	.	.	.	.	.	.	.	.	.	.	.	.	.	.	.	.	.	.	.	.	.	.	+	.	.	.	.	.	.	.	2	3	.	.
Crassocephalum crepidioides	.	.	.	.	.	.	.	.	.	.	.	.	.	.	.	.	.	.	.	.	.	.	.	.	.	.	.	.	.	.	.	+	+	.	.
Ixeris debilis	.	.	.	.	.	.	.	.	.	.	.	.	.	.	.	.	.	.	.	.	.	.	.	.	.	.	.	.	.	.	.	+	+	.	.
Vicia hirsuta	.	.	.	.	.	.	.	.	.	.	.	.	.	.	.	.	.	.	.	.	.	.	.	.	.	.	.	.	.	.	.	+	+	.	.
Setaria glauca	.	.	.	.	.	.	.	.	.	.	.	.	.	.	.	.	.	.	.	.	.	.	.	.	.	.	.	.	.	.	.	.	+	.	+
Gnapha affine	.	.	.	.	.	.	.	.	.	.	.	.	.	.	.	.	.	.	.	.	.	.	.	.	.	.	.	.	.	.	.	.	+	+	.
Setaria viridis	.	.	.	.	.	.	.	.	.	.	.	.	.	.	.	.	.	.	.	.	.	.	.	.	.	.	.	.	.	.	.	.	+	.	+
Echinochloa crus-galli	.	.	.	.	.	.	.	.	.	.	.	.	.	.	.	.	.	.	.	.	.	.	.	.	.	.	.	.	.	.	.	+	.	.	.
Polygonum longisetum	.	.	.	.	.	.	.	.	.	.	.	.	.	.	.	.	.	.	.	.	.	.	.	.	.	.	.	.	.	.	.	+	.	+	.
Pennisetum alopecuroides	.	.	.	.	.	.	.	.	.	.	.	.	.	.	.	.	.	.	.	.	.	.	.	.	.	.	.	.	.	.	.	.	.	.	2
Arthraxon hispidus	.	.	.	.	.	.	.	.	.	.	.	.	.	.	.	.	.	.	.	.	.	.	.	.	.	.	.	.	.	.	.	.	.	.	4
Arundinella hirta	.	.	.	.	.	.	.	.	.	.	.	.	.	.	.	.	.	.	.	.	.	.	.	.	.	.	.	.	.	.	.	.	.	.	1
Hydrocotyle maritima	.	.	.	.	.	.	.	.	.	.	.	.	.	.	.	.	.	.	.	.	.	.	.	.	.	.	.	.	.	.	.	.	.	.	2
Specularia perfoliatum	.	.	.	.	.	.	.	.	.	.	.	.	.	.	.	.	.	.	.	.	.	.	.	.	.	.	.	.	.	.	.	.	.	.	1

Character Species of Camellietia japonicae

Ligustrum japonicum	3	1	+	+	+	.	(+)	+	+	.	1	1	1	+	+	+	+	+	+	2	+	+	+	.	1	+	.	.	.	.	.	.	.	.	.
Pittosporum tobira	2	.	+	+	+	.	(+)	.	.	(+)	+	.	.	+	(+)	.	.	.	1	.	+	+	(+)	.	+	.	.	.	.	.	3	.	.	.	.
Euonymus japonicus	.	+	.	.	+	(+)	.	.	+	.	.	+	2	(+)	+	+	+	+	.	.	.	.	.	.	.	.	.	.	.	.	.	.	.	.	.
Rhaphiolepis umbellata	+	.	+	2	.	.	+	1	.	+	2	+	.	2	.	.	.	.	.	.	1	.	.	.	1	.	.	.	.	.	.	.	.	.	.
Viburnum awabuki	.	(+)	(+)	+	.	.	.	(+)	.	.	.	.	1	1	.	.	.	.	.	.	.	.	.	.	.	.	.	.	.	.	1	.	.	.	.
Camellia japonica	.	.	.	.	.	.	.	.	.	+	.	.	.	.	.	+	.	1	.	.	(+)	.	.	.	.	.	.	.	+	.	.	.	.	.	.
Ophiopogon japonicus	.	.	.	+	.	.	.	.	.	+	.	.	.	.	.	+	.	.	.	.	.	.	.	.	.	.	.	.	.	.	.	.	.	.	.
Elaeagnus pungens v. *caesp.*	.	.	.	.	.	.	.	.	.	.	.	.	.	.	.	.	.	+	.	.	.	.	.	.	.	.	.	.	.	.	.	.	.	.	.

Companion Species

Paederia scandens var. *mairei*	.	+	+	.	1	(+)	(+)	+	+	.	+	+	+	.	+	.	.	+	.	.	(+)	.	.	.	.	.	+	.	.	.	.	.	.	.	.
Oxalis corniculata	.	.	.	.	+	.	.	.	.	.	.	.	.	.	.	.	.	.	.	.	+	.	+	.	.	.	.	+	.	.	+	.	+	+	+
Vitis angustifolia	.	.	.	.	.	.	.	.	.	.	.	.	.	.	.	.	.	.	1	.	+	r	.	.	.	.	.	.	+	+	.	.	+	.	+
Solidago altissima	.	.	.	.	.	.	.	.	.	.	.	.	.	.	(+)	.	.	.	.	.	+	.	.	.	.	.	.	.	.	.	.	+	2	1	2
Artemisia princeps	.	.	.	.	.	.	.	.	.	.	.	.	.	.	.	+	.	.	.	.	.	.	.	.	.	.	.	.	3	.	.	.	2	1	+
Equisetum arvense	.	.	.	.	+	.	.	.	.	.	.	.	+	.	.	.	.	.	+	.	+	.	.	.	.	.	.	.	.	.	.	+	.	.	+

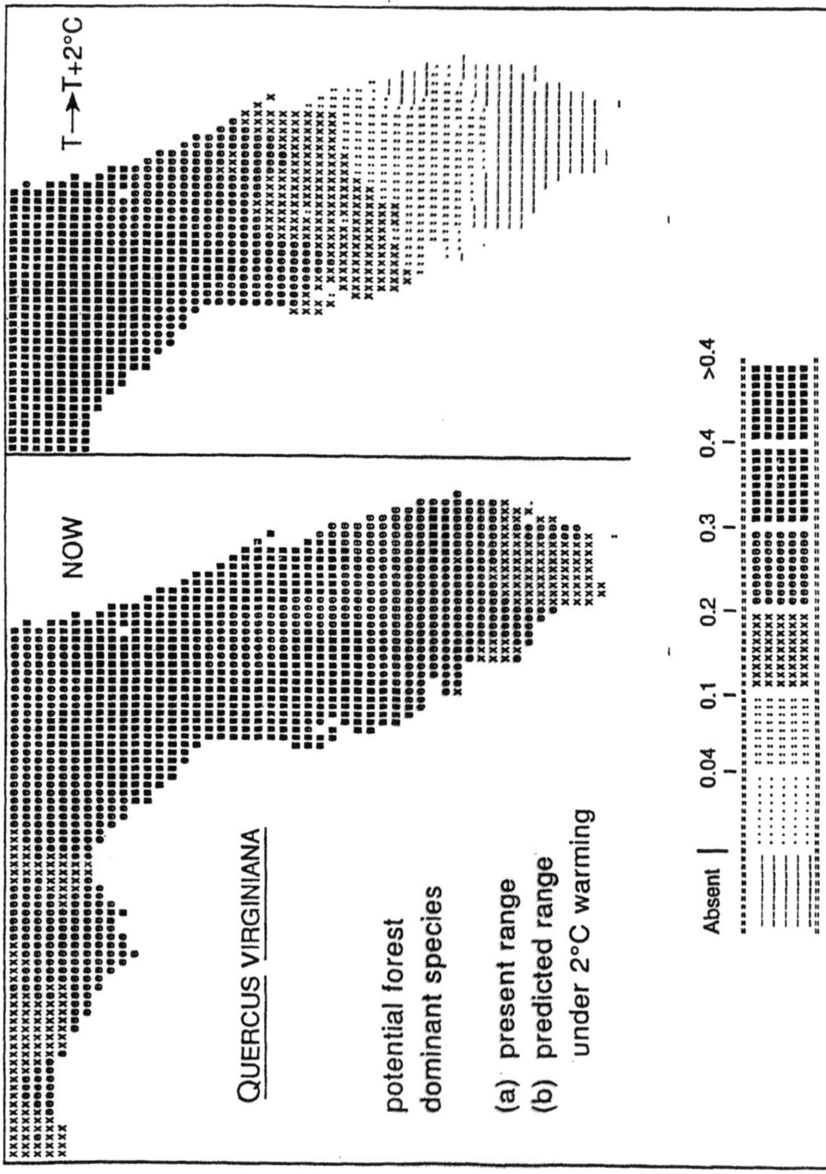

Figure 4: Potential change in the range in Florida of *Quercus virginiana* under 2°C climate warming

Shading represents "fitness" levels generated by computer algorithm, based on proximity to closest climatic limits

Continent	Tropical Forest	Subtropical Forests	Warm-Temperate Forests	Temperate/Tropical Montane Forests	Boreal Forest
Natural Types					
East and SE Asia	Dipterocarpaceae forest *Celtis* forest	Lauraceae forest Fagaceae forest	evergreen Fagaceae forest *Chamaecyparis* etc.	deciduous Fagaceae forest	*Abies*, *Picea*, *Pinus* forests
South America	Leguminosae and mixed forests	Lauraceae forest Mesic evergreen Chaco forest	*Araucaria* and *Podocarpus* forests	*Nothofagus* forest	- - - - -
Africa (E side)	Leguminosae, esp. Caesalp. forests *Croton-Ficus* forest	Lauraceae forest *Newtonia* forest	*Podocarpus* forest	*Hagenia-Hypericum* montane forest Ericaceae scrub	- - - - -
Substitute Vegetation					
	secondary forest with tropical pioneer species		secondary forest with temperate species (e.g. deciduous *Quercus*)		*Betula*, *Populus* woods, etc.
	tropical climber-dominated mantle communities		mantle communities with warm-temperate climbers and bushes		
	tropical weeds		mixed tropical and temperate weeds		temperate weeds
Coastal Vegetation					
	mangrove forest		salt marsh		
	tropical coastal vegetation		temperate/boreal coastal vegetation		

Table 5: Forest zones and substitute communities in East Asia and analogous areas

Conclusion

So far, in eastern North America, there seems to be no consistent evidence of northward expansion by warm-temperate evergreen broad-leaved tree species. This may be due in part, however, to two situations largely unique to southeastern North America:

1) geographic coincidence of the northern boundaries of many evergreen broad-leaved species with the change from coastal-plain sand to piedmont clay substrates; and

2) the combination of relatively high mean winter temperatures but unusually low, albeit infrequent extremes (see Box 1995a), which may periodically "knock back" potential northward migrants.

Various evergreen broad-leaved tree species survive as ornamentals in towns and around houses further north, especially *Magnolia grandiflora*, *M. virginiana*, and sometimes even *Quercus virginiana*, but they do not occur in natural populations. There is a definite northward expansion by the exotic East Asian evergreen *Ligustrum sinense*, which has taken over most floodplains in the Southern piedmont over the last few decades. East Asian evergreens, however, have thrived on the Southern piedmont since last century (such as *Lonicera japonica*, which is deciduous in Japan!), so the recent invasion by *Ligustrum sinense* may be a natural expansion into an open niche rather than a response to higher temperatures.

On the other hand, this is not to suggest that northward expansion will not occur. There is some evidence that some subtropical species may be expanding northward in warm-temperate Florida, perhaps also into southern Georgia. There also seems to be some evidence in Japan that bird-dispersed evergreen broad-leaved trees may be expanding or at least becoming more abundant in parts of central and northern Japan.

Evergreen broad-leaved species are numerous in East Asia, so one should expect at least some of them to spread easily into deciduous forest areas under warmer conditions. Species with less cold tolerance, especially Lauraceae such as *Persea, Cinnamomum*, and *Neolitsea*, may spread especially along coastal areas and locally in other areas with cold moderated by proximity to water bodies. On the other hand, *Quercus* may expand more in inland and other drier areas. Shrubby species, most of which have fleshy fruits, can spread easily, including *Elaeagnus* spp., *Fatsia japonica, Dendropanax trifidus*, *Eurya* spp., *Nandina domestica*, *Aucuba japonica*, *Ardisia* spp., and the palm *Trachycarpus fortunei* in particular. Some of these, growing inside evergreen conifer plantations (*Cryptomeria japonica*) or stands of dense, tall bamboo (*Phyllostachys heterocycla* var. *pubescens*), also withstand outside air temperatures below the usual evergreen-limiting value of -15°C and may represent relicts from the earlier Jomon period, which was about 3°C warmer than now (Fujiwara 1981). In areas with few evergreen broad-leaved species within the deciduous forest region, such as Europe, planted exotic evergreens can spread more easily into deciduous forests, forming weedy groves or other patches.

Responses to climatic warming will be seen in various processes and at various scales, including global changes, responses of non-plant organisms which may affect vegetation, and local changes in urban areas. Responses by plants and vegetation, however, may take considerable time. Insect and other pest populations, on the other hand, have much shorter lifespans and will adapt genetically much faster to changed conditions than will most plants, especially the longer-living (dominant) plants which provide the main structure of ecosystems. Climatic changes will drastically shift the competitive balance between plants and pest populations in favor of the latter, permitting enormous herbivory and pathological damage to vegetation during the transition period. It is not understood at all whether vegetation recovery from this transient damage will require decades, a few centuries, or several to many millennia.

References

Bartram W. 1792 (facs. 1973): Travels through North and South Carolina, Georgia, East and West Florida. Savannah: Beehive Press (facs. of London eds.)

Bourdeau P. F. and H. J. Oosting (1959): The maritime live oak forest in North Carolina. Ecology 40: 148-152.

Box E. O. (1988): Some Similarities in the Climates and Vegetations of Central Honshu and central eastern North America. In: Contributions to Knowledge of the Flora and Vegetation of Central Japan (A. Miyawaki and E. Landolt, eds.). Veröff. Geobot. Inst. ETH, Stiftung Rübel (Zürich) 98: 141-168.

Box E. O. (1994): Eastern North America: Natural Environment and Sampling Strategy. In: Vegetation in Eastern North America (A. Miyawaki et al., eds.) pp. 21-57. University of Tokyo Press.

Box E. O. (1995a): Global and local climatic relations of the forests of East and Southeast Asia. In: Vegetation Science in Forestry (E. O. Box et al., eds.), Handbook Vegetation Sci., vol. 12/1, pp. 23-55. Dordrecht: Kluwer.

Box E. O. (1995b): Handbook for the International Phytogeographical Excursions in the Southeastern USA. Sponsored by Internat. Assn. for Vegetation Science, produced by the author (paperback). 122pp (main) and 170pp (including Florida).

Box E. O. (1995c): Global Potential Natural Vegetation: Dynamic Benchmark in the Era of Disruption. In: Toward Global Planning of Sustainable Use of the Earth -- Development of Global Eco-engineering (Sh. Murai, ed.), pp. 77-95. Amsterdam: Elsevier.

Box E. O. (1997): Bioclimatic position of evergreen broad-leaved forests. In: Island and High-Mountain Vegetation: Biodiversity, Bioclimate and Conservation, Proceedings, IAVS Symposium Tenerife, pp. 17-38. Universidad de La Laguna, Serie Informes no. 40, Tenerife (Canary Islands), Spain.

Box E. O. and D. W. Crumpacker (1993): A Model for Predicting Climate-Induced Vegetation Change in Florida and Adjacent Areas. In: Proceedings, IGBP Internat. Symposium on Global Change, Waseda University, Tokyo, 519-535.

Box E. O., D. W. Crumpacker and E. D. Hardin (1993): A climatic model for location of plant species in Florida, U.S.A. J. Biogeography 20: 629-644.

Box E. O. and K. Fujiwara (1988): Evergreen broad-leaved forests of the southeastern United States: preliminary description. Bull. Inst. Environm. Sci. Technol., Yokohama Natl. Univ. 15: 71-93.

Box E. O., Song Y.-Ch., A. Miyawaki and K. Fujiwara (1991): An evergreen broad-leaved forest in transitional eastern China. Bull. Inst. Environm. Sci. Technol., Yokohama Natl. Univ. 17: 63-84.

Box E. O., Ch.-H. Chou and K. Fujiwara (1998): Richness, climatic position, and biogeographic potential of East Asian laurophyll forests, with particular reference to examples from Taiwan. Bull. Inst. Environm. Sci. Technol., Yokohama Natl. University 24: 61-95.

Christensen N. L. (1988): Vegetation of the Southeastern Coastal Plain. In: North American Terrestrial Vegetation (M. G. Barbour and W. D. Billings, eds.) pp. 317-363. Cambridge University Press.

Daubenmire R. F. (1990): The *Magnolia grandiflora - Quercus virginiana* forest of Florida. Amer. Midl. Nat. 123: 331-347.

Fujiwara K. (1981-86): Nihon-no jouryoku shoujou shinrin-no gunraku taikei (Phytosociological investigation of the evergreen broad-leaved forests of Japan). 4 parts. Bull. Inst. Environm. Sci. Technol., Yokohama Natl. Univ. 7: 67-133; 8: 121-150; 9: 139-160; 13: 99-149 (in Japanese).

Fujiwara K. (1986): Jouryoku Shoujou Shinrin (Evergreen Broad-Leaved Tall Forest.) In: Nippon Shukusei Shi (Japan Vegetation Record, 10 vols.) (A. Miyawaki, ed.), vol. 7: Kanto, pp. 129-145. Tokyo: Shibundo. (In Japanese)

Fujiwara K. and E. O. Box (1994): Evergreen Broad-Leaved Forest Region of the Southeastern United States. In: Vegetation in Eastern North America (A. Miyawaki et al., eds.) pp. 273-312. University of Tokyo Press.

Gray A. (1846): Analogy between the flora of Japan and that of the United States. Amer. J. Sci. Arts II(2): 135-136.

Greller A. M. (1980): Correlation of some climatic statistics with distribution of broad-leaved forest zones in Florida, USA. Bull. Torrey Botan. Club 107: 189-219.

Greller A. M. (1989): Correlation of warmth and temperateness with the distributional limits of zonal forests in eastern North America. Bull. Torrey Botan. Club 116: 145-163.

Hara M. and Ch. Yonebayashi (eds.) (1997a): Lucidophyllous Forests in Southwestern Japan and Taiwan. Natural History Research, Special issue no. 4. Chiba: Natural History Museum.

Hara M. and Ch. Yonebayashi (eds.) (1997b): Shouyou Jurin no Seitai-Gaku (Lucidophyll Forest Ecology). Chiba Central Museum. (in Japanese).

Hou H.-Y. (1983): Vegetation of China with reference to its geographical distribution. Ann. Missouri Bot. Garden 70: 509-548.

Hübl E. (1988): Lorbeerwälder und Hartlaubwälder (Ostasien, Mediterraneis und Makaronesien). Düsseldorfer Geobotan. Kolloq. 5: 3-26.

Kira T. (1969): (What is lucidophyll forest?) In: (Culture in the Laurel Forest Zone) (S. Ueyama, ed.), pp. 43-84. Chuuou-kouron-sha, Tokyo. (In Japanese)

Kira T. (1977): A Climatological Interpretation of Japanese Vegetation Zones. In: Vegetation Science and Environmental Protection (A. Miyawaki et al., eds.) pp. 21-30. Tokyo: Maruzen.

Kira T. (1991): Forest ecosystems of East and Southeast Asia in a global perspective. Ecol. Res. (Japan), 6: 185-200. Reprinted in: Vegetation Science in Forestry (E. O. Box et al., eds.) pp. 1-21. Dordrecht: Kluwer.

Klötzli F. (1984): The position of Fagaceae and Myrtaceae on the Pacific mountains. Erdwiss. Forsch. 18: 337-354.

Klötzli F. (1988): On the global position of the evergreen broad-leaved (non-ombrophilous) forest in the subtropical and temperate zones. In: Contributions to Knowledge of the Flora and Vegetation of Central Japan (A. Miyawaki and E. Landolt, eds.). Veröff. Geobot. Inst. ETH, Stiftung Rübel (Zürich) 98: 169-196.

Laurel Forest Culture Study Group (1986): (Laurel Forest Culture Trip: Visiting Yunnan, China.) Miyasaki: Jouyou-Shirin Bunka-o Kangaeru-Kai (Laurel Forest Culture Study Group). 96 pp. (in Japanese). (esp. Kawano, T.: Vegetation Outline of Yunnan Region, pp. 22-51).

Miyawaki A. (1967): (Vegetation of Japan, Compared with Other Regions of the World.) (Encyclopedia of Science and Technology). Vol. 3. Gakken, Tokyo. (in Japanese).

Miyawaki A. (1993):Restoration of Native Forests from Japan to Malaysia. In: Restoration of Tropical Forest Ecosystems (H. Lieth and M. Lohmann, eds.), pp. 7-28. Dordrecht: Kluwer.

Miyawaki A. and K. Fujiwara (1988): Restoration of natural environment by creation of environmental protection forests in urban areas. Bull. Inst. Environm. Sci. Technol., Yokohama Natl. Univ. 15: 95-102.

Miyawaki A., K. Fujiwara and M. Ozawa (1993): Furusato-no ki ni yoru furusato-no mori zukuri (Native forests from native trees). Bull. Inst. Environm. Sci. Technol., Yokohama Natl. Univ. 19: 73-107.

Myers R. L. and J. J. Ewel (1990): Ecosystems of Florida. Orlando: Univ. of Central Florida Press.

Olson D. F. (1983): Temperate Broad-Leaved Evergreen Forests of Southeastern North America. In: Temperate Broad-Leaved Evergreen Forests (J. D. Ovington, ed.) pp. 103-105. Amsterdam: Elsevier.

Ovington J. D. (ed.) (1983): Temperate Broad-Leaved Evergreen Forests. "Ecosystems of the World" series Vol. 10. Amsterdam: Elsevier.

Quarterman E. and Ch. Keever (1962): Southern mixed hardwood forest: climax in the southeastern coastal plain, U.S.A. Ecol. Monogr. 32: 167-185.

Rübel E. (1930): Pflanzengesellschaften der Erde. Bern: Verlag Hans Huber.

Song J.-S., S. Nakanishi and Sh. Itow (1990):Phytosociological studies on the laurel-leaved forest in Korea: I. Laurel-leaved forests on Cheju Island. J. Phytogeogr. & Taxon. 38: 127-136 (in Japanese).

Song Y.-Ch. (1986): Broad-leaved evergreen forests in central Japan in comparison with eastern China. Veröff. Geobot. Inst. ETH, Stiftung Rübel (Zürich)98:197-224.

Song Y.-Ch. (1995): On the global position of the evergreen broad-leaved forests of China. In: Vegetation Science in Forestry (E. O. Box et al., eds.), Handbook Vegetation Sci., vol. 12/1, pp. 69-84. Dordrecht: Kluwer.

Suzuki K. (1982): Distribution of evergreen and summergreen broad-leaved forest in Japan. Bull. Inst. Environm. Sci. Technol., Yokohama Natl. Univ. 8: 151-163.

Suzuki T. (1952): Tou-A no Shin-Rin Shokusei (Forest Vegetation of East Asia). Kokon Shoin, Tokyo. (in Japanese).

Suzuki T. (1953): The forest climaxes of East Asia. Jap. J. Bot. 14: 1-12.

Tagawa H. (1997): Worldwide distribution of evergreen lucidophyll oak-laurel forests. Tropics (Jap. Soc. Trop. Ecol.) 6(4): 295-316.

Tagawa H. and M. Hotta (eds.) (1997): Special issue on "Co-existence of the World Humid Evergreen Forest Ecosystems and People." Tropics (Jap. Soc. Trop. Ecol.) 6(4): 283-477.

Wang Chi-Wu (1961): The Forests of China, with a survey of grassland and desert vegetation. Maria Moors Cabot Foundation, publ. no. 5. Cambridge (Mass.): Harvard University.

Wells B. W. (1942): Ecological problems of the southeastern United States coastal plain. Botan. Review 8: 533-561.

Wildpret de la Torre W. and V. E. Martin Osorio (1997): Laurel forest in the Canary Islands: biodiversity, historical use, and conservation. Tropics (Jap. Soc. Trop. Ecol.) 6(4): 371-382.

Wu Zheng-Yi and committee (eds.) (1980): Zhongguo Zhibei (Vegetation of China). Science Press, Beijing. 1375 pp + 339 B/W photos (in Chinese, with Latin-Chinese species lists; no page index).

Yim Y.-J., E. S. Jeon and N. J. Lee (1996): (Distribution of evergreen broad-leaved woody plants in Korea.) (Ecosystems) 1: 3-17 (in Korean).

Plenary discussion

Statement: A. Borhidi

It was shown in detail that different types of vegetation occur according to the site position of the trees. But laurophyllous and sclerophyllous forests or semideciduous tropical forests are differentiated very much according to the structure. For instance it is very important to see not only the tree species but also the presence and absence of lianas or different types of epiphytes. So I think it is very important to study this also.

Statement: F. Klötzli

I was very happy you were able to fill all the room that our good colleague Song Yongchang would have filled, and I'm very happy you were able to show such an overview of all this. I think this gives us a very good basis because you have shown the classical areas of evergreen broad-leaved forests. I have just a few remarks. Of course, if we could have gone on into the Himalayas we would have seen altitudinal belts which are absolutely comparable to your zonal belts including *Castanopsis*, *Quercus* and Lauraceae, all the laurineous species and so on. The same I would be able to show for Mexico in the more moist areas. What I rather object to is to bring in the sclerophyllous part. I would say the difference is so high in the climate diagram, in the structure and the physiology of the woody species that I would treat the sclerophyllous evergreen woodlands separately. That's our outlook because we live near the sclerophyllous biome. For us it is absolutely another world and we would never take them together, although in tropical mountains they can overlap because of mist effects. Then I would like to have your opinion in the workshops as how to treat the limiting areas at 38°/39° latitudes. This is an area where you should be able to observe any changes; more than in the altitudinal belts the installation of seedlings in such limiting areas would be obvious. Another aspect I mentioned a few days ago has to be considered if we try to give a forecast on the changes, and that's the pathogenic moment. The virulence of certain pathogens will be far higher, so if we have no idea on the effects of such fungi or so, it is difficult to say what will really happen because it might just wipe out certain species unless they get adapted.

Recent shifts in vegetation boundaries of deciduous forests,
especially due to general global warming, pp. 301–316
edited by F. Klötzli and G.-R. Walther

Grassland-forest boundaries in southern Brazil

Valério De Patta Pillar[1] and Fernando Quadros[2]
[1] Departamento de Botânica
[2] Departamento de Plantas Forrageiras e Agrometeorologia,
Universidade Federal do Rio Grande do Sul,
Porto Alegre, RS 90046-900, Brazil

Abstract

We discuss the apparent inconsistency between climate and vegetation in southern Brazil, especially in Rio Grande do Sul, a transitional zone with elements of the broad-leaved subtropical forest, mixed Araucaria-broadleaf forest and coastal rain forest vegetation types dominating in the north, and the campos grassland and shrublands dominating in the south and southwest. We show that on the large spatial scale, seasonal and annual climate fluctuations cause traumatic water deficits that may prevent the dominance of forest in some regions. This contradicts large scale vegetation models based on climate averages predicting that forest should prevail. On the small spatial scale the water budget is affected by local heterogeneity related to geomorphic and soil features (slope, aspect, drainage). We review evidence and conjecture that interactions of water budget with vegetation type and fire and grazing regimes may explain the resilience of a relict vegetation pattern.

Introduction

Since long ago vegetation scientists have been puzzled by the fact that grassland dominates in a region of South America under a macroclimate that apparently would be adequate to support forests (e.g. Klein 1975, Box 1986). The same has been said for the vegetation of Rio Grande do Sul (Lindman 1900, Rambo 1956), the southernmost state in Brazil, between latitudes 27° to 33° S. The annual precipitation record in the state shows an ascending south-north gradient from 1235mm to 2162mm (on the northeastern plateau) and at all sites it is greater than the annual potential evapotranspiration (Mota et al. 1970). These records apparently indicate absence of zonal water deficits.

Large scale vegetation models of Holdridge (1947) and Box (1981) predict forest vegetation under these conditions. However, on large and small scales grassland, shrubland and forest mingle in a mosaic as the potential vegetation (Teixeira et al. 1986). This is consistent with the fact that the region is at a transitional zone between the subtropical forest and coastal rain forest vegetation types, which are more dominant north of the state, and the campos and pampa grassland formations, which are more dominant to the south and southwest (Cabrera 1971, Schultz 1957).

The grassland-forest transitional region is interesting for studying vegetation dynamics under local and global scale environmental change, notwithstanding the fact that, in the limit, transitions could be found anywhere at any scale, but some transitions may be more obvious than others.

We elaborate hypotheses in this paper and attempt to explain grassland-forest vegetation patterns on large and small spatial scales in southern Brazil. We review evidence, discuss past and present day patterns, and consider their connections to climate, fire and grazing regimes.

Forests and campos

The vegetation of Rio Grande do Sul has been classified in phytogeographic regions by Reitz et al. (1983). Vegetation types are also described in Klein (1975) and Waechter et al. (1984). Based on these and existing maps (Hueck & Seibert 1972, Teixeira et al. 1986) we can define the following general types (Figure 1):

(1) Atlantic coastal rain forest, which shows its southernmost occurrence on the northern quaternary coastal plain and the eastern steep slopes and valleys of the northern plateau of the state, with several species of Lauraceae (*Ocotea catharinensis*, *Nectandra* spp.), Myrtaceae (*Myrcia* spp.), *Ficus organensis*, *Alchornea triplinervia* and *Euterpe edulis* (palm) as characteristic tree species.

(2) Broad-leaved subtropical (laurophyllous) forest of the Paraná and Uruguay river basins, up to elevations of 500-800m, occurring in the north and northwest portions of the state, with elements penetrating along the Ibicuí and medium and lower Jacuí river basins, characterized by winter deciduous emergent trees such as *Apuleia leiocarpa*, *Parapiptadenia rigida*, *Peltophorum dubium*, *Enterolobium contortisiliquum*, *Cordia trichotoma*, *Cabralea cangerana*, by non-deciduous non-emergent trees such as *Nectandra* spp., *Ocotea* spp. and *Patagonula americana* and small trees as *Actinostemon concolor*, *Sorocea bonplandii* and *Trichilia claussenii*.

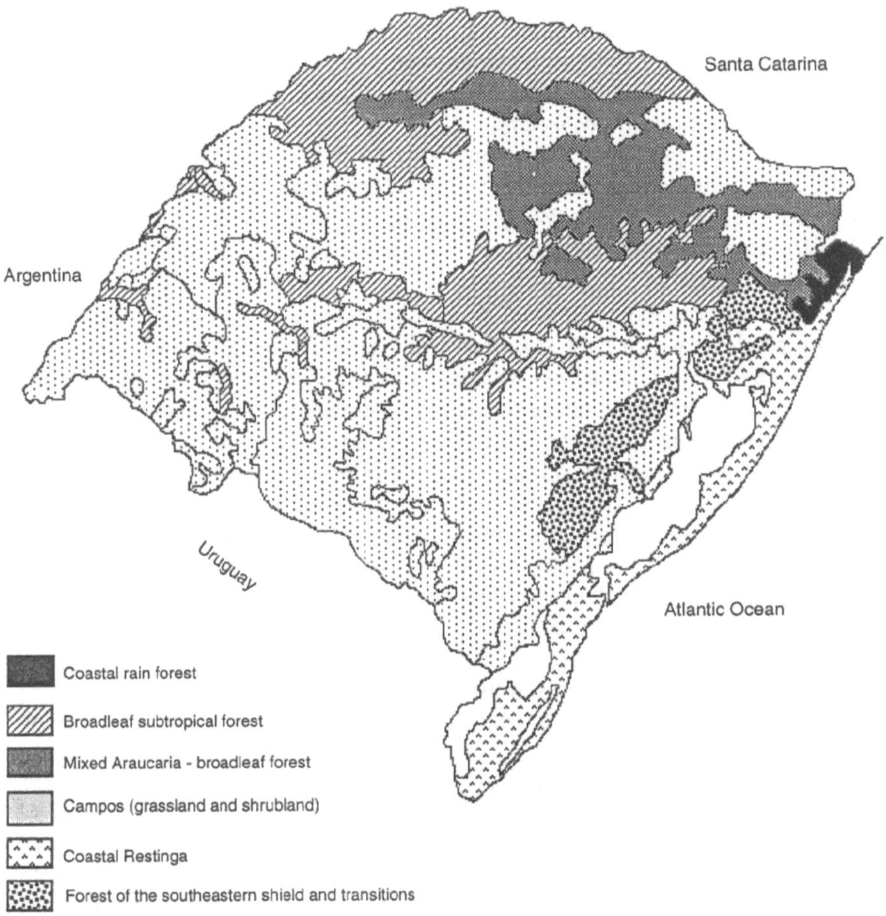

Figure 1: Map of the potential vegetation in the state of Rio Grande do Sul, Brazil. Adapted from Hueck & Seibert 1972, Klein 1975 and Teixeira et al. 1986. Subdivision within types reflect physiognomic variation.

(3) Mixed Araucaria - broad-leaved (laurophyllous) forest, with *Araucaria angustifolia* as the emergent tree and at a lower stratum species of Lauraceae (*Ocotea pulchella, O. puberula, Cryptocarya aschersoniana, Nectandra lanceolata, N. grandiflora, N. megapotamica*), Myrtaceae (*Myrcia bombycina, Myrceugenia euosma, Gomidesia sellowiana, Psidium cattleyanum, Myrcianthes gigantea*), Aquifoliaceae (*Ilex paraguariensis, Ilex* spp.), Sapindaceae (*Matayba elaeagnoides, Cupania vernalis*), Winteraceae (*Drimys brasiliensis*), Podocarpaceae (*Podocarpus lambertii*) and Leguminosae (*Mimosa scabrella*).

(4) Acacia-Prosopis parkland, a very small area on the westernmost portion of the state, containing typical floristic elements of the Chaco and Espinal formations (Cabrera 1971), such as *Prosopis nigra, P. algarobilla, Acacia caven, Parkinsonia aculeata* (Leguminosae), *Acanthosyris spinescens* (Santalaceae), *Sapium glandulatum* (Euphorbiaceae).

(5) The southeastern shield presents a complex pattern of forest, shrubland and campo patches, which have not been well studied; on the eastern slopes and valleys forest cover is denser, while in the interior of the region open forests, shrubland and campo dominate, containing elements that are also found in the other forest types, such as *Dodonaea viscosa, Schinus molle, Lithraea brasiliensis, Scutia buxifolia, Calyptranthes concinna, Gomidesia sellowiana, Ocotea pulchella, Matayba elaeagnoides, Podocarpus lambertii, Ilex paraguariensis, Araucaria angustifolia*.

(6) "Restinga" on the 600km long coast, with vegetation typical of the different successional stages on dunes and on lake shores.

(7) Grassland and shrubland (campos) dominate mostly on the southwest, central areas between the Jacuí river and the southeastern shield, associated with the mixed Araucaria - broad-leaved forest on the north and northeast, and along the coastal plain, composed of many species of Poaceae (about 800 species, mainly *Paspalum, Axonopus, Andropogon, Aristida, Schizachyrium, Eragrostis, Piptochaetium*), Compositae (*Baccharis, Vernonia, Eupatorium, Gamochaeta, Chaptalia, Senecio*), Cyperaceae (*Eleocharis, Rhynchospora*), Leguminosae (*Desmodium, Vicia, Trifolium, Galactia, Stylosanthes*), Rubiaceae (*Borreria, Relbunium*), Umbelliferae (*Eryngium, Centella*), among others. The term "campo" is used to refer to physiognomies that vary from true grassland to shrubland and frequently is used in the plural as "campos".

Authors have distinguished general types of campos on the basis of forage quality and geographical distribution (Gonçalves 1990, Barreto & Boldrini 1990). The campos in the southwestern region tend to present better forage quality ("fine campos"), in general dominated by prostrate or low tussock grasses, especially of *Paspalum* and *Axonopus* among C4, summer flowering species and of *Briza, Bromus, Piptochaetium* and *Stipa* among C3, spring flowering species, and by legumes of *Desmodium*, *Trifolium* and *Adesmia*. The campos in the other regions are the so-called "gross campos", characterized by a lower stratum with the previously referred species and an upper stratum where predominate tall tussock grasses of *Andropogon*, *Aristida*, *Erianthus*, *Hypoginium*, *Schizachyrium* and *Trachypogon*, all C4 species, which during the growing season accumulate biomass that remains attached to the plants, increasing flammability when they are dry. However, most of the differences seem to be determined by the grazing regime, since grazing exclusions or low intensity grazed areas in both the fine and gross campo are dominated by tall tussock grasses that were mentioned and, with some variation depending on the region, *Eryngium horridum* and shrubs of *Baccharis*, *Campomanesia*, *Eupatorium*, *Pteridium*, *Senecio* and *Vernonia*. Some of the variation may be related to soil conditions. In this respect, *Aristida*, *Piptochaetium* and *Eryngium horridum* are found mostly on drier, upper slopes, while *Andropogon lateralis*, *Baccharis trimera*, *Schizachyrium microstachium* are found mostly on moister sites (Pillar et al. 1992). There are clear geographically determined patterns in the dominance of some conspicuous species, e.g. *Eupatorium buniifolium* is a typical shrub of the campos on the southwest and *Aristida jubata* is typical of some campos on dry sites on the northern plateau and on patches of sandy ill-developed soils in the southwest.

The forest vegetation of the state of Rio Grande do Sul has been under strong human interference mainly after the 19th Century's intense colonization by Europeans. Historical records and present day mature or recovering, secondary forest fragments are indicative of the potential vegetation in some areas (see map in Figure 1). Forest dominates as potential vegetation in large areas in the northern half of the state, mainly on the slopes and valleys along the east and south edges of the northern basaltic plateau and on the slopes and valleys of the Uruguay river basin. Forest also dominates on some parts of the southeast granite shield, mainly on the eastern slopes. Though less dominant, riparian forests and smaller patches ("capões") on moist sites on slopes are observed throughout typical campo areas in other parts of the state, containing floristic elements of the aforementioned forest types.

The campos vegetation presently covers about 120'000km^2 in the state of Rio Grande do Sul, which corresponds to about half of the state's land mass. In the 17th Century, during the establishment of the Jesuit missions on the eastern side of the Uruguai river, cattle and horses were introduced (Kern et al. 1991) and large ranches ("vacarias") were run by natives, including campos in the Araucaria forest region. A century later, immigrants from the Azores and other parts of Brazil colonized campo areas, mainly along the medium and lower Jacuí river valley and on the southeastern shield, with a relatively slight effect on forests comparing to the colonization of the northern half of the state a century later. Since then the campo vegetation has been the major support for a cattle ranching economy. Fire is regularly used in late winter when ranch management is such that excessive biomass accumulates during the growing season, being senescent by the next spring. In the last three decades a large proportion of areas with campo vegetation has been allocated to cropland. Sometimes crops are cultivated on a rotational basis in which the natural grassland recovers during the few years the land is left fallow but grazed.

Factors affecting the patterns

Climate

The climate in southern Brazil is mainly a Cfa type in Köppen's classification, with a small portion with a Cfb type on the higher elevation plateau. In the state of Rio Grande do Sul the mean annual temperatures range from 15 to 20°C, with mean temperatures in the coldest month ranging from 11 to 14°C, when frost is frequent, but snow rarely observed and only in the high elevation areas. Seasonal and annual fluctuations in precipitation cause notable water deficit in the summer in some areas (Buriol et al. 1979, see map in Figure 2). The water budget figures shown in the map are based on series of meteorological observations (>30 years). Instead of using the normal (average) values of precipitation and estimated evapotranspiration, the method computes the water budget at each time step and, on the basis of the series of water budget figures, generates probabilities for different levels of water excess and deficit at the different sites and months of the year, giving a more realistic picture of the soil moisture conditions faced by vegetation. The map in Figure 2 indicates the water deficit levels that may occur, at a 0.2 probability, on the different regions of the state of Rio Grande do Sul. The northeastern part of the state is the least affected by water deficits, while areas in the southwest and east are the most affected.
The computation of water budget in Buriol et al. (1979) does not take into account surface and subsurface runoff, percolation and outflows of water, which are impossible to estimate in such a large scale study and are associated with geomorphic and soil related features (slope, soil texture, depth of impermeable layer), aspect, vegetation cover (Miller 1977). Therefore, we can consider that water deficits can be aggravated or moderated by these factors at the smaller spatial scales. Under a given regional climate regime, the local heterogeneity determines water budgets that may prevent the development of forest on some portions of the landscape (mostly ridges and upper slopes), while facilitating forest development on other portions (valleys). It is assumed that plants of the campos are more tolerant to water deficit than plants of the forests (Pillar & Boldrini 1996). Information is yet scanty regarding the degree to which these water deficits would

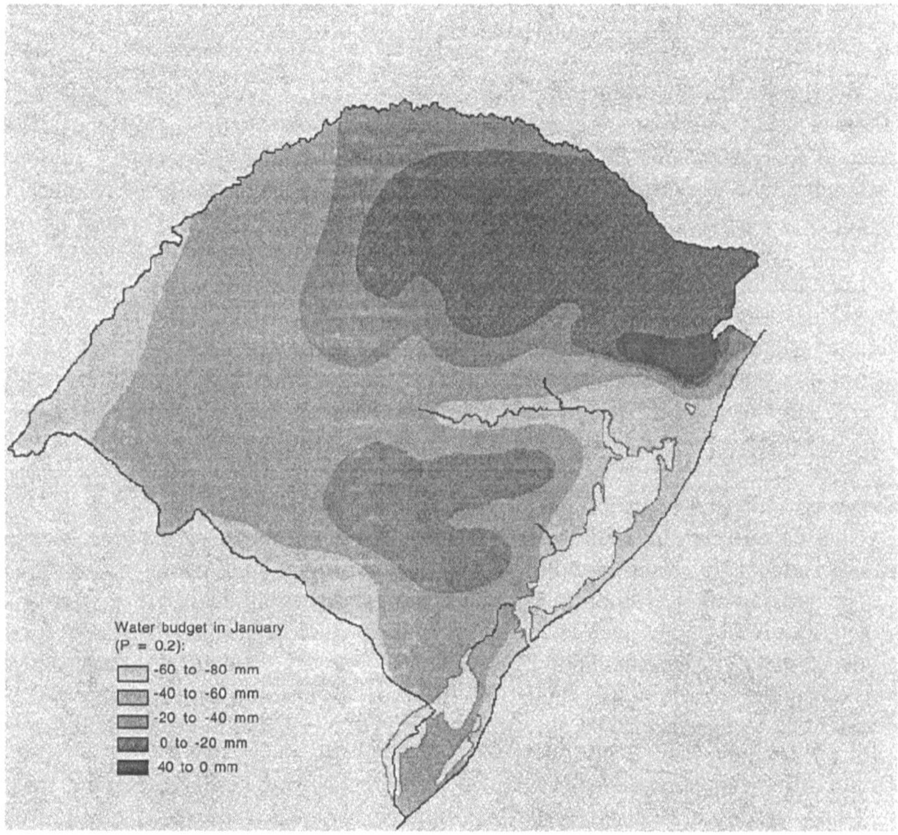

Figure 2: Water budget in January, in the state of Rio Grande do Sul, Brazil. There is a probability of 0.2 that in a given year the indicated water budget will occur. Negative values imply water deficit, positive values imply water excess, considering a soil water storage capacity of 100mm (from Buriol et al. 1979).

be directly detrimental to the development of forest tree saplings. However, there is a reasonable coincidence between the zones with more water deficit (Figure 2) and potential vegetation cover with campos (Figure 1), except in the area of campos associated with mixed Araucaria - broad-leaved forest, which we will discuss later. The coincidence is an evidence that water deficits may have prevented directly or indirectly (by influencing fire regimes) the dominance of forest in some regions of the state. Local heterogeneity also may explain the occurrence of forest in regions where the regional climate would indicate campo.

Past climate and vegetation shifts

There are evidences that the climate in South America during the last glacial period were in general cooler than today, but moisture patterns showed strong regional differentiation (Markgraf 1989). More specific evidence of past climate changes is found in pollen records in Rio Grande do Sul (Lorscheitter 1992) and Santa Catarina (Behling 1993, 1995). The climate was colder and drier until 10'000 years B.P., warmer and drier from 10'000 to 2900 years B.P. and cooler and moister from 2900 to present (Behling 1995). Behling's data from the Araucaria region in Santa Catarina (Figure 3) indicates that until around 2600 years B.P. the vegetation was much more campo-like than at present, which confirms the hypothesis advanced by Klein (1975) that the campos within the Araucaria forest region are relicts of a drier climate period.

Grazing and fire regimes

By the time European settler's cattle herds and horses started to graze in the campos and pampa regions in South America, grazing animals of the native fauna were small, mainly deer, rhea and small rodents. These could subject the grassland vegetation only to low and localized grazing pressure. There is, however, fossil evidence that large grazing mammals of Equidae (*Equus*), Camelidae and Cervideae existed in the region until 8000 B.P. (Kern 1994, Bombin & Klamt 1975). As conjectured by Schüle (1990), since the collision of North and South America in the Upper Pliocene, about 3 million years B.P., until the end of the Pleistocene, the South American vegetation may have been considerably damaged by these invading large Laurasian ungulates and other megafauna elements. If this is true, intensive grazing is not far away in the evolutionary history of the present day flora of the campos.

A quite interesting question is to know if fire has been always a high frequency element of the natural campos ecosystem or if it has intensified after the arrival of humans in the region around 12-13 thousand years B.P. (Kern et al. 1991). The phytopalaeofire history in southern Brazil is not well known, but recent findings in the mixed Araucaria - broad-leaved forest region in Paraná (Behling 1996), which shows a similar mosaic of forest and campos as in Rio Grande do Sul, indicate that the frequency of palaeofires increased at the beginning of the Holocene, likely anthropogenic at that time. This finding further supports the idea that the pre-Colombian anthropogenic influence on the vegetation, including dissemination of naturally occurring plants, should not be overlooked (Schüle 1990, Dean 1995). Also, Behling found that the highest frequency of fires was during the late

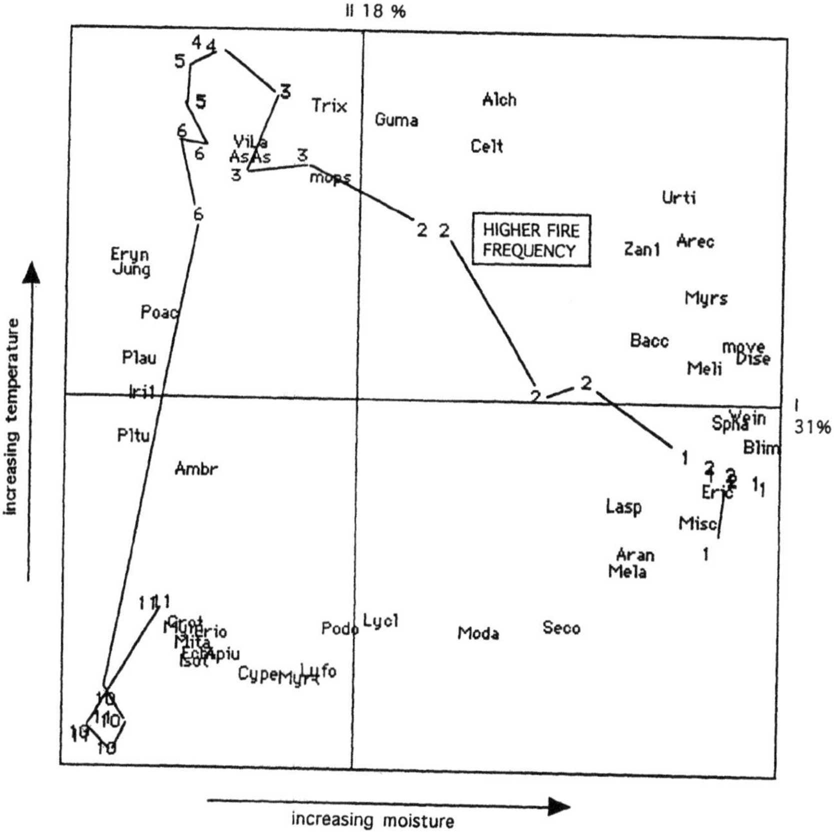

Figure 3: Ordination of the pollen record in 30 sections of a core taken at Serra do Rio do Rastro, Santa Catarina, Brazil.

Data is from Behling (1993). The numbers identify radiocarbon estimated age in thousand years B.P. (from 800 to 11013 are connected in sequence). Ordination method is principal coordinates analysis, applied to euclidean distances, with pollen counts square root transformed and standardized by the total count in each core section. Species most correlated (|r| > 0.6) to the ordination axes are plotted (correlations rescaled accordingly are the coordinates). Species labels: Alch Alchornea, Ambr Ambrosia-type, Apiu Apium, Aran Araucaria angustifolia, Arec Arecaceae, AsAs Asteraceae subf. Asteroideae undiff., Bacc Baccharis-type, Blim Blechnum cf. B. imperiale, Celt Celtis, Crot Croton, Cype Cyperaceae, Dise Dicksonia sellowiana, Echi Echinodorus, Eric Ericaceae, Erio Eriocaulon/Paepalanthus, Eryn Eryngium-type, Guma Gunnera manicata, Iri1 Iridaceae (type I), Isot Isoëtes, Jung Jungia-type, Lasp Lamanonia cf. L. speciosa, Lycl Lycopodium clavatum-type, Lyfo Lycopodiaceae (foveolate), Mela Melastomataceae, Meli Meliosma, Misc Mimosa cf. M. scabrella, Mita Mimosa cf. M. taimbensis, Moda Moritzia dasiantha, Myri Myriophyllum, Myrs Myrsine, Myrt Myrtaceae, Plau Plantago cf. P. australis, Pltu Plantago cf. P. turficola, Poac Poaceae, Podo Podocarpus, Seco Sebastiana commersoniana, Spha Sphagnum, Trix Trixis, Urti Urticales, ViLa Vicia/Lathyrus, Wein Weinmannia, Zan1 Zanthoxylum (type I), mops Pteridophyta (monolete psilate), move Pteridophyta (monolete verrucate). Climate interpretation is based on Behling (1995). Fire regime is inferred based on data from a similar area in Paraná (Behling 1996).

Holocene, coinciding with the migration of the mixed Araucaria - broad-leaved forest from the valleys into the areas with campos. Fire could have been a driven factor in the Araucaria forest expansion (Soares 1990).

The campos, even the areas dominated by shrubs, are fire prone if sufficient senescent biomass has accumulated. The rate of senescence of older leaves in grasses increases during dry spells. Furthermore, senescent leaves of tall tussock grasses in general remain attached to the plants, increasing flammability of the whole plant (Bond & Wilgen 1996). Accumulation of biomass, and therefore flammability levels, are related to grazing regimes (Bond & Wilgen 1996). Ranchers usually adjust the stocking rate of pastures on the campos on the basis of the carrying capacity during the winter. The consequence is more accumulation of ungrazed biomass during the main growing season in the spring and summer in the campos presenting more seasonality in the biomass production, which are the ones in the colder winter areas, associated to the mixed Araucaria - broad-leaved forest.

The broad-leaved subtropical forest and the coastal rain forest types are not fire prone, i.e. usually do not burn. Young trees of these and other forests types cannot tolerate fire, since fire tolerance is related to the protection to the cambium that develops as the bark becomes thicker. References indicate that in general tree species are sensitive to fire when the plants are younger up to 2 m height or with a bark thickness less than 10 mm (Peterson & Ryan 1986, Uhl & Kauffman 1990). *Araucaria angustifolia* seedlings and young trees cannot tolerate fire (Veblen et al. 1995). However, there is evidence that the regeneration of *Araucaria angustifolia* is dependent on regular disturbances, mainly fire. Human induced light fires in old pine stands may in fact increase regeneration (Soares 1990). Veblen et al. (1995) give similar indications for the forest in Southern Chile and Argentina. Fires in the mixed Araucaria- broad-leaved forest can be ignited by lightning, based on a long term study of forest fire occurrence in the central Paraná state, in which lightning was responsible for 20% of the fires (Soares 1990, Soares & Cordeiro 1974). If natural fires are possible in the pine forest, would they be in the campos? This question may not be relevant for campos in the pine forest region, since when a fire starts in the forest it is probable that, if fuel is accumulated, the fire will spread to adjacent campo, no matter where it started. But, outside the pine forest region, knowing whether or not natural fires were possible is critical to understand the pattern dynamics of campo and non-fire prone forest types. Although there is no formal study on this topic in the region, at present the occurrence of natural fires starting on grassland areas is very rare (see also references in Bond & Wilgen 1996:218). However, this is not an indication that natural fires in grassland were also rare at pre-Colombiam times; some evidence of palaeofires was detected since 35'000 B.P. in data from a mountain region in São Paulo, which was dominated by campos until the end of the Pleistocene (Behling 1997). It is quite obvious that grazing intensity of grassland areas increased after

the introduction of domestic cattle, resulting in a reduction of the number of areas with enough fuel accumulation for extended naturally ignited fires. Furthermore, when grassland is left ungrazed for a long time fuel accumulates and any fire tends to be much more intense, being capable of burning plant types that would not burn under a low intensity fire, such as forest trees invading grassland.
Data of long term field studies evaluating the effects of (the absence of) fire and grazing on the dynamics of boundaries between campos grassland and forest are yet to be made available by ongoing research at long term ecological research sites near Porto Alegre and Santa Maria. Data from exclosures (Boldrini & Eggers 1996) and our own observations of areas in which grazing and fire were excluded, some for more than 10 years, indicate that in some of them the grassland vegetation tended to evolve to shrubland and to a forest type vegetation but not in others. Also, observations of excluded areas at grassland-forest boundaries indicated that in general forest tended to invade adjacent grassland. Although the data is scanty and in some areas where informal observations were made the history of disturbances is unknown, we can conjecture that under suppression of fire and grazing the grassland-forest boundaries would tend to shift up (mostly in the elevation gradient) to a new equilibrium state determined by the soil water budget and by the new type of vegetation cover. The reversed effect is expected only upon imposing an intense grazing or fire regime, and may take longer. Also, there is a clear effect of cattle trampling hindering tree regeneration within forest patches.
We could conjecture that in the mixed Araucaria - broad-leaved forest region natural or anthropogenic fires have determined a mosaic of forest and grassland, notwithstanding the fact that in that region periods of soil water deficit are rare in the present climate, and therefore the extent of areas that would be too limiting for forest is smaller. Green's (1989) simulation model predicts a mosaic pattern in the case when two fire prone vegetation types (which is the case here) mingle with regard to environmental requirements (moisture is adequate for both grassland and forest). That is, fire may have been the main factor explaining grassland-forest patterns in the mixed Araucaria- broad-leaved forest region.

Final remarks

Whether the environmental limit for plant types is set by water budget or by fire may not change the resulting pattern in the areas with greater water deficits. If fire prone plants characterize grassland, and grassland is typical of drier sites, a forest plant may not occur on drier sites because at younger age cannot withstand droughts, or because at some stage is fire sensitive, or because of both. For one reason or another the resulting vegetation cover may be a campo. However, in the moister areas at present covered with a mosaic of fire prone Araucaria forest and campos, fire may be the major factor explaining the resilience of relict campos; the water budget - fire mechanism may be relevant only on the drier portions of the landscape which are unsuitable for Araucaria plants. Although empirical evidence is not clear, long term suppression of grazing and fire in the campos in both areas may determine local grassland-forest boundaries shifts (up in the elevation gradient) to a new equilibrium state determined by the soil water budget and by the new type of vegetation cover.

We therefore conclude that the hypothesis is tenable that on the large spatial scale seasonal and annual climate fluctuations cause water deficits that prevent the dominance of forest in some regions. On the small spatial scale, water deficit is aggravated or moderated by local heterogeneity related to geomorphic and soil features. Interactions of water budget with current vegetation cover, fire and grazing disturbances explain observed vegetation patterns and grassland - forest boundaries.

Acknowledgements

Financial support from CAPES (Brazil) to the authors is gratefully acknowledged. Thanks are extended to H. Behling for allowing the use of his pollen data and to L. Orlóci for helpful suggestions on the manuscript.

References

Behling H. (1993): Untersuchungen zur spätpleistozänen und holozänen Vegetations- und Klimageschichte der tropischen Küstenwälder und der Araukarienwälder in Santa Catarina (Südbrasilien). Dissertationes Botanicae Band 206. J. Cramer, Berlin.

Behling H. (1995): Investigations into the Late Pleistocene and Holocene history of vegetation and climate in Santa Catarina (S Brazil). Vegetation History and Archaeobotany 4: 127-152.

Behling H. (1996): Late Quaternary vegetation, climate and fire history of the Araucaria forest and campos region from Serra Campos Gerais, Paraná State (South Brazil). Review of Palaeobotany and Palynology (in press).

Behling H. (1997): Late Quaternary vegetation, climate and fire history from the tropical mountain region of Morro de Itapeva, SE Brazil. Palaeogeography, Palaeoclimatology, Palaeoecology (in press).

Barreto I. L. & I. I. Boldrini (1990): Aspectos físicos, vegetação e problemática das regiões do Litoral, Depressão Central, Missões e Planalto do Rio Grande do

Sul, Brasil. In: J. P. Puignau: Introducción, conservación y evaluación de germoplasma forrajero en el Cono Sur, p.199-210. IICA-PROCISUR, Montevideo.

Boldrini I. I. & Eggers L. (1996): Vegetação campestre do sul do Brasil: resposta e dinâmica de espécies à exclusão. Acta Botanica Brasilica 10: 37-50.

Bombin M. & E. Klamt (1975): Evidências paleoclimáticas em solos do Rio Grande do Sul. Anais do XXVIII Congresso Brasileiro de Geologia, P. Alegre, p.183-193.

Bond W. J. & van Wilgen B. W. (1996): Fire and Plants Chapman & Hall, London.

Box E. O. (1981): Macroclimate and Plant Forms: An Introduction to Predictive Modelling in Phytogeography. Junk, The Hague.

Box E. O. (1986): Some climatic relationships of the vegetation of Argentina, in global perspective. Veröff. Geobot. Inst. ETH 91: 181-216.

Buriol G. A., A. V. Saccol, V. Estefanel, A. B. Heldwein & F. M. Schneider (1979): Distribuição geográfica das disponibilidades hídricas do solo possíveis de ocorrerem no estado do Rio Grande do Sul. Revista do Centro de Ciências Rurais 9: 111-169.

Cabrera A. L. (1971): Fitogeografia de la Republica Argentina. Sociedad Argentina de Botanica, Buenos Aires.

Dean W. (1995): With broadax and firebrand: The destruction of the Brazilian Atlantic Forest. University of California Press, Berkeley.

Gonçalves J. O. N. (1990): Informações básicas sobre solos, clima, vegetação, áreas agroecológicas homogêneas e centros de pesquisa na região Sul do Brasil. In: J. P. Puignau: Introducción, conservación y evaluación de germoplasma forrajero en el Cono Sur, p.187-198. IICA-PROCISUR, Montevideo.

Green D. G. (1989): Simulated effects of fire, dispersal and spatial pattern on competition within forest mosaics. Vegetatio 82: 139-153.

Holdridge L. R. (1947): Determinations of world formations from simple climatic data. Science 105: 367-368.

Hueck K. & Seibert P. (1972): Vegetationskarte von Südamerika. Mapa de la Vegetación de America del Sur. Gustav Fischer, Stuttgart.

IPAGRO (1989): Atlas agroclimático do Estado do Rio Grande do Sul. IPAGRO, Seção de Ecologia Agrícola, Porto Alegre.

Kern A. A. (1994): Antecedentes indígenas. Porto Alegre, Ed. da Universidade.

Kern A. A., A. Jacobus & P. Ribeiro (1991): Arqueologia pré-histórica do RS. Mercado Aberto, Porto Alegre.

Klein R. M. (1975): Southern Brazilian phytogeographic features and the probable influence of upper Quaternary climatic changes in the floristic distribution. Boletim Paranaense de Geociências 33: 67-88.

Lindman C. A. M. (1900): Vegetationen i Rio Grande do Sul (Sydbrasilien). Stockholm, Nordin & Josephson. Portuguese version by A. Löfgren, Porto Alegre, Livraria Universal, 1906, also facsimilar in C. A. M. Lindman & M. G. Ferri, A Vegetação no Rio Grande do Sul (Brasil Austral). Itatiaia/USP, Belo Horizonte/São Paulo, 1974.

Lorscheitter M. L. (1992): Pollen registers of the South and Southeast regions of Brazil during the last 40000 years. In: L. D. de Lacerda, B. Turcq, B. Knoppers & B. Kjerfve (eds.) Paleoclimatic changes and the carbon cycle, p.55-61, Univ. Federal Fluminense, Niterói.

Markgraf V. (1989): Paleoclimates in Central and South America since 18000 BP based on pollen and lake-level records. Quaternary Science Reviews 8: 1-24.

Miller D. H. (1977): Water at the surface of the earth; an introduction to ecosystem hydrodynamics. Academic Press, New York.

Mota F. S. da, C. O. Goedert, N. F. Lopes, J. R. B. Garcez & A. S. Gomes (1970): Balanço hídrico do RS. Pesquisa Agropecuária Brasileira 5: 1-27.

Peterson D.L. & K. C. Ryan (1986): Modelling postfire conifer mortality for long range planning. Environmental Management 10: 797-808.

Pillar V. De P. & I. Boldrini (1996): Lindman e a ecologia da vegetação campestre do Rio Grande do Sul. Ciência & Ambiente 13: 87-97.

Pillar V. De P., A. V. A. Jacques & I. Boldrini, I. I. (1992): Fatores ambientais relacionados à variação da vegetação de um campo natural. Pesquisa Agropecuária Brasileira 27: 1089-1101.

Rambo B. (1956): A fisionomia do Rio Grande do Sul, 2nd ed. Livraria Selbach, Porto Alegre.

Reitz R., R. M. Klein & A. Reis (1983): Projeto madeira do Rio Grande do Sul. Sellowia 34-35: 1-525.

Schüle W. (1990): Landscapes and climate in prehistory: interactions of wildlife, man and fire. In: J. G. Goldammer (ed.), Fire in the Tropical Biota, p. 273-318, Springer-Verlag, Berlin.

Schultz A. R. (1957): Some phytogeographical and phytological data from Rio Grande do Sul, Brazil. Vegetatio 7: 355-360.

Soares R. V. (1990): Fire in some tropical and subtropical South American vegetation types: an overview. In: J. G. Goldammer (ed.), Fire in the Tropical Biota, p. 63-81, Springer-Verlag, Berlin.

Soares R. V. & L. Cordeiro (1974): Análise das causas e épocas de ocorrência de incêndios florestais na região centro-paranaense. Floresta 5: 46-49.

Teixeira M. B., A. B. Coura Neto, U. Pastore & A. L. R. Rangel Filho (1986): Vegetação. In: Levantamento de recursos naturais. IBGE, Rio de Janeiro, v. 33, p. 541-632.

Uhl C. & J. B. Kauffman (1990): Deforestation, fire susceptibility, and potential tree responses to fire in the eastern Amazon. Ecology 71: 437-49.

Veblen T.T., B. R. Burns, T. Kitzberger, A. Lara & R. Villalba (1995): The ecology of the conifers of southern South America. In: Enright & Hill (eds) Ecology of the Southern Conifers, p.120-155. Melbourne University Press, Melbourne.

Waechter J. L., L. A. Cestaro & S. T. S. Miotto (1984): Vegetation types in the ecological station of Aracuri, Esmeralda, RS, Brazil. Phytocoenologia 12: 261-9.

Plenary discussion

Statement: F. Klötzli

I just want to say one thing and that is a technical term: Patana grassland (extensive discussion and references in: W.L. Werner (1984) Die Höhen- und Nebelwälder auf der Insel Ceylon (Sri Lanka) pp. 136-143. In: W. Rauh (Ed.): Tropische und subtropische Pflanzenwelt 46, Akad. Wiss. u. Lit. Mainz Wiesbaden Steiner, 200pp. + 1 Karte), Patana grassland in Sri Lanka and southern India. If you look at literature on grasslands on Horton Plains in Sri Lanka, you find exactly the same discussion, because it is the same pattern. And people have been wondering why even in montane rainforest areas, grassland and forest isles co-exist. So I'm very glad you were able to find a picture which in my eyes has a high probability the way you explain it. I would maybe put more weight on the texture of the soil and the fact that physiological dryness is playing a decisive role. Under certain circumstances the plant has a great difficulty to take up the necessary water because the texture doesn't allow it. So I'm very happy with your picture, and especially also with the palaeoclimatic outlook and the influence of grazing animals I would strongly suggest to try to get the links with the characteristics of Patana grassland.

Statement: A. Borhidi

My impression is that *Araucaria* must have been more extended in the former times because in the grasslands you have shown the composition is the same as it is under the Araucaria forest as well. *Baccharis* has the same role in the vegetation as *Vaccinium* in the northern hemisphere which is practically a needle-leaved forest subshrub, for instance in Mexico and Cuba and even in Florida. And even the other species of *Vernonia*, *Eupatorium* and so on they are really the genera which are abundantly represented in the central American Pine forests. So it is the same phenomenon that exists e.g. in Cuba or Mexico if you see montane grassland with 1500mm precipitation. These are composed of the same grassland species which are in the other groundvegetation of the Pine forest. So I would suggest to study more on the historical basis and not only on the pollen diagrams but also on the cultural historical data.

Recent shifts in vegetation boundaries of deciduous forests,
especially due to general global warming, pp. 317–331
edited by F. KLÖTZLI and G.-R. WALTHER

Laurophyllisation of deciduous *Nothofagus*-Forests in Southern Chile

Dorothea FRANK[1] and Manfred FINCKH[2]
[1] Lehrstuhl Biogeographie, Universität Bayreuth, D-95447 Bayreuth, Germany.
[2] Proyecto CEE, Universidad de La Frontera, Casilla 54d, Temuco, Chile

Abstract

Stratified measurements of frequency of the major tree and liana species have been used to interpret the vegetation dynamics of deciduous *Nothofagus obliqua* forests in the Central Valley of south-central Chile. In terms of both cover and frequency, secondary forests are dominated by *Nothofagus obliqua* in association with evergreen and laurophyllous-indifferent species. Laurophyllous trees of Valdivian origin, such as *Aextoxicon punctatum,* dominate all strata of the few remaining old growth forests, most of them showing regeneration patterns typical of shade tolerant late-sucessional species. Evergreen tree species from Central Chile which reach their southern limit in the study area are most frequent in early successional stages on "zonal" sites, although their original niches in this region were “extrazonal” thermophilous sites. The current dominance of the deciduous *Nothofagus obliqua* appears to be a consequence of anthropogenic destruction of stands during the last century. Long periods without disturbance would cause these forests revert to evergreen broad-leaved rain forests. This laurophyllisation process appears to be controlled by land-use practice and intensity rather than by global warming. Consequently, studies on laurophyllisation should pay particular attention to land-use history and changing patterns of land-use.

Introduction

The Central Valley of the IX. Region of Chile is a transition zone between the evergreen-sclerophyllous forests of mediterranean Central Chile and the Valdivian rainforests of perhumid Southern Chile. During the last 150 years the previous, more or less continuous, forest ecosystems of this ecotonal zone have been destroyed by extensive burning and logging, and primary forests have almost disappeared outside the national parks. The extant secondary forests in this ecotonal zone of Southern Chile are dominated by deciduous *Nothofagus obliqua* with regenerating laurophyllous species in their understories. This fact has provoked a long-lasting debate about how these forests should be classified. Most classifications and vegetation maps classify the forests of this transition zone as deciduous (Schmithuesen 1956, Hueck 1966, Gajardo 1994), but the composition of their understories seems to support the hypothesis that laurophyllous species are invading these, apparently deciduous, forests. Contemporary phytosociology allows both the classification of vegetation units and analysis of kinetics of vegetation dynamics. The objective of this study is to investigate the vegetation dynamics of these *Nothofagus obliqua* forests by analyzing their species composition and structure. Veblen et al. (1979) studied *Nothofagus obliqua* forests in the Andean precordillera and emphasised the natural disturbance regime of this region with its high frequency of catastrophic events caused by volcanic eruptions, fires or earthquakes. Near Temuco, however, similar *Nothofagus obliqua* forests are found on old and stable Tertiary land surfaces, far from the volcanic disturbances of the Andes. Our hypothesis is that the dominance of *Nothofagus obliqua* is man-made and due to land-use and that, without anthropogenic disturbance, laurophyllisation will lead to evergreen broad-leaved forests.

Study area

The study area is situated directly north-west of Temuco, the capital of the IX. Region of Chile, at about 38°40' S and 72°40' W. It occupies an area of ca. 500 km^2 and is characterized by a mountain chain formed by continental-volcanic materials which ascends to 770m a.s.l. and a lower part with gently rolling hills formed from marine sediments. The lower parts (below ca. 300m) are covered by old red clay soils (Ultisols), and are separated by a heterogeneous transition zone from the upper parts (above ca. 500m) which have young volcanic ash soils (Andosols).

The climate of the study area is intermediate between perhumid Southern Chile and the mediterranean climate of the Central Zone. Temuco has an annual rainfall of about 1300mm with a mean temperature of 11.6°C (Walter 1990). Most of the precipitation falls during the mild winter and there is typically an arid period of about two months in summer (van Husen 1967). The concurrence of relatively cool air temperatures (due to the proximity of the Humboldt Current) and high insolation (due to the subtropical latitude) favours high photosynthetic rates and low respiratory losses that result in highly productive forests (Schwabe 1956).

At lower altitudes, the study area is dominated by deciduous *Nothofagus obliqua* forests corresponding to the Nothofago-Perseetum lingue (Schmithüsen 1956, Oberdorfer 1960). Above ca. 500m, the forest remnants are classified as Nothofago-Dasyphylletum (Finckh 1995) and contain species such as *Nothofagus alpina*, *Dasyphyllum diacanthoides* and *Myrceugenia planipes*. Neither these "highland forests" growing on young volcanic ash soils nor the myrtle forests growing on hydromorphic soils along the rivers are considered further in this paper.
The remaining forest patches are surrounded by combinations of three types of land-use: extensive subsistence agriculture of the Mapuche Indians, intensive agriculture associated with large farms, and forest monocultures of introduced *Pinus* and *Eucalyptus* species. Most of the remaining native forests are secondary, but two remnants of more or less primary forest vegetation still exist in the mountainous part of the study area, about 10km north of Temuco. The "Reserva Nacional Cerro Ñielol" (80ha, ca. 100 - 300m) and "Rucamanque" forest reserve (450ha, ca. 200 - 550m). Each of the two forest remnants represents a different type of the formerly widespread forests of the area. The Reserva Nacional Cerro Ñielol (henceforth called "Ñielol") represents a forest type which has, to some extent, warmer and drier "extrazonal" site conditions, whereas the site at "Rucamanque" provides the cooler and more humid conditions characteristic of the "zonal" type of these forests. "Rucamanque" has lower human impact than "Ñielol", with 45% of its area typical of primary forest.

Methods

Black and white aerial photographs (FONDEF 1994, 1:20.000) were used to map the actual forest vegetation of the study area. Information about land-use history was obtained from aerial photographs dating from 1945, 1961 and 1988 and from bibliographic sources (e.g. Berninger 1929). The results were digitized and processed with a Geographic Information System (ATLAS-GIS).
Subsequently, we checked the composition of the forest vegetation of 93 stands, using modified Braun-Blanquet (1964) methodology. The vertical structure of the stands was recorded in seven different strata with three tree layers (T1 > 20m, T2 20m - 10m, T3 10m - 5m), a shrub layer (S 5m - 0.5m), a field layer (F < 0.5m) and two layers for epiphytes and lianas (L1 lianas in T1 - T3, L2 lianas in S and F). These structural divisions allow the regeneration dynamics of the different species to be analyzed. The cover of each species in the 7 different vertical layers was estimated by the Braun-Blanquet-Scale (5 = 75 % - 100 %; 4 = 50 % - 75 %; 3 = 25 % - 50 %; 2 = 5 % - 25 %; 1 = 1 % - 5 %; + < 1 %). The frequency of each species in each layer was calculated as a percentage of the number of stands. We defined five stand-types for statistical analysis: young/low secondary forests not reaching the T1-layer, older/taller secondary forests with canopy reaching the T1-layer, (more or less) intact old

growth forests and extrazonal stands of secondary and primary forest containing *Cryptocarya alba* (a mediterranean element indicating warmer and drier site conditions). Due to small numbers of stands, secondary forests containing *Cryptocarya alba* were not separated into low and tall secondary forest.
Statistical analyses were confined to liana and tree species capable of reaching heights of more than 10m that also reached at least 15 % frequency in one of the layers of at least one stand type. The relative importance of the trees was estimated by averaging the median values of each Braun-Blanquet score in each stand-group.
Climatic measurements were obtained from three stations in open areas.

Results

Species composition and structure

Figure 1 maps the distribution of forest vegetation in the study area, and includes both native forests and plantations of introduced species. The former continuous forest of the study area is now highly fragmented with less than 1 % of the total area still covered by old growth forest; ca. 6 % with deciduous *Nothofagus obliqua* forest and 9 % covered by degraded shrublands.
The selected tree species and lianas have been classified according to ecological types and life forms (Figure 2). Figure 3 shows the mean frequency of the selected tree species in the various layers and forest types. In secondary forests, deciduous *Nothofagus obliqua* has the highest frequency of any tree. He occurs with high frequency in all vertical strata of secondary forests less than 20m high, including those with *Cryptocarya alba* (the extrazonal type). In these latter forests, *Nothofagus obliqua* has its highest frequency values in the T3 layer, indicating successful regeneration of this deciduous tree. The taller secondary forests show a similar pattern, with the peak frequency of *Nothofagus obliqua* in the T1-layer and decreased frequencies in the lower strata. In the two primary forest-types, *Nothofagus obliqua* shows a different pattern, and appears mostly as an emergent in the top layer (T1) but is almost absent from all other strata, indicating an almost complete inhibition of *Nothofagus obliqua* regeneration in both primary forest types.
The pattern of frequencies of the other tree species in the different forest types show marked differences to that of *Nothofagus obliqua*. The high frequencies of "evergreen-mediterranean" trees in the understory of the extrazonal secondary forests and less tall secondary forests indicate successful regeneration under the canopy of *Nothofagus obliqua*. *Peumus boldus* and *Cryptocarya alba* reach the main canopy of the primary forests of the warmer and drier sites but are almost absent from the zonal primary forests. *Citronella mucronata* and *Maytenus boaria* seem to be restricted to disturbed secondary forests in the study area, and *Podocarpus saligna* shows a similar pattern.

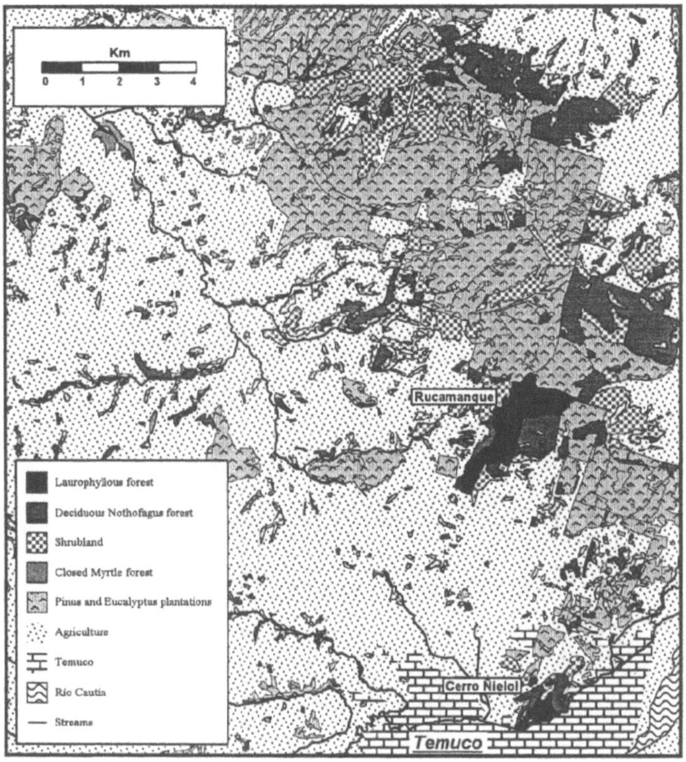

Figure 1: Vegetation map of the study area

Deciduous tree species:	**Evergreen tree species:**	**Laurophyllous-indifferent tree species:**	**Laurophyllous-Valdivian tree species:**
Nothofagus obliqua	*Cryptocarya alba* *Peumus boldus* *Citronella mucronata* *Maytenus boaria* *(Podocarpus saligna)*	*Persea lingue* *Laurelia sempervirens*	*Aextoxicon punctatum* *Eucryphia cordifolia* *Laureliopsis philippiana* *Lomatia dentata* *Gevuina avellana* *Weinmannia trichosperma*
Deciduous lianas:	**Geophytic lianas:**	**Laurophyllous-indifferent lianas:**	**Laurophyllous-Valdivian lianas:**
Muehlenbeckia hastulata *Rubus constrictus*	*Bomarea salsilla* *Dioscorea auriculata*	*Boquila trifoliolata* *Cissus striata* *Lapageria rosea* *Lardizabala biternata*	*Elytropus chilense* *Hydrangea serratifolia* *Luzuriaga radicans* *Mitraria coccinea* *Pseudopanax valdivianum*

Figure 2: Species classification according to ecological types and life forms

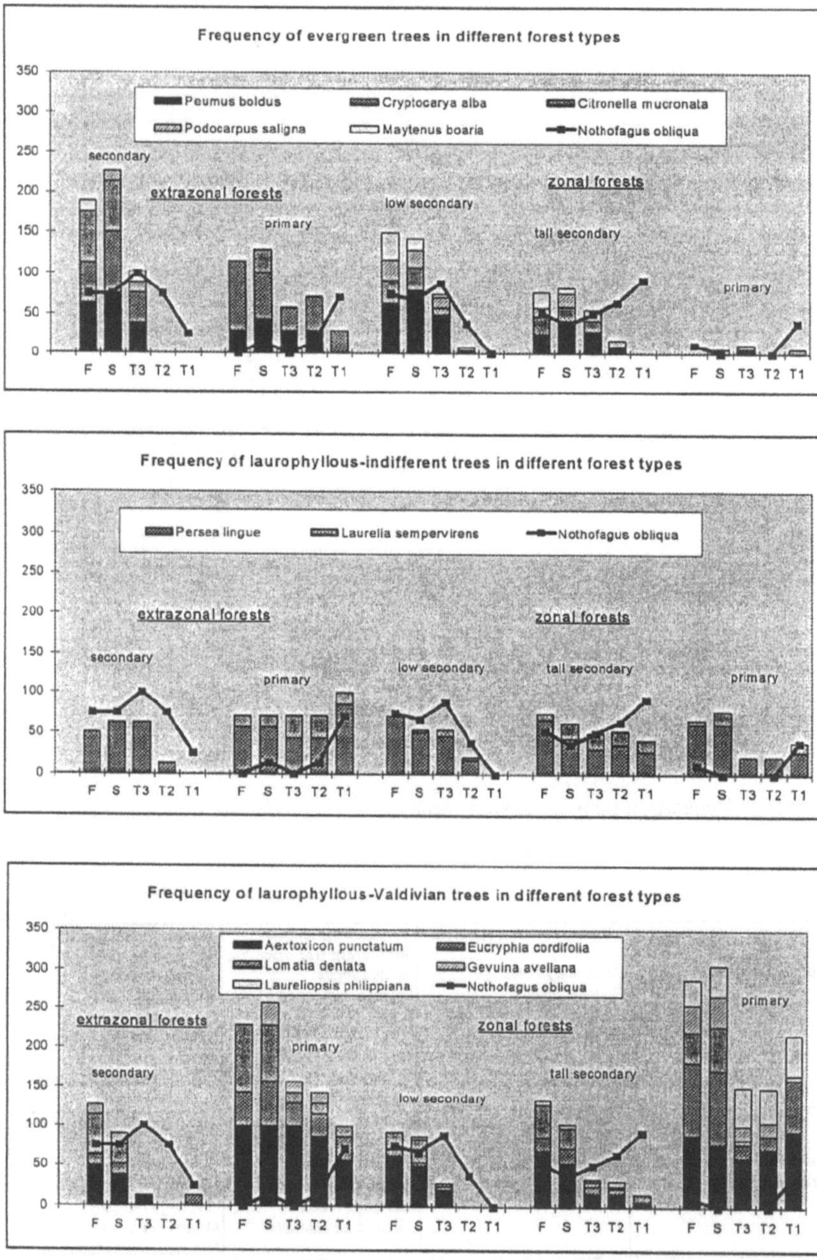

Figure 3: Mean frequency of selected tree species in different forest types

The laurophyllous-Valdivian tree species are rarely present in the canopy of the secondary forests, but their high frequency in the understory of these stands indicates that they are regenerating. *Aextoxicon punctatum* appears with frequencies of ≥ 50 % in the field and shrub layers of deciduous *Nothofagus obliqua* forests indicating a strong regeneration in the shady understory. The less frequent species, *Lomatia dentata* and *Gevuina avellana,* show similar patterns.
In the zonal primary forests, *Aextoxicon punctatum* has very high frequencies in all layers. *Eucryphia cordifolia* and *Laureliopsis philippiana* are restricted to zonal primary forests and, together with *Aextoxicon punctatum*, form the topmost layer of these stands. *Aextoxicon punctatum* is also the most frequent tree in the extrazonal forests and, together with *Persea lingue*, *Lomatia dentata* and the evergreen *Peumus boldus,* dominates the tree layers of this forest-type. *Persea lingue* and *Laurelia sempervirens* behave differently from the laurophyllous-Valdivian tree species, and we have termed them "laurophyllous-indifferent". Both seem to have a wide ecological amplitude, establishing in insolated sites together with *Nothofagus obliqua*, but showing continuous regeneration in primary forests, normally in small or medium sized gaps. Consequently, we conclude that mixed evergreen-laurophyllous forests are the endpoint of forest succession on undisturbed warmer sites, while pure laurophyllous forests appear to be the "climax" vegetation of zonal stands.
The mean coverage of the tree species in the five forest-types (Figure 4) shows a similar, but much more pronounced, pattern than that in Figure 3. In extrazonal secondary forests the evergreen-mediterranean species have about 20 % coverage, much less than *Nothofagus obliqua* (ca. 60 %), but in the extrazonal primary forests *Nothofagus obliqua* has only 18 % coverage while *Peumus boldus* and *Cryptocarya alba* have a canopy coverage of ca. 25 %. Although the evergreen species have ca. 20 % coverage in the younger secondary forests, the regeneration dynamics of these evergreen species is completely different in the zonal forests where they almost disappear with ongoing succession.
The laurophyllous-Valdivian species start with a coverage of less than 10 % in the extrazonal and zonal secondary forests, but reach absolute dominance (60 %, 120 %) in the respective primary forests with *Aextoxicon punctatum* becoming by far the most important species of old-growth forests. Again the two laurophyllous-indifferent species, *Persea* and *Laurelia,* show less variation, but their coverage maxima in extrazonal primary and tall zonal secondary forests indicate optima under medium shade.
The frequencies of the selected lianas show a similar pattern to those of trees comparing *Nothofagus obliqua* forests with old growth stands (Figure 5). Geophytic and deciduous lianas are widespread in *Nothofagus obliqua* forests, but are almost absent from extrazonal primary forests and completely absent from zonal old growth forests. The introduced deciduous climber *Rubus*

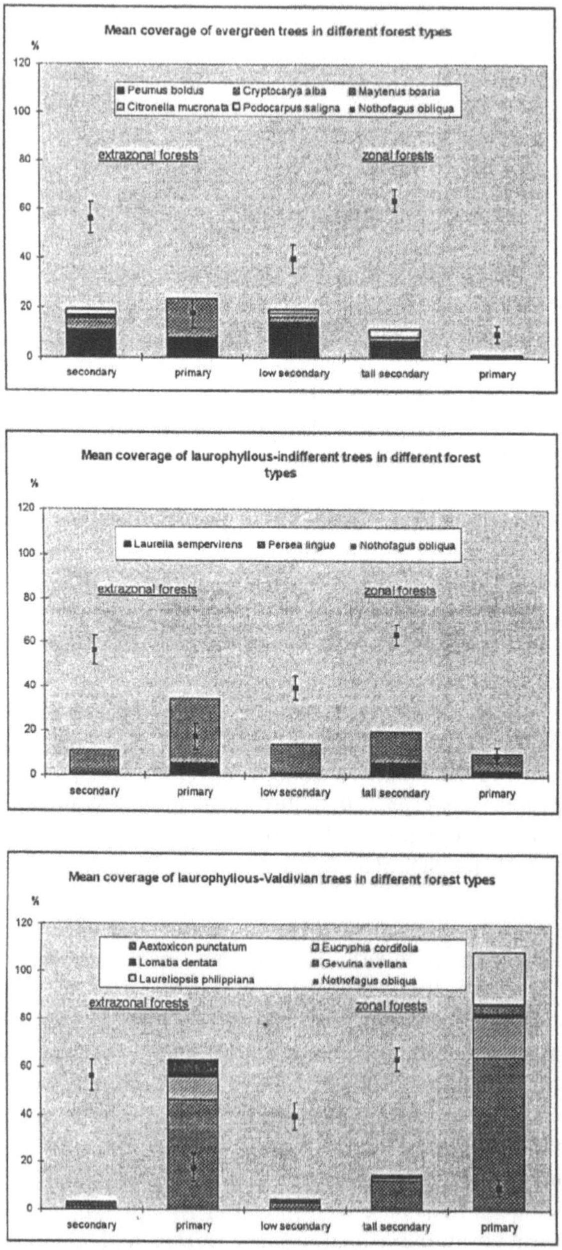

Figure 4: Mean frequency of selected liana species in different forest types

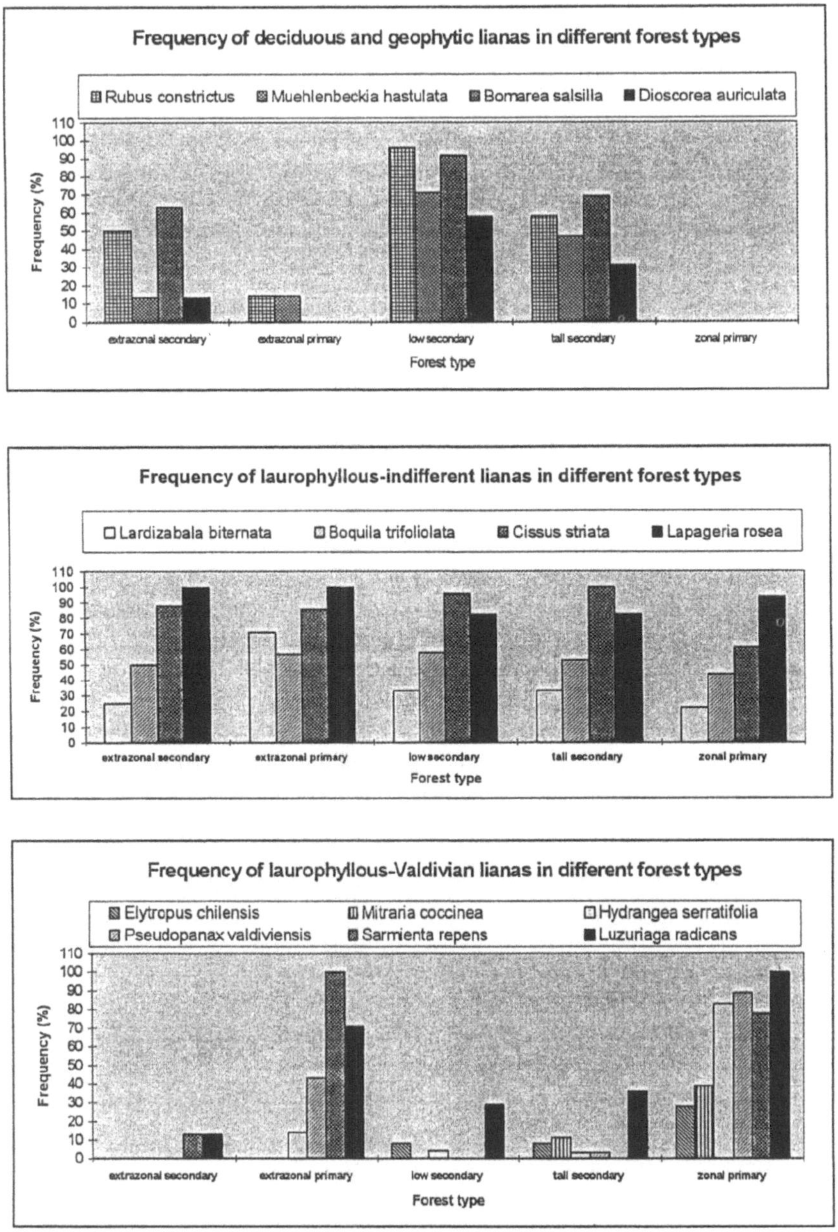

Figure 5: Mean frequency of selected liana species in different forest types

constrictus behaves exactly like the native species *Muehlenbeckia hastulata* and the geophytic vines *Bomarea salsilla* and *Dioscorea auriculata*.
Laurophyllous-indifferent lianas (*Lardizabala biternata*, *Boquila trifoliolata*, *Cissus striata* and *Lapageria rosea*) behave like the laurophyllous indifferent trees: they appear in primary and secondary forests, but with slightly higher frequencies in secondary and extrazonal stands.
The laurophyllous-Valdivian lianas show the opposite pattern, behaving exactly like the laurophyllous-Valdivian trees - with high species numbers and frequencies in zonal primary forests, reduced diversity in intact thermophilous forests and low frequencies or complete absence in secondary forests dominated by *Nothofagus obliqua*.

Discussion

There are few studies of forest dynamics in the ecotonal zone of south central Chile. Veblen et al. (1979) analyzed the stand structure of a similar *Nothofagus obliqua*-dominated old-growth forest in an area with much higher annual precipitation (about 3.000mm) near Villarrica (ca. 40°S). They found no recent regeneration of *Nothofagus obliqua* and continuous regeneration maintaining a steady-state population of *Aextoxicon punctatum*, both observations coinciding with ours in showing laurophyllisation with increasing stand age.
Veblen et al. (1979) found intermittent regeneration of *Persea lingue*, *Eucryphia cordifolia* and *Laurelia sempervirens* and suggested that these species regenerated only in sizeable canopy gaps (30 - 50m^2), although *Persea lingue* (> 10cm dbh) was found to be randomly distributed throughout the stand. Our data show *P. lingue* to be present and regenerating in all vertical strata of old-growth forests (with mean frequencies of 30 - 60 % and > 10 % cover) irrespective of external disturbance. In our study area, *Eucryphia cordifolia* behaves as a late successional species with optimal coverage and ongoing regeneration in old-growth stands. The high number of saplings resulting from abundant root suckering of *Eucryphia* may have biased the results of Veblen et al. (1979) which were based on stem numbers. Nevertheless, like *Aextoxicon punctatum*, these species will be favored in gaps caused by fallen trees.
Our data indicate that *Gevuina avellana* is a late successional species as it attains higher frequencies and coverage only under closed canopies. *Lomatia dentata* is most typical of the main canopy of extrazonal old-growth forests, whereas *Peumus boldus* persists in old-growth forests on zonal sites, and plays an important role on extrazonal sites. The range of both species extends up to the relict *Aextoxicon* forests in the coastal range of North-Central Chile at about 32°S.
The species composition and stand structure of the forest fragments in our study area may be summarized as follows. *Nothofagus obliqua* dominates the

secondary forests, accompanied by evergreen and laurophyllous-indifferent species, whereas in the primary forests different laurophyllous tree species build a closed canopy of 30 - 35m height. Laurophyllous trees of Valdivian origin (e.g. *Aextoxicon punctatum* and *Eucryphia cordifolia*) dominate all vertical strata, indicating favorable conditions for their regeneration. Most of the laurophyllous trees show regeneration patterns typical of shade tolerant late-sucessional species, although *Laurelia sempervirens* and *Persea lingue* show a wide successional amplitude. Evergreen tree species from Central Chile that are near their southern limit in this area show other regeneration patterns. In zonal stands they are most common in earlier succession stages, less frequent in taller secondary forests, and almost absent from the primary forests. In this region, these species originally inhabited extrazonal thermophilous sites with relatively shallow soils. In secondary forests on extrazonal sites, *Nothofagus obliqua* is the first to establish, followed by evergreen trees as *Peumus boldus*, *Cryptocarya alba* and *Citronella mucronata* and, much later, by laurophyllous trees. Intact thermophilous forests have laurophyllous canopies with emergent *Nothofagus obliqua*, a reduced diversity of laurophyllous species, and increased representation of *Lomatia dentata*, *Persea lingue* and *Laurelia sempervirens*. The evergreen trees *Peumus boldus*, and *Cryptocarya alba* form part of the canopy and *Citronella mucronata* survives in the understory with higher frequencies than in the zonal primary forests.

In the study area, any natural reafforestation has followed catastrophic disturbance, typically clear-felling or fires during the last hundred years. In the first post-disturbance period there is a massive cohort establishment of *Nothofagus obliqua* which, after several decades, is typified by a bell-shaped distribution of stem diameters of this tree (Veblen et al. 1979). The establishment of evergreen and (after some delay) laurophyllous tree species occurs after the deciduous canopy is closed, inhibiting the further regeneration of *Nothofagus obliqua* in the understory. Finally, and after long disturbance-free periods, the forest shifts towards a laurophyllous canopy with deciduous emergents - the typical stand structure of the current primary forest fragments. Further absence of disturbance would result in the dominance of *Aextoxicon punctatum* with a considerable admixture of *Persea lingue*, *Laurelia sempervirens*, *Eucryphia cordifolia* and *Laurelia phillippiana* in the main canopy and a high diversity of laurophyllous lianas - i.e. a forest with a structure and species composition typical of the temperate Valdivian rain forests (Gajardo 1994).

The various classifications of the vegetation of southern Chile often disagree about the limits of the different forest types. The "Nothofago-Perseetum lingue" of Oberdorfer (1960) coincides with the mid-successional stages of the "zonal" forests that are widespread in the study area, whereas the thermophilous subassociation, Nothofago-Perseetum boldetosum, corresponds to the mid-successional stages on "extrazonal" sites. Arroyo et al. (1996) excluded our study area from the Valdivian rainforest, assigning the northern limit of this vegetation type in the Central Valley at 41° S. Other authors have also defined individual subareas of deciduous *Nothofagus obliqua* forest in the Central Valley between 40° and 37°S. (Schmithüsen 1956, Quintanilla 1974, Gajardo 1994), despite the fact that most of these forests appear to be anthropogenic and of secondary origin. The zonal old growth forest fragments are dominated by laurophyllous trees, most of them characteristic species of Valdivian rainforests. Laurophyllous lianas and epiphytic ferns contribute to the high species diversity and most of the species have a marked southern temperate (Gondwanan) origin (Finckh 1995). Consequently these forests must be classified as temperate laurophyllous rainforests.

The question as to whether the vegetation shift from deciduous to laurophyllous forests is due to global warming or due to successional processes can be answered only by examining climatic data and the ecophysiology of the trees. Figure 6 shows an annual record (1.7.1996 to 30.6.1997) of monthly maximum and minimum air temperatures measured in the open at a height of 1.5m (means from three stations in the study area).

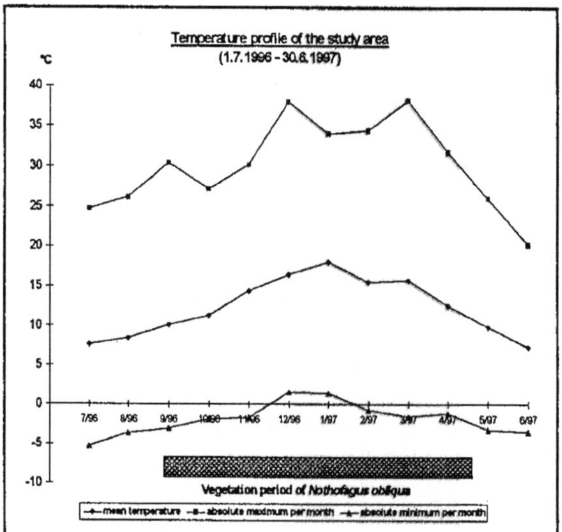

Figure 6: Temperature profile of the study area (1.7.1996 - 30.6.1997)

Maximum temperatures at the soil surface will be considerably greater than those measured at 1.5 m height (up to 40° C). Seedlings of many Valdivian laurophyllous species (e.g. *Eucryphia cordifolia, Weinmannia trichosperma* and *Laureliopsis philippiana*) cannot survive heating and the corresponding desiccation (Donoso 1993). Minimum temperatures of < 0°C occurred from February until mid-November (Figure 6). Frosts may occur at any time of year and are frequent from September to mid-November, the early part of growing season. Seedlings of many laurophyllous species are susceptible to frost, especially in late spring, so that it may be difficult for them to establish in the open (Alberdi 1996). In contrast, *Nothofagus obliqua* sprouts vigorously from stumps left after stand destruction. These sprouts are less susceptible to temperature extremes than seedlings and tolerate minimum temperatures down to -15°C (Golte 1983, Read & Brown 1996).

The temperatures in the open are markedly different from those within the stand. Within stands of *Nothofagus obliqua,* the temperature extremes are less pronounced, with lower maxima and higher minima than in the open, outside the stand. After leaf expansion (mid September) and during the whole vegetative period (until the end of April), temperatures within the stand of *Nothofagus obliqua* normally neither exceed 30°C nor fall below 2°C. This amelioration of temperature within stands is well-known and especially significant in regions with frequent frosts and pronounced summer drought, such as the study area. In such conditions, the deciduousness of *Nothofagus obliqua* provides the best strategy for early successional stages (Donoso 1993, Alberdi 1996); neither the seedlings of the evergreen tree species nor those of the laurophyllous trees are able to establish in the open. The sclerophyllous evergreen species appear earlier in the succession than the broad-leaved evergreen trees, but are less adapted to the shaded old-growth conditions in which the Valdivian laurophyllous species find their synecological optimum.

These man-made differences between temperatures in the open and within stands far exceed the changes predicted for global warming. Consequently, the effects of global warming on laurophyllisation are likely to be minor .

Conclusions

Species distribution, vegetation structure and patterns of regeneration in secondary *Nothofagus obliqua* forests and old-growth stands suggest that the laurophyllisation of the *Nothofagus obliqua* forests is a consequence of successional dynamics. After natural or man-made disturbance, especially after forest fires, *Nothofagus obliqua* appears as a pioneer species recolonizing devastated areas. Once the canopy (dominated by *Nothofagus obliqua*) is closed, temperature extremes are less pronounced and the shade tolerant laurophyllous species start to regenerate inside these stands. This process of laurophyllisation further changes the microclimatic conditions, preventing frost in the understory and maintaining high and more stable humidities throughout the year. This is reflected in the appearance of a large number of indicator species associated with high humidity and typical of Valdivian rainforests (e.g. *Hymenophyllaceae*) in the old-growth stands that are now quite scarce in the study area. The current climatic differences between areas in the open and within the stands are much greater than any potential differences due to changes in the global climate.

In the study area, the balance between forests dominated by *Nothofagus obliqua* and those undergoing laurophyllisation depends on the frequency and intensity of man-made fires and other disturbances. Fossil pollen records from the more humid Lakes Region, 400km south of the study area, indicate a continuous incidence of man-made fires since the arrival of the Palaeo-Indians at about 10.000 years B.P. Tree ring analyses from the "primary" Rucamanque forest give us an estimated stand age of about 350 years (A. Vogel, pers. comm.). Huge cleared areas apparently were present at the beginning of the seventeenth century and were subsequently recolonized by forest. Similar stand devastation ca. 300 years ago was described by Lusk (1996) for *Fitzroya cupressoides* forests in Southern Chile. These catastrophic disturbances might be related to the general uprising of the Mapuche Indians against the Spanish colonizers. The use of fire to clear land for agriculture or silviculture increased again during colonization between approx. 1880 and 1940 and is still common in the study area, accounting for forests that are now dominated by *Nothofagus obliqua*. Currently, fire frequency is decreasing due to improved fire control. Cattle grazing in forests and extraction of wood for fuel is frequent among small farmers and wood is still the major domestic fuel for Temuco's 250.000 inhabitants. The intensity and frequency of disturbance regulates the laurophyllisation process - consequently we postulate that the laurophyllisation of these forests is controlled by land-use practice and intensity. Research on laurophyllisation should pay special attention to land-use history and changing land-use patterns and the remaining laurophyllous "climax" forests should be strategic priorities for nature conservation.

Acknowledgements

We would like to thank in particular Peter Bannister (University of Otago) for idiomatic revision and critical comments on the first draft of this paper. Research was funded by European Comunity.

References

Alberdi M. (1996):Ecofisiología de especies lenosas de los bosques higrófilos templados de Chile: Resistencia a la sequía y bajas temperaturas. In: J. Armesto et al. (eds.): Ecología de los Bosques Nativos de Chile, pp. 279-300. Editorial Universitaria, Santiago de Chile.

Arroyo M.T.K. et al. (1996): Relaciones fitogeográficas y patrones regionales de riqueza de especies en la flora del bosque lluvioso templado de Sudamerica. In: J. Armesto et al. (eds.): Ecología de los Bosques Nativos de Chile, pp. 71-99. Editorial Universitaria, Santiago de Chile.

Berninger O. (1929): Wald und offenes Land in Süd-Chile seit der spanischen Eroberung. Geographische Abhandlungen 3: 1-130. Stuttgart.

Braun-Blanquet J. (1964): Pflanzensoziologie. 3th ed. Springer, Berlin-Wien-New York.

Donoso C. (1993): Bosques templados de Chile y Argentina. Editorial Universitaria, Santiago de Chile.

Finckh M. (1995): Die Wälder des Villarrica-Nationalparks (Südchile) - Lebensgemeinschaften als Basis für ein Schutzkonzept. Dissertationes Botanicae 259. Cramer, Berlin-Stuttgart.

Gajardo R. (1994): Sistema básico de clasificación de la vegetación nativa chilena. Editorial Universitaria, Santiago de Chile.

Golte W. (1983): Verbreitung und ökologische Grundlagen der laubwerfenden *Nothofagus*-Arten im südlichen Andenraum.- In: W. Lauer (ed.) Beiträge zur Geoökologie von Gebirgsräumen in Südamerika und Eurasien, pp. 9-51. Wiesbaden.

Hueck K. (1966): Die Wälder Südamerikas. G. Fischer, Stuttgart.

Lusk C. (1996): Gradient analysis and disturbance history of temperate rain forests of the coast range summit plateau, Valdivia, Chile. Rev. Chil. Hist. Nat. 69: 401-411.

Marticorena C. & Quezada M. (1985): Catálogo de la flora vascular de Chile. Gayana 42: 1-157.

Oberdorfer E. (1960): Pflanzensoziologische Studien in Chile. Flora et Vegetatio Mundi 2. Weinheim/Bergstr.

Quintanilla V. (1974): Les Formations végétales du Chili tempéré.-Doc. de Cartogr. Ecol. 14: 33-80.

Read J. & Brown M. (1996): Ecology of Australian *Nothofagus* Forests. In: T. Veblen et al. (eds.): The Ecology and biogeography of *Nothofagus* Forests, pp. 131-181. Yale University Press, New Haven-London

Schmithüsen J. (1956): Die räumliche Ordnung der chilenischen Vegetation. Bonner geogr. Abh. 17: 1-89.

Schwabe G. (1956): Die ökologischen Jahreszeiten im Klima von Mininco (Chile). Bonner geogr. Abh. 17: 139-182.

Veblen T. T. et al. (1979): Tree regeneration strategies in a lowland *Nothofagus*-dominated forest in south-central Chile. J. of Biogeography 6: 329-340.

Walter H. (1990): Vegetation und Klimazonen. 6th ed. Ulmer, Stuttgart.

Workshops

Workshop I (Head: FUJIWARA, Rapporteur: BRUELHEIDE)

Laurophyllisation: Monitoring in a global context ("Global Change")

Questions:

1) Is monitoring of vegetation change in ecotonal areas advisable, necessary and/or important?
2) Are ecotonal areas suitable for that purpose or should monitoring be restricted to core areas?
3) If so, how should permanent plots be installed?

Report: H. BRUELHEIDE

First Question:

Before addressing this point I have a premise. We first discussed and agreed that it is very important to make use of all information that is already available. Before establishing new plots we should make use of previously established plots. This would allow us to use historical documents such as old relevés and old mappings in a similar way as has already been done in Switzerland (Klötzli et al. 1996, Verh. Ges. Ökol. 26:537-550).

We first tried to clarify what we were talking about. How are ecotonal areas exactly defined? We will use the term „ecotone" on a broad geographical scale as the transition between biomes, which are the main ecological zones of the earth (= "Zono-Ökotone" sensu Walter). The biomes correspond to the main groups of vegetation formations. Although ecotones are encountered between all biomes, we focused on those which are located in a steep climatic gradient between adjacent biomes. Ecotonal areas of this kind present a quickly changing environment from one biome to the next. Why have we concentrated on these areas? We expect these areas to be the most responsive to climatic change. The longer the ecological gradient, the more time vegetation needs to react to climatic change. Therefore, we agreed to answer the first question clearly with "yes". We all expect vegetation changes to occur as a result of climatic changes and we expect them to occur all over the world. Consequently, monitoring is advisable. I think most ecologists will agree that there is a demand for monitoring.

This answers the question of why we should monitor. We examined a further question: Where should we monitor? There are a lot of ecotones with steep climatic gradients. Which biomes should be chosen for investigation and to establish permanent plots? The most worthwhile ecotones will be those adjacent to a biome that is clearly limited by temperature, as is the case of the ecotone between broad-leaved evergreen and deciduous forests. We have thus far mainly concentrated on this transition zone, as we did at this symposium during the last few days, but we would also like to point out some alternative regions that could be worth further investigation (see Steffen W.L. &

Shvidenko A.Z. (eds.) (1996):IGBP Northern Eurasia study: prospectus for an integrated global change research project. IGBP Report 37: 1-91. Stockholm.). One is the timberline, especially where human influence is minimal. If timberlines with negligible human influence exist, they should be included in a global monitoring program. Although we found it important to exclude human impact, we agreed that this would be impossible. In this case, human influence should be identified and quantified as far as possible in order to be able to distinguish between changes due to a change of land use and changes due to climatic change (see also p.22 f. for transects under human influence in Koch G.W.; Scholes R.J., Steffen W.L., Vitousek P.M. & Walker B.H. (1995): The IGBP terrestrial transects: science plan. IGBP Report 36: 1-54. Stockholm).
Another transition zone worth studying could be the lower boundary of altitudinal belts. One example is the transition zone between the submontane and the montane belts, which can be encountered in many regions.

Discussion of 1st Question:

Statement: F. KLÖTZLI

There are a few new ideas, especially regarding timberlines. I would say Japanese timberlines and North American timberlines are less influenced because there is no cattle up there. There is no Alp management. In our areas it would be tedious to put up such a thing on anthropogenic (sub-)alpine pastures.

Statement: K. FUJIWARA

Therefore we ask to isolate anthropogenic impacts near and on the timberline.

Statement: A. BORHIDI

I would like to add: Amongst timberlines there is also the xeric timberline, not only the temperature induced imberline which should be considered and investigated.

Answer: H. BRUELHEIDE

With respect to vegetation changes at the timberline, we have to consider temperature and precipitation, of course. The causes of timberlines are difficult and complex. There are various climatic factors which can influence the timberline. In addition to minimum temperature and precipitation in summer, the length of the vegetation period etc. may be important. We started our discussion by determining that the best knowledge about vegetation shifts and moving boundaries has been gathered at the northern timberline in Scandinavia (see Jäger 1995: Geobotany. Progr. in Bot. 56: 396-415). We found these results very stimulating for research work and consequently included timberlines in our list.

Statement: V. PILLAR

There is one thing we did not mention directly: that is that climatic models predict larger temperature changes in higher latitudes. So the changes in lower latitudes will be slower.

Question: G. GIANONI

What about the priorities?

Answer: H. BRUELHEIDE

We have concentrated on ecotones between deciduous and broad-leaved evergreen forests because this symposium has focused on that point.

Statement: N. ZAZANASHVILI

There are "different classes" of ecotones. The zone between deciduous forests and subtropical belts can be determined as "macroecotone" and laurophyllous species may be the most sensible to the climate changes in this zone. The exposition of their dynamic trends can be considered as the best indicators of climate changes. The dynamic of these species (and some other indicators) better reveal in the ecotones of the "lower classes" ("regional" or "meso" ecotones), as in the subalpine belt (timberline). This process can be called the "double ecotone" effect. I am sure that a sufficient investigation of the subalpine belt in the Colkhis will show an interesting and actual picture of the dynamics of laurophyllous species.

Answer: H. BRUELHEIDE

There are other reasons why we have concentrated on this ecotone. We think that the biome with broad-leaved evergreen vegetation can easily be described by a climatic envelope. We can assume that there is a distinct limitation by minimum temperatures and a requirement for summer precipitation. If we expect a change in winter temperature, this would affect the broad-leaved evergreen species. Therefore, these species may serve as indicators for winter temperature, perhaps for summer precipitation as well.

Second Question:

We understood this question to be a choice between the core and the edge. The edge corresponds to an ecotone in a steep climatic gradient located between two biomes. We expect the edge to be the more dynamic area with more rapid and clear changes. Therefore, we suggest concentrating on the boundary areas of biomes. However, we found it useful to have monitoring plots in the core as well, which means in the centre of the range of a biome. Although the core area is likely to be more static, we need it for reference or as a control. Just imagine the situation in which a rise in temperature has caused the permanent plots near the boundary to change, evergreen species have invaded the plots and we have no plots in other regions that have remained unchanged.

Third Question:

It was easy to agree on permanent plots. We are all vegetation scientists and have a high opinion of permanent plots. It is clear that they should be marked properly. We did not discuss whether to choose the localities systematically, randomly or subjectively. We agreed to arrange them on transects that should be placed across vegetation boundaries. We discussed at length which size of plot would be most appropriate to record the change. Valerio Pillar suggested applying a multiscale approach, because we not only have to consider different strata, but also different objects. Therefore, we suggest simultaneous monitoring according to different scales. For monitoring forests, a plot size of

10 x 10 m to 20 x 20 m was suggested. It would be important to have a constant size in all the places where we place the transects. We should record the complete species composition. We should apply a cover estimation scale (e.g. Londo, Braun-Blanquet). We also suggest including structural data because we may experience no change in species composition, but changes in membership to different strata. For example, understorey species may climb up the trees. One suggestion was to make crown projections when recording the plots. Other structural data could also be included. Furthermore, special species should also be monitored in more detail, for example broad-leaved evergreen species. These species should be monitored using biometrical methods including flowering, fructification etc. We should also mark single individuals for biometrical records.

Since this can be a huge amount of work in a 20 x 20 m plot, it would be more appropriate - according to our multiscale approach - to have small plots within the large plots which allow us to observe changes on different scales (nested plots or nested transects).

The next point concerns climatic measurements. We consider it very useful to have weather stations near the transects. This would ensure that we have real evidence for climatic change in the context of a vegetation change without having to take it for granted. Furthermore, we think it would be possible to record at least temperature and precipitation for every plot.

Finally, we suggest making use of vegetation mapping. Mapping is related to the multiscale approach. Even if we are able to establish a huge amount of permanent plots, there is still the chance that we may fail to trace a change although it is evident in the surrounding area. Mapping the surrounding area on a relatively small scale would also help us relate changes to remote sense monitoring.

Discussion of 2nd & 3rd Question:

Statement: K. FUJIWARA

Prof. Nakhutsrishvili mentioned to make relevés with a vegetation profile.

Statement: G. CARRARO

I don't agree that the relevé areas must be always the same because in a lot of cases there are some species which need a bigger area: to describe a type of vegetation with shrub you need $100 m^2$ and for a large coniferous forest you need a larger relevé area. I don't think that it is so important to have a regular form and area.

Statement: V. PILLAR

I can say something about this: to monitor vegetation you can not establish a priori an appropriate size because who knows how it would be in ten or twenty years. So the size you think is good now may not be good in ten/twenty years. So this is very true in monitoring grasslands which go towards scrub or even forest. This is why we suggest a multiscale approach in which you put side by side smaller plots which could be enlarged simply by pulling them together. So you increase the size at anytime and you don't loose the information. If you decide to increase the size in twenty years you have the information of the quadrats that are adjacent. So you can easily do a multiscale analysis of the process.

Statement: F. KLÖTZLI

I'm very much in favour of these thoughts. In case you should have time to give a profile then it should be at least tree high or two or three hights. And then I would like to come back to the problem of the structure data: It may also occur that structures of trees may change. So for instance the best known here in our country is the structure of spruce: it is not the same in the continental and/or subalpine areas compared with that of montane or lowland areas. It changes its structure because of severe mechanical influence especially snow pressure. So it would be worth it to take care of these data as well if you have time enough because these structures they might change with climate change.

Workshop II (Head: BORHIDI, Rapporteur: DOYLE)

Research priorities in ecotonal areas from evergreen broad-leaved to deciduous forests in Europe

Questions:

1) Which areas in Europe might be suitable for monitoring ongoing changes?
2) Are laurophyllous plants (exotics and endemics) suitable for monitoring?
3) Which research priorities are considered to be (top) important to assess any effects of Global Change on vegetation?

Report: G. DOYLE

We engaged in a general discussion of the problems addressed during the Symposium. In summarising the impact of climate change we recognised that future change might relate to (a) elevated temperatures in either the summer or winter periods; (b) changes in water regimes, with some areas becoming wetter, some drier, and others experiencing seasonal shifts; and (c) changes in light levels available to vegetation, owing to increasing or decreasing cloud cover.

These climate changes would then be reflected in altered plant growth and in vegetation shifts. These impacts would be reflected in changes in the horizontal patterns of vegetation boundaries and in the vertical structure of vegetation types. They could also impact by altering current species distributions, through increases or decreases in the altitudinal limits of sensitive species, and movements in distribution boundaries along oceanic/continental gradients.

We were asked to make a comment on those sites or regions that might prove to be appropriate for investigation and monitoring. We considered that the submontane to montane boundaries would be ideal for these purposes, since they remain relatively natural. Sites at lower elevations tend to be severely influenced by man and his animals and might prove to be impossible to monitor. In approaching such boundary situations, it would be essential to follow both 'top down' and 'bottom up' approaches, investigating upwards and downwards shifts in species and vegetation boundaries. During the Symposium, there appeared to be a concentration on the upward movement of species, and less consideration of the movement of montane species to lower elevations.

We were also asked to consider which species would be most appropriate in monitoring boundary shifts related to climate changes. We considered some species to be obvious candidates. We thought that *Ilex aquifolium* would be an ideal species, since it has both an altitudinal limit and an eastern limit along the oceanic/continental gradient. As a consequence, this species could be used to test altitudinal shifts and geographical shifts from West to East as oceanicity extends. Ivy (*Hedera helix*) was discussed at some length during the Symposium, and would appear to be an ideal candidate species. It has both an altitudinal limit and an eastern limit. It has structural features that vary with oceanicity, climbing high up in trees in oceanic regions and remaining at ground level in more continental areas. In addition, the phenological characteristics of *Hedera* vary, depending on the relative oceanicity/continentality in any one area. Other potential candidate species included *Ruscus* species, *Viburnum tinus* and *Daphne laureola*.

We discussed in detail the potential use of exotic species as monitors for climatic change. Arising from presentations and discussions during the Symposium, it was obvious that there was a general lack of knowledge concerning the physiological reactions of exotic species to climate. As a consequence, it might be possible to erroneously conclude that an expansion in the distribution of an exotic species was a reaction to climate change. The example of expansion of *Rhododendron* in Ireland might be considered a reaction to altered weather patterns, but is clearly not so since expansion pre-dated recent climate change and is related to the biology of the species and the availability of sites suitable for invasion (Doyle, this volume). There is considerable work yet to be completed in unravelling the impacts of climate from changes mediated by altered land-use and management practices. With the exotic species, experiments should be designed that would distinguish between climatic effects and the reactions to disturbance and grazing impacts. It is apparent, however, that exotic species might create biological problems. The expansion of *Rhododendron* in Ireland is such an example. It has become a pest, acting as a weed and excluding native species and inhibiting regeneration of natural oak forest. An outsider visiting the Ascona region might agree that the local forest would be radically changed in a relatively short time owing to the expansion of exotics, but would have difficulty in predicting whether climate change, alteration in forest management practices, the innate biological characteristics of the exotics or a combination of these factors was responsible for the expansion. Such observations demonstrate the necessity for a carefully designed research program that would provide a sound understanding of the biology and ecophysiology of exotic species that have expanded in their distribution in the recent past. Despite these reservations, some exotic species, such as *Laurus nobilis*, might prove to be useful monitors.

Following our discussion we identified a list of research priorities.

1. There should be a sound geographical understanding of ecotonal areas, related to a broad appreciation of species and vegetation distributions from the Atlantic to Continental regions of Europe.

2. Mapping of vegetation boundaries would be essential since these would be used to monitor change.
3. Mapping the distribution limits of sensitive species would be equally essential.
4. Establishment of permanent plots in appropriate areas would allow effective monitoring.
5. Exclosure and enclosure studies should be established: exclosures would help to unravel the impacts of climate and grazing; enclosures might be used in site manipulation studies.
6. Population dynamics investigations of sensitive species would be important. Comparison of population dynamics at the centre of distribution of species and at their climatic limits would be especially instructive.
7. Investigation of the ecophysiology of sensitive species, especially at their distribution limits, would be an important part of such a research program.
8. Examination of allelopathic characteristics of species showing expansion of their distribution might help in explaining their movement into new territories.

Statement: A. BORHIDI

Within the mapping of the limit of species ecological limits are included as well not only the edaphical or climatical one.

Workshop III (Head: GIANONI, Rapporteur: CONEDERA)

Importance of laurophyllisation for Swiss forest policy

Questions:

1) Is “Laurophyllisation“ a suitable indicator to detect and characterize “Global Change“ in Switzerland?
2) If such monitoring is accepted, how should we procede?
3) Are there any consequences in Swiss forest policy in the near future, and can laurophyllous species be integrated in our forests? (cf forest policy in the EU and its consequences in Cantabria, clearing indigenous forests for the benefit of exotics, e.g. Eucalypts)

Report: M. CONEDERA

We tried to turn the problem because we were not so happy with the term of the question because of linguistic, political and scientific reasons. (Comparison with forest decline). We are dealing with laurophyllisation as a fact. We want to start with investigations on the physiology, on colonization dynamics and on external factors which are i.e. forest management and sylvicultural treatments. Neophytes (*Robinia*, *Pinus strobus*), autochthonous thermophilous species as well as the climate could also be included into the study.

We would like to start with two things at the same time: One is the understanding of the processes, literature study of the state of the art and so on; the other one is monitoring. For this task we also suggest the multiscale approach.

We do believe that it is very important to look at laurophyllous forests just as normal forests. This means protection function first and then economic and social functions. And we do think that this is possible. Maybe in the future laurophyllisation could be a touristic attraction. Another very important thing that Ivo Ceschi mentioned is the legal definition for new colonized areas. We have to define what is forest in the sense of the Swiss forest law. He told us of existing problems with palms: are they to be considered as forest or not? A very important aspect if we consider the related value of the land.

Question: G. DOYLE

In your contribution the other night in Locarno it was clear that the utilisation of the forests locally has changed remarkably. Its economic value has declined, it is no longer used as a resource. So you are going to design a program to accomodate this change in landuse so that ecotourism or whatever becomes the main value for these areas.

Answer: I. CESCHI

Yes, I would say so. The new forest management has to be adapted to the changing situation. We start from the function. Every forest has any of the main functions which is protection, production and ecological and recreational value. But everybody can agree that at different points we have different ratios of these main functions. In one place you have the protection value which is the most important, in others you have the recreational value which is the most important value. So we have to define where we have protection forest, where we have recreational forest and where we have to produce timber. Now forest production is declining and so we have to redefine the function and to manage the forest according to the redefined functions. So we have to separate these different management strategies, that's the challenge we have. What does laurophyllisation to this change? I think it could be an advantage that we can show that we have something very dynamic going on in our forests and interesting trees are coming in. And so we can show something innovative to what our visitors are not used to see. So far we have had no problem, let's say no sylvicultural problem, but so far we can't exploit this situation.

http://www.birkhauser.ch

Baur, B., Erhardt, A., Institut für Natur-, Landschafts- und Umweltschutz der Universität Basel (NLU) / **K. C. Ewald,** ETH Zürich, Departement Wald- und Holzforschung / **Freyer, B.,** Forschungsinstitut für Biologischen Landbau, Frick

Ökologischer Ausgleich und Biodiversität
Grundlagen zur Beurteilung des Naturschutzwertes ausgewählter landwirtschaftlicher Nutzflächen

Die Intensivierung der Anbau- und Bewirtschaftungsmethoden führte in den vergangenen 50 Jahren zu einer Verarmung der Biodiversität in mitteleuropäischen Landwirtschaftsgebieten. Durch landeseigene Gesetze und mit der Ratifizierung der Konvention über die biologische Vielfalt (Agenda 21) ist die Schweiz verpflichtet, Massnahmen zur Erhaltung und Förderung der Biodiversität zu treffen. Als eine dieser Massnahmen wurden ökologische Ausgleichsflächen eingeführt, mit denen die natürliche Artenvielfalt im Landwirtschaftsgebiet gefördert werden sollte. Die Bewirtschafter erhalten für bestimmte Arten von ökologischen Ausgleichsflächen finanzielle Entschädigungen, während der Unterhalt von anderen ökologischen Ausgleichsflächen und Landschaftsstrukturelemente vom Bund nicht unterstützt wird.

Anhand von veröffentlichten wissenschaftlichen Studien wird die Bedeutung der verschiedenen Arten von ökologischen Ausgleichsflächen und Landschaftsstrukturelementen für die Erhaltung und Förderung der Biodiversität untersucht und mit der heutigen Entschädigungspraxis verglichen.

Mit dieser Arbeit wird die Grundlage für eine objektive Beurteilung und Überprüfung der bestehenden Ausgleichsbeiträge geschaffen. Änderungen in den finanziellen Entschädigungen werden vorgeschlagen, damit die natürliche Arten- und Strukturvielfalt im Landwirtschaftsgebiet besser gefördert werden kann.

"Die vorliegende Arbeit bildet einen wichtigen Mosaikstein im besseren Verständnis und der besseren Bewertung ökologischer Ausgleichsflächen. Die Literatur ist in kürzester und prägnantester Form zusammengefasst. Von grossem Nutzen für einen Anwender sind die thematische Untergliederung und die Schlussfolgerungen in Form von Empfehlungen."
Ph. Roch, Direktor Bundesamt für Umwelt, Wald und Landschaft

Baur, B.
Ökologischer Ausgleich und Biodiversität
1997. 102 Seiten. Softcover
ISBN 3-7643-5802-5
SPPU Themenhefte

BioSciences with Birkhäuser

For orders originating from all over the world except USA and Canada:

Birkhäuser Verlag AG
P.O. Box 133
CH-4010 Basel / Switzerland
Fax: +41 / 61 / 205 07 92
e-mail: orders@birkhauser.ch

For orders originating in the USA and Canada:

Birkhäuser Boston, Inc.
333 Meadowland Parkway
USA-Secaucus, NJ 07094-2491
Fax: +1 / 201 348 4033
e-mail: orders@birkhauser.com

Birkhäuser

http://www.birkhauser.ch

Lautenschlager-Fleury, D. und E.

Die Weiden von Mittel- und Nordeuropa

Der Bestand vieler Weidenarten ist in den letzten Jahrzehnten zurückgegangen und einige Spezies gelten heute bereits als gefährdet. Das Verständnis und die Kenntnis der Weidenarten ist notwendig, um die Vielfalt dieser Pflanzen zu erhalten und ihren Bestand zu sichern.

Das vorliegende Buch liefert dazu einen wesentlichen Beitrag. Ausführliche Bestimmungsschlüssel und detaillierte Artbeschreibungen, die durch zahlreiche Zeichnungen und Fotografien der Blätter, der Kätzchen und der Zweige mit Knospen veranschaulicht werden, ermöglichen es Fach-Botanikern wie Weidenliebhabern die Weiden und Weidenbastarde von Mittel- und Nordeuropa während des ganzen Jahres einfach und sicher zu bestimmen.

Darüber hinaus bietet es Informationen über die Standorte, die Lebensräume und die ökologische Bedeutung der einzelnen Weidenarten und ihren sinnvollen Einsatz im Grünverbau.

Dagmar Lautenschlager-Fleury wurde im Jahr 1955 an der Universität in Basel im Fach Botanik promoviert und arbeitete als Lehrerin am Mädchen-Gymnasium in Basel.

Ernst Lautenschlager war Biologie-Assistent am Mathematisch-Naturwissenschaftlichen Gymnasium in Basel. Seit seiner Pensionierung im Jahr 1974 stehen die Weiden im Zentrum seiner Interessen.

Dr. Dagmar und Ernst Lautenschlager-Fleury sind Ehrenmitglieder der Basler Botanischen Gesellschaft und Autoren verschiedener Bücher und zahlreicher Publikationen über Weiden.

Basierend auf dem Werk „Die Weiden der Schweiz und angrenzender Gebiete" umfaßt dieses Buch die Weidenarten von ganz Mitteleuropa und Schwedisch-Lappland, wodurch ein Vergleich alpiner und nordischer Weidenspezies möglich ist.

Lautenschlager-Fleury, D. und E.
Die Weiden von Mittel- und Nordeuropa
1994. 172 Seiten, ca. 200 s/w Abb.
und zahlreiche Strichzeichnungen.
Gebunden.
ISBN 3-7643-2994-7

BioSciences with Birkhäuser

For orders originating from all over the world except USA and Canada:

Birkhäuser Verlag AG
P.O. Box 133
CH-4010 Basel / Switzerland
Fax: +41 / 61 / 205 07 92
e-mail: orders@birkhauser.ch

For orders originating in the USA and Canada:

Birkhäuser Boston, Inc.
333 Meadowland Parkway
USA-Secaucus, NJ 07094-2491
Fax: +1 / 201 348 4033
e-mail: orders@birkhauser.com

Birkhäuser

MIX
Papier aus verantwortungsvollen Quellen
Paper from responsible sources
FSC® C105338

If you have any concerns about our products,
you can contact us on
ProductSafety@springernature.com

In case Publisher is established outside the EU,
the EU authorized representative is:
Springer Nature Customer Service Center GmbH
Europaplatz 3, 69115 Heidelberg, Germany

Printed by Libri Plureos GmbH
in Hamburg, Germany